Advances in Asphalt Pavement Technologies and Practices

Advances in Asphalt Pavement Technologies and Practices

Editors

Amir Tabakovic
Jan Valentin
Liang He

MDPI • Basel • Beijing • Wuhan • Barcelona • Belgrade • Manchester • Tokyo • Cluj • Tianjin

Editors
Amir Tabakovic
Netherlands Organisation for Scientific Research (TNO) and Delft University of Technology (TUDelft)
The Netherlands

Jan Valentin
Faculty of Civil Engineering, Czech Technical University in Prague
Czech Republic

Liang He
School of Civil Engineering, Chongqing Jiaotong University
China

Editorial Office
MDPI
St. Alban-Anlage 66
4052 Basel, Switzerland

This is a reprint of articles from the Special Issue published online in the open access journal *Applied Sciences* (ISSN 2076-3417) (available at: https://www.mdpi.com/journal/applsci/special_issues/asphalt_pavement_technologies).

For citation purposes, cite each article independently as indicated on the article page online and as indicated below:

LastName, A.A.; LastName, B.B.; LastName, C.C. Article Title. *Journal Name* **Year**, *Volume Number*, Page Range.

ISBN 978-3-0365-5493-8 (Hbk)
ISBN 978-3-0365-5494-5 (PDF)

Contents

About the Editors

Amir Tabakovic

Dr. Amir Tabakovic is a Marie Skłowoska-Curie Fellow at the Netherlands Organisation for Applied Scientific Research (TNO—Netherlands), a visiting research fellow at the Delft University of Technology (TU Delft—Netherlands) and a visiting assistant professor at the UCD School of Civil Engineering (Ireland). Amir is also a founding member and director of a start-up company called Revolution Road Ltd., which focuses on the development and promotion of innovative sustainable technologies for asphalt pavement design, construction, and maintenance. Amir holds a BEng, an MPhil in Mechanical Engineering from Technological University Dublin (Ireland), and a PhD in Civil/Mechanical Engineering from University College Dublin (UCD - Ireland). Amir has over 18 years of experience in asphalt pavement research. His research focuses on the development of bio-inspired systems for asphalt pavements, bio-binders, and self-healing systems for asphalt pavements. He has published widely on this subject and he also works with the International RILEM committee to develop international standards for evaluating self-healing asphalt. His career goal is to introduce bio-inspired technology into asphalt pavement design and maintenance processes, with a focus on sustainability and the circular economy.

Jan Valentin

Jan Valentin (Ph.D.) has been the deputy head of the Department of Road Construction at the Faculty of Civil Engineering of the Czech Technical University in Prague for the last 13 years. At the same time, he has been working for 20 years as an expert in various positions within the multinational company HOCHTIEF (currently as sustainability manager of HOCHTIEF CZ). In the past 10 years he has also led several national research projects and two international (European) projects—CoRePaSol and FunDBitS—within the CEDR research and development calls. Since 2022, he has been the coordinator of the European project RECONMATIC within Horizon Europe, focused on whole life-cycle waste management in combination with digitalization and automatization. Jan holds a master's degree and a PhD degree from the Czech Technical University in Prague. Professionally, his main focus is on asphalt pavement technologies, including aspects of recycling, functional testing of asphalt mixtures and binders. In recent years, he has also increasingly focused on aspects of sustainability and circular economy with an emphasis on the issue of increasing recycling. He is the author or co-author of a number of publications, two patents, more than 20 utility models and many proven technologies implemented in the Czech Republic. He represents the Czech Republic in the European Asphalt Pavements Association.

Liang He

Liang He is a professor at Chongqing Jiaotong University since 2020, and the director of China-Poland Joint Center for Sustainable Pavement Innovative Materials. His general research interests are in the areas of advanced road maintenance methods, sustainable pavement materials, and multiscale simulation of asphalt material behaviour. In the past 5 years, he has led 6 national scientific research projects, including 4 projects funded by the National Natural Science Foundation of China. He has published more than 50 high-level journal papers at home and abroad, and authorized

5 national invention patents. He is a member of the Scientific Advisory Committee of Archives of Civil Engineering, Youth Editorial Board of Journal of Traffic and Transportation Engineering, and Executive Editorial Board of Materials Reports. He is also a co-chair of the organization committee the 1st International Workshop on the Use of Biomaterials in Pavements. http://teacher.cqjtu.edu.cn/heliangf1/zh_CN/index.htm.

Preface to "Advances in Asphalt Pavement Technologies and Practices"

The MPDI Applied Sciences Special Issue, "Advances in Asphalt Pavement Technologies and Practices", aimed to compile and curate the latest state-of-the-art research in the field of asphalt pavement technology worldwide. The special issue is a joint effort between the MPDI Applied Sciences Editorial Office and three Guest Editors: Dr. Amir Tabaković, Dr. Jan Valentin and Dr. Liang He. The selected papers went through a rigorous MDPI Applied Sciences review process. All authors of the accepted papers have made special efforts to increase the quality of the technical content via a rigorous review process. The special issue comprises 19 papers describing a unique and innovative research work focusing on the current challenges facing road owners and the asphalt industry.

The published papers cover research on new materials and new technologies for asphalt pavement design, construction, and maintenance, such as the development of a unique self-healing asphalt technology, conductive alginate capsules that encapsulate rejuvenators, and an application of android sensors and Google Earth for pavement maintenance. Another paper uses a random forest model to predict the location of potential damage on asphalt pavements. A number of papers investigate specific distress problems in asphalt pavements, such as the correlation between pavement distress and the roughness measurement, the low-temperature performance of a porous asphalt mixture, and the influence of the crumb rubber asphalt mixture on a multiscale. There is also a paper that evaluates and corrects the method of the asphalt pavement rutting performance prediction model based on the RIOHTrack long-term observation data. This Special Issue also benefits from a review paper on the influential factors and evaluation methods of the performance of grouted semi-flexible pavement (GSP). In a special effort to tackle climate change and to introduce the sustainability of asphalt pavement, two papers investigate recycling aged asphalt using a hard bituminous binder for hot-mix recycling, whereas another paper discusses the use of shredded cigarette filters as stabilizing fibres in stone mastic asphalts. The Special Issue also includes an editorial where Guest Editor Dr. Tabaković presents the current challenges facing the asphalt industry with the depletion of the petroleum-based asphalt binder (bitumen), but he also discusses solutions to this challenge, including the utilisation of biomass for the production of bio-based rejuvenators and bitumen modifiers.

The Guest Editors of this Special Issue would like to give their appreciation to the authors of the papers for their efforts in contributing good quality research, and to the staff in the Editorial Office of MDPI Applied Sciences, especially Jim Wang, for their assistance during the course of organizing and publishing this issue.

Last, but not least, the Guest Editors would like to thank the dedicated reviewers who contributed their valuable time to review the papers.

Amir Tabakovic, Jan Valentin, and Liang He
Editors

applied sciences

MDPI

Editorial

Bio-Binder—Innovative Asphalt Technology

Amir Tabaković [1,2,3]

1 Centre for Research in Engineering Surface Technology (CREST), FOCAS Research Institute, Technological University Dublin, 13 Camden Row, Kevin Street, D08 CKP1 Dublin, Ireland; amir.tabakovic@tudublin.ie

2 School of Civil Engineering, University College Dublin, D04 K3H4 Dublin, Ireland

3 Civil Engineering and Geosciences, Delft University of Technology, 2628CN Delft, The Netherlands

Received: 10 November 2020; Accepted: 30 November 2020; Published: 3 December 2020

The global road network spans 16.3 million km [1], of which 5 million km is in the EU. These road networks fulfil major economic and social goals by facilitating the movement of goods and people throughout the EU, and are therefore of the utmost importance to the economic and social life of the EU [2]. National governments invest heavily in their road networks, e.g., in 2014, EUR 53.33 billion was invested in the development and maintenance of the EU road network [3]. Each year, the world produces 1.6 trillion tonnes of asphalt [4], of which 218 million tonnes is produced in the EU [5]. The average cost of asphalt in the EU is EUR 200 per tonne. These figures show that the construction and maintenance of road networks are a significant cost to the tax payer. These costs are set to rise further as the sources of bitumen (a product of crude oil) diminish [6,7]. Improved road design and enhanced road materials offer the road industry the potential for improved efficiency and financial savings. The challenge is to develop road materials and construction methods which will improve the environmental cost and reduce the economic cost of road construction. Unlike other construction materials, road materials have developed minimally over the past 100 years [8], but since the 1970s, the focus has been on more sustainable road construction materials, e.g., recycled asphalt pavements [9]. Recycling asphalt involves removing old asphalt, and mixing it with new (fresh) aggregates, binder and/or rejuvenator [9,10]. The primary purpose of the recycling process is to restore the original molecular structure of the aged bitumen in order to extend the lifespan of the asphalt pavement (road). The road lifespan is extended by adjusting the properties of the asphalt mix, i.e., reducing its stiffness [11]. Some commercially available rejuvenating agents include Modeseal R20, Reclamite, Paxole 1009, Cyclepave and ACF Iterlene 1000. Two recent studies by Garcia et al. [12] and Su et al. [13] demonstrated that food by-products (i.e., vegetable oil) can also be used as rejuvenators and offer an alternative to crude oil. Other recent road material innovations, i.e., warm asphalt mixtures, have had unanticipated environmental impacts, e.g., contaminating soil and groundwater [14]. There is growing pressure on the road industry to reduce its environmental impact. Self-healing asphalt has potential in this regard, but it has not yet delivered in terms of improved sustainability [15], or demonstrated its reduced cost and environmental impact [16].

Bitumen, which has traditionally been used as a binder within asphalt mix, is a product of crude oil. The production of crude oil is in decline [7], and the environmental and financial costs of extraction are on the increase [6,7]. There is a need to identify alternatives to bitumen—preferably cheaper and more sustainable alternatives. Researchers have considered replacing bitumen with a plant-based alternative [17], but diverting plants from food resources to asphalt would be unacceptable given the increased pressure on food resources globally. Researchers [18,19] have studied the utilization of waste cooking oil as a binder rejuvenator in the asphalt recycling process. Lignin, a by-product of wood processing, was identified by Van Vliet et al. as a suitable source of biomass [20] for the production of bitumen [21], but the uncertain future of wood processing plants makes this potentially an unreliable/unsustainable material source.

Microalgae have been considered as an alternative source of bio fuel [22], but also have potential as an alternative bitumen or asphalt binder [23]. Audo et al. [23] demonstrated that microalgae can be converted into crude oil, which is an ideal material for the production of bio-binder. The bio-oil produced has an equivalent energy value to fossil fuels [24], making it suitable for bio-fuel but also possibly as an asphalt binder source [23]. Maximum biocrude yields in the range of 40–50% have been reported [23], though they are in an initial stage of development. The BioRePavation project [25] demonstrated that microalgae oil can be used as an asphalt binder in the construction of new asphalt mixtures and in the asphalt recycling process.

This is a promising development in terms of the improvement of the sustainability of microalgal cultivation [26]. The high uptake capacity of microalgae for nitrate and phosphate, coupled with its sustained photosynthetic activity, is the basis of its recognized potential in treating wastewater [27]. Pal et al. [28] have shown that this approach is especially relevant in arid areas, where food and water resources are scarce and microalgae can play an indispensable role in nutrient and carbon recovery from waste resources, as well as in the treatment/use of waste or other marginal water resources. The microalgae obtains nutrition from the dissolved nitrogen and phosphorus found in the wastewater (which would otherwise be a pollution threat) [29]. This process will convert the nutrients found in wastewater into an environmentally friendly fertilizer. Microalgae that proliferate in wastewater facilities are not suitable for human consumption [29], but are ideal for the production of products to be implemented in road construction, such as asphalt binder and bitumen rejuvenators [23]. This presents a unique opportunity to create a circular economy, whereby microalgae can be cultivated within the wastewater treatment process and later processed to extract the oil from the algae. The oil can be further used in the production of the asphalt bio-binder, and other co-products of the process can be used as animal feed or land fertilizer. The utilization of bio-oils in bitumen and asphalt production presents a timely opportunity for the asphalt industry to simultaneously improve its sustainability record and reduce its negative impact on the environment.

Funding: This research received no external funding.

Acknowledgments: This Special Issue would not be possible without the contributions of all authors and reviewers. Special thanks to Jim Wang for his continuous support with organising this Special Issue.

Conflicts of Interest: Author declares no conflict of interest.

References

1. OECD. *Road Traffic, Vehicles and Networks. 2013 in Environment at a Glance 2013: OECD Indicators*; OECD Publishing: Paris, France, 2013.
2. Vita, L.; Marolda, M.C. *Road Infrastructure—the Backbone of Transport System*; EU Directorate General for Research and Sustainable Surface Transport: Brussels, Belgium, 2008.
3. Federation, E.U.R. *Roads Statistics: Year Book 2017*; European Union Road Federation Place Stéphanie 6/B B-1050: Brussels, Belgium, 2017.
4. EAPA. *The Asphalt Paving Industry, A Global Perspective*; EAPA: Brussels, Belgium, 2011.
5. EAPA. *Key Figures of the European Asphalt Industry in 2014, in Asphalt in Figures*; EAPA: Brussels, Belgium, 2016.
6. Leggett, J. Big Oil's Looming Bubble. In *New Internationalist*; McGowen House: Northampton, UK, 2014; pp. 20–21.
7. Worth, J. Ending the oil age. In *New Internationalist*; McGowen House: Northampton, UK, 2014; pp. 12–16.
8. Tabaković, A.; Braak, D.; Gerwen, M.; Copuroglu, O.; Post, W.; Garcia, S.J.; Schlangen, E. The compartmented alginate fibres optimisation for bitumen rejuvenator encapsulation. *J. Traffic Transp. Eng.* **2017**, *4*, 347–359. [CrossRef]
9. Tabaković, A. Recycled Asphalt (RA) for Pavements. In *Handbook of Recycled Concrete and Demolition Waste*; Pacheco-Torgal, F., Tam, V.W.Y., Labrincha, J.A., Ding, Y., de Brito, J., Eds.; Woodhead Publishing: Sarston, UK, 2013; pp. 394–423.
10. Tabakovic, A.; Gibney, A.; McNally, C.; Gilchrist, M.D. The Influence of Recycled Asphalt Pavement on the Fatigue Performance of Asphalt Concrete Base Courses. *ASCE J. Mater. Civ. Eng.* **2010**, *22*, 8. [CrossRef]

11. Brownridge, J. The role of an asphalt rejuvenator in pavement preservation: Use and need for asphalt rejuvenation. In Proceedings of the International Conference on Pavement Preservation, Newport Beach, CA, USA, 13–15 April 2010.
12. Garcia, A.; Jelfs, J.; Austin, C.J. Internal asphalt mixture rejuvenation using capsules. *Constr. Build. Mater.* **2015**, *101*, 309–316. [CrossRef]
13. Su, J.F.; Qiu, J.; Schlangen, E.; Wang, Y.Y. Investigation the possibility of a new approach of using microcapsules containing waste cooking oil; in-situ rejuvenation. *Constr. Build. Mater.* **2015**, *74*, 83–92. [CrossRef]
14. Trombulack, S.C.; Frissell, C.A. Revbiew of ecological effects of roads on terestrial and aquatic communities. *Conserv. Biol.* **2000**, *14*, 13.
15. Tabaković, A.; Post, W.; Cantero, D.; Copuroglu, O.; Garcia, S.J.; Schlangen, E. The reinforcement and healing of asphalt mastic mixtures by rejuvenator encapsulation in alginate compartmented fibres. *Smart Mater. Struct.* **2016**, *25*, 084003. [CrossRef]
16. Anderson, F.A. Final Report on the Safety Assessment of Melamine/Formaldehyde Resin. *J. Am. Coll. Toxicol.* **1995**, *14*, 373–385.
17. Garcia, A.; Austin, C.J.; Jelfs, J. Mechanical properties of asphalt mixture containing sunflower oil capsules. *J. Clean. Prod.* **2016**, *118*, 9. [CrossRef]
18. Saha, R.; Melaku, R.S.; Karki, B.; Berg, A. Effect of Bio-Oils on Binder and Mix Properties with High RAP Binder Content. *J. Mater. Civ. Eng.* **2020**, *32*, 04020007. [CrossRef]
19. Rodrigues, C.; Capitão, S.; Picado-Santos, L.; Almeida, A. Full Recycling of Asphalt Concrete with Waste Cooking Oil as Rejuvenator and LDPE from Urban Waste as Binder Modifier. *Sustainability* **2020**, *12*, 8222. [CrossRef]
20. Gellerstedt, G.; Henriksson, G. Chapter 9—Lignins: Major Sources, Structure and Properties. In *Monomers, Polymers and Composites from Renewable Resources*; Belgacem, M.N., Gandini, A., Eds.; Elsevier: Amsterdam, The Netherlands, 2008; pp. 201–224.
21. Vliet, D.; Slaghek, T.; Giezen, C.; Haaksman, I. Lignin as green alternative for bitumen. In Proceedings of the 6th Euroasphalt and Eurobitumen Congress, E&E Congress 2016, Prague, Czech Republic, 1–3 June 2016.
22. Murphy, F.; Devlin, G.; Deverell, R.; McDonnell, K. Biofuel Production in Ireland—An Approach to 2020 Targets with a Focus on Algal Biomass. *Energies* **2013**, *6*, 6391–6412. [CrossRef]
23. Audo, M.; Paraschiv, M.; Queffélec, C.; Louvet, I.; Hémez, J.; Fayon, F.; Lépine, O.; Legrand, J.; Tazerout, M.; Chailleux, E.; et al. Subcritical Hydrothermal Liquefaction of Microalgae Residues as a Green Route to Alternative Road Binders. *ACS Sustain. Chem. Eng.* **2015**, *3*, 8. [CrossRef]
24. Umakanta, J.; Das, K.C. Comparative Evaluation of Thermochemical Liquefaction and Pyrolysis for Bio-Oil Production from Microalgae. *Energy Fuels* **2011**, *25*, 5472–5482.
25. French Institute of Science and Technology devoted to Transport, P.a.N.I. BioRePavation. 2020. Available online: http://biorepavation.ifsttar.fr/ (accessed on 20 October 2020).
26. Kesaano, M.; Sims, R.C. Algal biofilm based technology for wastewater treatment. *Algal Res.* **2014**, *5*, 231–240. [CrossRef]
27. Craggs, R.; Sutherland, D.; Campbell, H. Hectare-scale demonstration of high rate algal ponds for enhanced wastewater treatment and biofuel production. *J. Appl. Phycol.* **2012**, *24*, 329–337. [CrossRef]
28. Pal, D.; Khozin-Goldberg, I.; Cohen, Z.; Boussiba, S. The effect of light, salinity, and nitrogen availability on lipid production by Nannochloropsis sp. *Appl. Microbiol. Biotechnol.* **2011**, *90*, 1429–1441. [CrossRef] [PubMed]
29. Wiley, P.; Elliott, C.; Thomas, H. Microalgae Cultivation Using Offshore Membrane Enclosures for Growing Algae (OMEGA). *J. Sustain. Bioenergy Syst.* **2013**, *3*, 18. [CrossRef]

Publisher's Note: MDPI stays neutral with regard to jurisdictional claims in published maps and institutional affiliations.

applied sciences

Article

The Effect of Conductive Alginate Capsules Encapsulating Rejuvenator (HealRoad Capsules) on the Healing Properties of 10 mm Stone Mastic Asphalt Mix

Amir Tabaković [1,2,*], Christopher Faloon [3] and Declan O'Prey [3]

1 Structural Reliability Section, Department of Infrastructure & Maritime, TNO, Stieltjesweg 1, 2628 CK Delft, The Netherlands
2 Department of Materials, Mechanics, Management & Design (3Md), Faculty of Civil Engineering and Geosciences, Delft University of Technology, Stevinweg 1, 2628 CN Delft, The Netherlands
3 Lagan Breedon House, Rosemount Business Park, Ballycoolin Road, D11 K2TP Dublin, Ireland; christopher.faloon@breedongroup.com (C.F.); declan.oprey@breedongroup.com (D.O.)
* Correspondence: a.tabakovic@tudelft.nl; Tel.: +31-15-27-81985

Featured Application: Conductive alginate capsules encapsulating a bitumen rejuvenator (HealRoad capsules) is a unique self-healing asphalt technology which reinforces the asphalt pavement by improving its tensile and compressive strength and enhances its autonomous healing properties. The concept combines two existing self-healing asphalt technologies (induction heating and rejuvenator encapsulation) to create a self-healing system which provides quick and effective asphalt pavement repair. The induction heating is used to repair asphalt damage (cracks and ravelling) while the rejuvenator is released to rejuvenate aged asphalt binder (bitumen).

Abstract: Conductive alginate capsules encapsulating a bitumen rejuvenator (HealRoad capsules) has demonstrated good healing abilities in pure bitumen and mortar mixes. HealRoad capsules can efficiently heal damage via induction heating. They also release the encapsulated rejuvenator, thereby rejuvenating aged bitumen. These findings indicate that HealRoad capsules and induction heating systems combined could represent a possible asphalt pavement maintenance method. This paper investigated the effect of HealRoad capsules on the mechanical performance of the 10 mm stone mastic asphalt mix and measured the damage repair (healing) efficiency of the capsules in an asphalt mix. The results indicate that in small amounts, >1%, HealRoad capsules do not degrade the mix performance (indirect tensile strength and rutting resistance) and in some cases, the HealRoad capsules actually improve mix performance, e.g., in terms of the indirect tensile strength ratio (water sensitivity). However, the HealRoad capsules are unable to stimulate induction healing due to the small volume of capsules within the mix. Further investigation demonstrated that increasing the capsules in the mix to >5% can stimulate induction heating effectively. However, it also indicated that a high content of HealRoad capsules reduces the asphalt mix strength. The study has shown that HealRoad capsules are an effective healing system for high bitumen content mixtures such as mortar mixtures but is an inefficient healing system for a full asphalt mix, such as the 10 mm stone mastic asphalt mix.

Keywords: asphalt pavements; self-healing; hybrid self-healing system; induction heating; indirect tensile strength; water sensitivity; wheel tracking rutting resistance

Citation: Tabaković, A.; Faloon, C.; O'Prey, D. The Effect of Conductive Alginate Capsules Encapsulating Rejuvenator (HealRoad Capsules) on the Healing Properties of 10 mm Stone Mastic Asphalt Mix. *Appl. Sci.* **2022**, *12*, 3648. https://doi.org/10.3390/app12073648

Academic Editor: Luís Picado Santos

Received: 7 March 2022
Accepted: 30 March 2022
Published: 5 April 2022

1. Introduction

For the past two decades, researchers have been intensively developing self-healing asphalt technology as a cheaper, safer and more sustainable method of maintaining and repairing asphalt roads [1–4]. The idea of the self-healing road stemmed from the concept of the "forever open road", motivated by the need to avoid the traffic disruption and associated risk of accidents caused by road maintenance activities on busy roads. The challenges for the road industry are to improve road material performance (make roads more durable), improve the sustainability of road construction methods (by introducing green technology) and improve the safety of the roads for both road users and road construction workers [5–7]. Self-healing asphalt technology has the potential to address several of the technical, economic, environmental and safety challenges currently facing the road industry. For example, self-healing asphalt technology has the potential to reduce the need for maintenance-related road closures, thereby reducing the risk to road maintenance workers, while simultaneously reducing the environmental impact of road construction and maintenance. It is estimated that self-healing asphalt technology could reduce energy consumption and CO_2 emissions by 30% over the lifetime of a road [8].

To date, researchers have tested three extrinsic self-healing methods for asphalt pavements, which are [1]: induction heating [1,9–12], microwave heating [4,13,14], rejuvenation (rejuvenator encapsulation) [2,15–18].

The rejuvenator encapsulation approach represents a more favourable method of asphalt self-healing as it is a passive healing system and does not require external action in order to initiate healing. The rejuvenation enables the aged bitumen to return to its original chemical, physical and mechanical properties. However, there are limitations with this method of self-healing because the healing rate is slow and healing efficiency low [2,18]. Researchers have demonstrated that the induction and microwave heating methods are efficient methods for healing asphalt damage (3 to 5 min) in comparison to rejuvenation [2,19]. However, a challenge with both methods is that the heating process (both induction and microwave) ages the bitumen, causing binder brittleness and the premature failure of the asphalt [9,20].

Hybrid self-healing systems provide a solution to the challenges associated with induction, microwave and encapsulated rejuvenation. The hybrid self-healing technology combines induction heating and encapsulated rejuvenation [2]. In this process, induction heating is used to repair the asphalt damage (and the cracks), while rejuvenation is used to replenish the aged asphalt binder (bitumen). Xu et al. [2] demonstrated that a hybrid healing system improves healing efficiency allowing for rapid damage repair and a simultaneous rejuvenation of the aged asphalt. However, a challenge with this approach is that the bituminous mix must be adjusted to accommodate both capsules and steel fibres. A solution to this challenge lies in combining the induction and rejuvenation into one product. Wan et al. [4,21] developed calcium alginate/nano-Fe_3O_4 composite capsules for controlled rejuvenator oil release using microwave heating. Wan et al. showed that the capsules demonstrated superior levels of healing with microwave heating due to the combined effect of thermal induction and rejuvenator healing. Wen et al. showed that hybrid Fe_3O_4–calcium alginate capsules encapsulating a rejuvenator have a higher self-healing capacity (8.6%) when compared to the mix containing pure calcium alginate capsules encapsulating a rejuvenator and 19.3% in comparison to a standard asphalt mix. The asphalt mix containing 2% of the capsules achieved up to 87.2% healing after 40 s of healing time.

The author (Tabaković et al. [3]) has developed a novel extrinsic self-healing asphalt technology: conductive alginate capsules encapsulating a bitumen rejuvenator (HealRoad capsules). The HealRoad technology combines two existing self-healing asphalt methods: (i) rejuvenator encapsulation and (ii) induction heating. Initial results demonstrated that the HealRoad capsules had sufficient thermal and mechanical strength to survive the asphalt mixing process and also that they sufficiently healed the asphalt bitumen damage [3]. Initial results also demonstrated that the capsules can improve the mechanical properties of the asphalt mix in terms of improving tensile strength and stiffness. With a capsule content of 20% in pure bitumen, the mix strength of the material increased to 118%, whereas for mortar bituminous mixtures (bitumen + fine aggregates − sand), the mix strength of the material increased by up to 67%.

This study evaluated the effect of HealRoad capsules in a 10 mm stone mastic asphalt (SMA) mix. The aim of the work was to investigate whether the HealRoad capsules can efficiently repair the damage in an asphalt mix (close cracks and restore asphalt mix strength). A second aim of the research was to ascertain how the inclusion of HealRoad capsules would affect the mechanical performance of the asphalt mix. Indirect tensile strength (ITS), water sensitivity and resistance to rutting tests were conducted to evaluate the effect of HealRoad capsules in the 10 mm stone mastic asphalt (SMA) mix.

2. Materials and Methods

2.1. Materials

2.1.1. The 10 mm Stone Matic Asphalt Mix

The 10 mm Stone Mastic Asphalt (SMA) mix was prepared in accordance with the Transport Infrastructure Ireland (TII) Specification for Road Works Series 900 guidance. The aggregate was an Irish Metasandstone PSV60. The filler was an imported limestone filler to EN. The bitumen was a polymer-modified 65/105-60 bitumen. Figure 1 shows the mix grading curve.

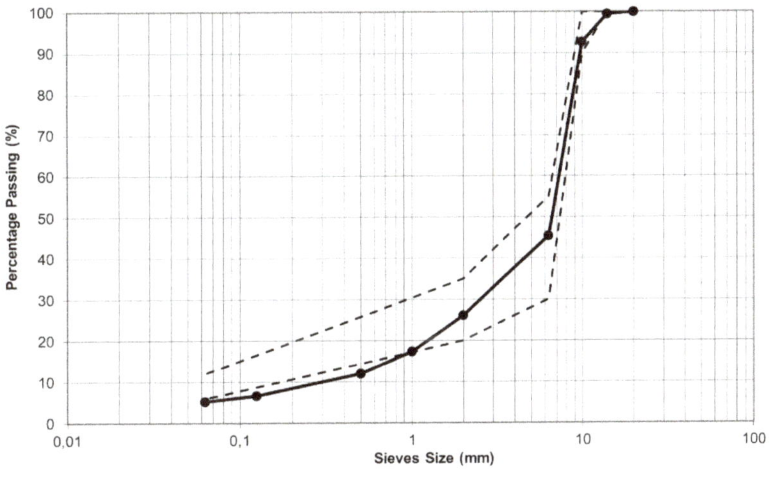

Figure 1. The 10 mm SMA grading curve.

Table 1 summarises the SMA mix constituents and shows their proportions in the mix, both with and without capsules. The HealRoad capsules were added in amount of 0%, 0.31%, 0.44%, 0.64%, 1.02% and 1.45%. These proportions reflect the following proportions of capsules in the capsule and bitumen portion of the mix: 0%, 5%, 7%, 10%, 15% and 20%. The HealRoad capsules are assumed to comprise the bitumen portion of the mix as their purpose is to initiate healing and rejuvenate the bitumen.

2.1.2. Capsule Design

The HealRoad capsules were prepared using a drop process [16,22] from an emulsion of rejuvenator suspended in a water solution of sodium alginate. To this aim a 6 wt % solution of sodium alginate in deionized water was prepared. At the same time, a 2.5 wt % poly(ethylene-alt-maleic-anhydride) (PEMA) polymeric surfactant solution was prepared by dissolving the copolymer in water at 70 °C and mixing it for 60 min. After the PEMA was dissolved in water, it was allowed to cool to room temperature (20 ± 2 °C) and was then combined with the rejuvenator. For this study, a vegetable oil of 0.9 g/cm^3 density at room temperature (20 ± 3 °C) was used, forming a bitumen healing agent solution, in a PEMA/rejuvenator with a 1/1.5 proportion by weight. The sodium alginate solution was mixed with iron powder (40 µm particle size), at 700 rpm for 1 h to allow the uniform dispersal of iron particles within the alginate mix. Following the initial investigation [3], an Alginate (Alg):iron powder (Fe) mix ratio was prepared in proportion by weight of dry constituents ratios: 20:80. After the sodium alginate and iron solution was fully mixed, the PEMA and rejuvenator (oil) solution was added to the alginate—iron powder solution mix in a ratio of 70% rejuvenator:30% sodium alginate [23]. The full capsule solution (20 L) was then mixed at 700 rpm for 20 min. More detail on capsule production can be found elsewhere [3]. All chemicals used in the production of the HealRoad capsules were purchased from the Merck Group—Sigma Aldrich, Wicklow, Ireland, except for the rejuvenator (vegetable oil) which was purchased from a local supermarket.

Table 1. Mix constituents proportions.

Mix Constituent	Mix Constituent Proportions					
Aggregates Bitumen Capsules	94.2 5.8 0	93.9 5.8 0.31	93.8 5.8 0.44	93.6 5.8 0.64	93.2 5.8 1.02	92.8 5.8 1.45
Total % of constituents in the mix	100	100	100	100	100	100
Capsules + bitumen percentage in the mix	5.80	6.11	6.24	6.44	6.82	7.25
Capsule percentage in capsule + bitumen mix proportion	0	5	7	10	15	20

2.1.3. Test Specimen Compaction

Cylindrical test specimens were compacted following EN 12697-31:2019. A Coopers Gyratory CRT-GYR-EN was used to compact the test specimens. Test specimens were compacted in dimensions of 100 mm diameter and 70 mm height. After 12 h of curing, specimens were cut, using a masonry saw to a diameter of 100 mm and a height of 32 mm.

2.2. Methods

The 10 mm SMA mix containing varying percentages of capsules were evaluated following the TII Series 900 Asphalt mix design. The aim was to optimize the 10 mm SMA mix containing HealRoad capsules, i.e., to determine the optimum capsule content for the 10 mm SMA mix.

2.2.1. Binder Drainage Test

The binder drainage test was conducted in accordance with EN12697-18 using the Schellenberg method. A sample of mixed 10 mm SMA material was placed in the glass jar and kept at 180 °C for 1 h. The contents of the jar were then emptied out and the jar reweighed to calculate the amount of binder remaining on the inside of the glass jar.

The principle behind the binder drainage test is to quantify the amount of material lost by drainage, i.e., the material that has adhered to the truck or mixer at the plant. Therefore, it is important to verify what effect the addition of capsules would have on the binder drainage of the 10 mm SMA mix. The equation employed for calculating the binder drainage is as follows:

$$BD = 100 \times \frac{[W5 - W3 - W6]}{[W4 - W3]} \tag{1}$$

where: BD = the drained material (%); $W3$ = mass of the empty beaker (g); $W4$ = mass of the beaker plus batch (g); $W5$ = mass of the beaker plus retained material after upturning (g); $W6$ = mass of the dried residue retained on the sieve (g).

2.2.2. Water Sensitivity Test—Indirect Tensile Strength Ratio (ITSR)

An indirect tensile strength ratio (ITSR) was used to evaluate the asphalt mix resistance to the moisture damage. The ITSR test was performed in accordance with European standard EN 12697-12. For each mix, six specimens were prepared. The specimens were then divided into two subsets of three. One set was stored in the laboratory at room temperature, 20 ± 3 °C, and the second set was placed under distilled water and subjected to a vacuum of 6.7 kPa for 30 min. After conditioning in vacuum, the second set was placed into a water bath at 40 °C for 72 h. Both sets of test specimens were then conditioned at a test temperature of 15 °C for two hours prior to testing. The dry set was conditioned in a temperature-controlled air chamber, and the wet set conditioned in a temperature-controlled water bath. A Controls 34T-107 compression testing machine with a Controls Digimax Plus (ver2.11-1) data acquisition instrument was employed to complete the indirect tensile strength test (ITST) in accordance with EN 12697-23.

2.2.3. Indirect Tensile Strength Test

The indirect tensile strength (ITS) test was conducted in accordance with EN 12697-23. The test specimens were conditioned in a temperature-controlled chamber at 15 °C for 2 h. The tests were conducted at 15 °C. The ITS test applies a vertical compressive strip load at a constant loading rate, in this case 50 mm/min, to the cylindrical specimen. The load is distributed over the thickness of the specimen through two loading strips at the top and bottom of the test specimen [24]. The specimens were loaded until the load value had fallen back to zero or the specimen had fully split into two. Use of the ITS test created two halves of a test specimen that could then be recombined and subjected to induction heating.

2.2.4. Wheel Tracking Test

The wheel tracking test is a test conducted on an asphalt material in order to establish the asphalt materials ability to resist the accumulation of permanent deformation under repeated loading, also known as rutting. The wheel tracking tests were conducted out in accordance with EN 12697-22. These tests were conducted on specimens oscillating under a wheel with a load of 700 N at a frequency of 26.5 cycles per minute for 10,000 cycles at the temperature of 45 °C. The specimen slabs had a depth of 40 mm and a length and width of 305 mm × 305 mm. The wheel tracking test was conducted on asphalt mixes containing 0%, 0.44% and 0.64% of HealRoad capsules.

2.2.5. Extrinsic Healing Regime

The aim of the healing test was to determine the optimum volume of HealRoad capsules in the mix in order to achieve full repair of the test specimen after the ITS test. The healing was performed using the induction heating system Abrell EKO 10/100C, PWR, CNTRL EKOHEAT® 10/100C, ES, with a solid-state induction power supply CE rated (input: WYE configured, 360–520 VAC, 50/60 Hz, 3-phase; output: 10 kW terminal, 50–150 kHz). A solenoid coil was used to apply the induction heating to the test specimen. Figure 2 shows the induction system set up. The heating (healing) was performed at 7.6 kW and 148 Hz induction machine energy output for a duration of 300 s (5 min). The temperature of the specimen during the healing period was measured using Eventek Infra Red Non-Contact Digital Laser Temperature gun with a temperature range between −50 °C and 420 °C.

The healing index (HI) was calculated employing the following rule [16,25]:

$$HI = \frac{ITS_h}{ITS_i} \times 100\ (\%) \tag{2}$$

where: ITS_i—initial ITS test result (MPa), ITS_h—ITS test result after healing (MPa).

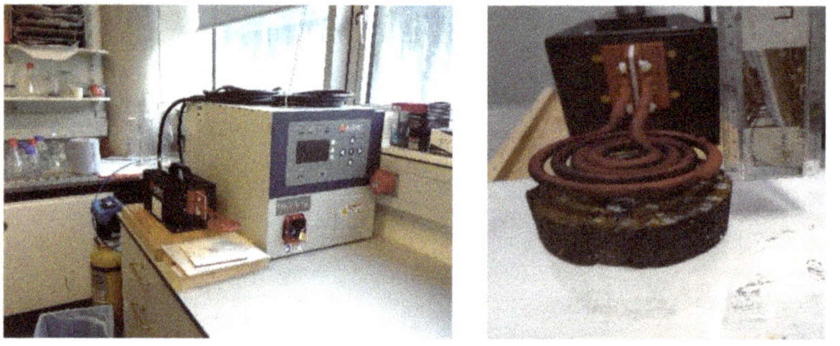

Figure 2. Ambrell Ekoheat 10 kW induction machine—laboratory induction apparatus.

3. Results

3.1. Binder Drainage Test

Table 2 summarizes the binder drainage test results of six 10 mm SMA mixes containing varying amounts of HealRoad capsules. The results show that the HealRoad capsules slightly increase binder drainage, but that they remain within acceptable limits. The mix containing 0.44% of capsules shows the same drainage level as the control mix (0% of capsules), whereas the mixes containing 0.64% and 1.45% of HealRoad capsules demonstrate an increased binder drainage level. This is probably due to the release of the rejuvenating oils caused by exposure to high temperatures of 180 °C.

Table 2. Binder Drainage.

Mix No.	Capsule Content in the Mix (%)	Capsule/Bitumen Ratio (%)	Mix Binder Drainage (%)
1	0	0	0.03
2	0.31	5	0.04
3	0.44	7	0.03
4	0.64	10	0.05
5	1.02	15	0.04
6	1.45	20	0.05

3.2. Indirect Tensile Strenght Test

Figure 3 shows the effect of the HealRoad capsule content in the 10 mm SMA mix. The results indicate that the mixes containing 0.44% and 0.64% of HealRoad capsules are the only mixes performing as well as the control mix. The results also indicate that increasing the capsule content above 1% results in a decreased ITS. These results indicate that in terms of 10 mm SMA mix strength performance, the optimal HealRoad capsule content is between 0.44% and 0.64%.

3.3. Water Sensitivity Test

Following the ITS test, it was decided to focus on mixtures containing 0.44% and 0.64% of HealRoad capsules to further investigate the effect on the 10 mm SMA mix. Figure 4 shows the indirect tensile strength ratio of the mixes. The results show that all mixes outperform the standard ITSR requirement of 80%. Furthermore, the results demonstrate that the HealRoad capsules have a positive effect on the water sensitivity of the 10 mm SMA mix. An increase of HealRoad capsule content in the mix, up to 0.64%, results in an increased ITSR. The ITSR value for both mixes containing HealRoad capsules is higher than 90%.

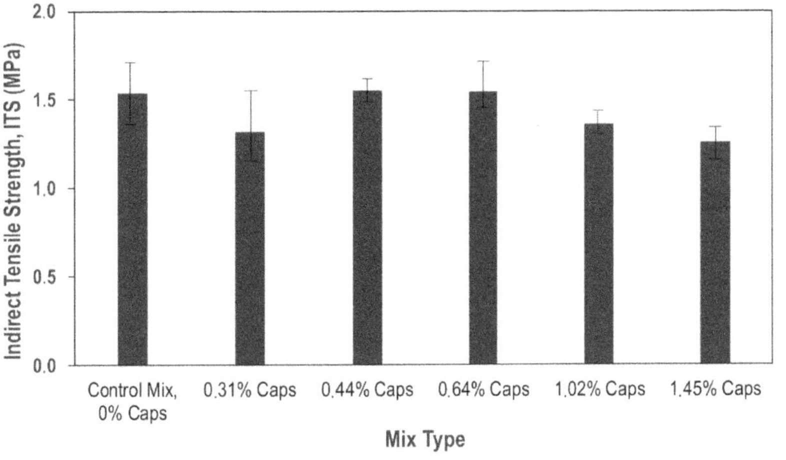

Figure 3. Indirect tensile strength test results.

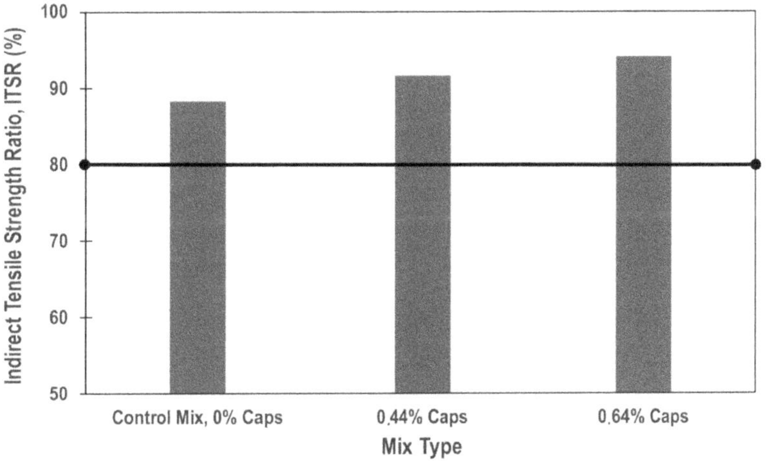

Figure 4. Water sensitivity test results—ITSR.

3.4. Wheel Tracking Test

The wheel tracking tests results are presented in Figure 5. The results show a very slight increase in rut depth with 0.4% of HealRoad capsules in the mix. The results also show that by increasing the HealRoad capsule content to 0.64%, the rut depth within the mix decreases by 15%, to below the control mix. These results indicate that the HealRoad capsules can improve the mix resistance to rut deformation.

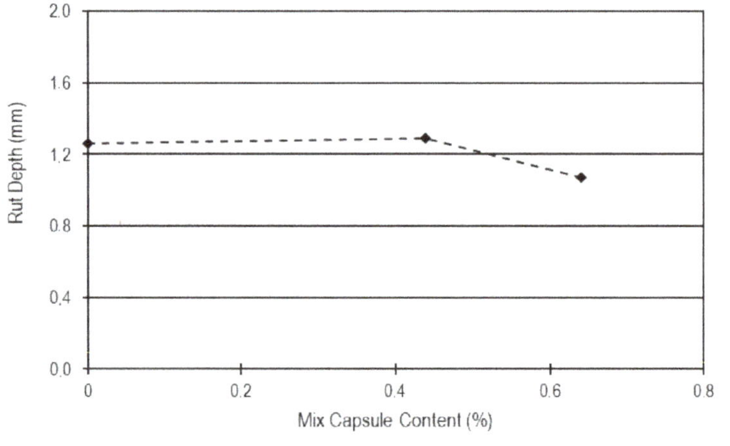

Figure 5. Wheel tracking results—resistance to rut deformation.

3.5. *Extrinsic Healing*

3.5.1. SMA Mix with Low Capsule Content

Following the ITS tests (see Section 3.2), cylindrical test specimens were subjcted to induction heating. The aim was to determine whether the HealRoad capsules within an asphalt mix can conduct induction energy and initiate asphalt healing. Unfortunately, none of the asphalt mixtures demonstrated healing capacity. This is because of the low volume and high dispersion rate of the HealRoad capsules within the mix, which resulted in poor conductivity of the induction energy. Figure 6 shows a cross section of test specimen containing the highest volume of HealRoad capsules—1.45%. The image shows that the HealRoad capsules are well embedded in the test specimen, however, such capsule dispersion causes the HealRoad capsules to have poor induction energy conductivity.

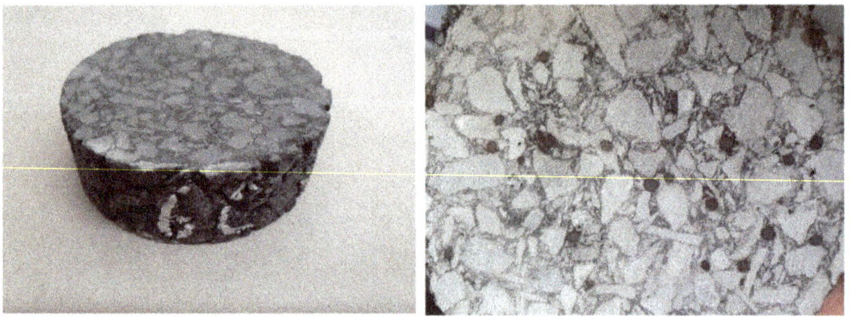

Figure 6. Image of 10 mm SMA—mix 6 containing 1.45% capsules in the mix.

3.5.2. High Capsule Content SMA Mix

In order to test the HealRoad capsules for induction heating in an asphalt mix, the amount of HealRoad capsules was increased to 5%, 10% and 20% (when capsule content is higher than 20%, the HealRoad capsules do not disperse evenly in the mix). The percentages of constituent material in each mix are summarized in Table 3. These mixtures were not investigated for standard mix performance, the aim was only to test whether the increased volume of HealRoad capsules had the capacity to initiate induction heating and self-heal. Figure 7 shows the images of 10 mm SMA mix containing varying percentages of HealRoad capsules, 0–20%. From the Figure 7c,d, it is evident that the mixes containing 10% and 20% of HealRoad capsules show evidence of corrosion. Another finding is that the mixes containing 20% of HealRoad capsules had fewer large aggregates in comparison to mixtures without HealRoad capsules. This could be an issue for the material strength as large angular aggregates, which determine the load carrying performance of a mix, are replaced by softer, round HealRoad capsules.

Table 3. The 10 mm SMA mix designs.

Aggregate Type	Constituent Content in the Mix (%)			
	Mix 1 (Control Mix)	Mix 2	Mix 3	Mix 4
14	4.02	4.01	3.79	3.34
10	47.21	44.60	42.10	37.11
6.3	7.55	7.14	6.74	5.92
Dust	32.10	30.33	28.63	25.23
Filler	3.30	3.12	2.95	2.60
Capsules	0.00	5.00	10.00	20.0
Binder		5.8		

(a) (b) (c) (d)

Figure 7. The 10 mm SMA mix containing varying amounts of capsules: (**a**) 0% capsules, (**b**) 5% capsules, (**c**) 10% capsules and (**d**) 20% capsules.

The specimens were subjected to the ITS test protocol described in Section 2.2.4 and test specimen healing was performed following the protocol described in Section 2.2.5. After the ITS test, the test specimens were tied with cable ties to join two test specimen parts and to allow test specimens without HealRoad capsules to heal. Figure 8 shows the tied test specimens.

Figure 8. The 10 mm SMA test specimen containing 5% capsules ready for induction healing.

The healing results show a positive response to induction heating with a test specimen temperature increase. As expected, test specimens containing higher amounts of HealRoad capsules were able to reach high temperatures, while test specimens containing 20% of HealRoad capsules reached a temperature of 80 °C. Figure 9 shows the temperature readings (measured using an infrared laser temperature gauge).

Figure 9. Temperature reading of the induction healing.

Figure 10 shows the test samples after the ITS test and before healing. The damage in the mixture containing 20% of HealRoad capsules was significant. We believe that this was caused by the high proportion of capsules in the mix, which replaced larger aggregates. Large aggregates (angular, flaky and elongated aggregates) in an asphalt mix form a load carrying skeleton which is responsible for pavement strength and for resistance to rutting [26]. Figure 11 shows the images of the same samples after healing. It is evident that mixtures with a high HealRoad capsule content—Mix 3 with 10% of capsules and Mix 4 with 20% of capsules—more efficiently repaired the damage, i.e., closed the cracks, in comparison to the control mix—Mix 1 containing 0% of capsules and Mix 2 containing 5% of capsules.

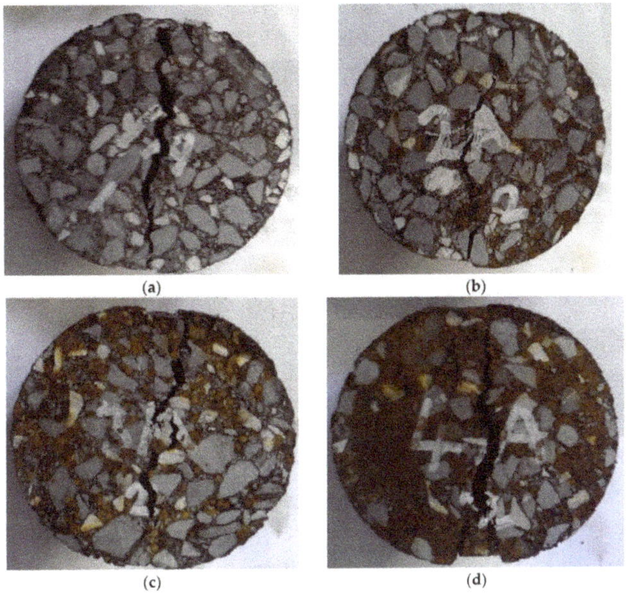

Figure 10. Images of test specimens after ITS test: (**a**) control mix—Mix 1, (**b**) Mix 2–5% capsules, (**c**) Mix 3–10% capsules and (**d**) Mix 4–20% capsules.

Figure 11. Images of test specimens after induction healing: (**a**) control mix—Mix 1, (**b**) Mix 2–5% capsules, (**c**) Mix 3–10% capsules and (**d**) Mix 4–20% capsules.

Figure 12 shows the initial ITS result for mixtures with a HealRoad capsule content of ≥5%. When compared with data from the initial tests, for mixes containing <1.45% of HealRoad capsules, the material strength was reduced as capsule content increased. In mixes containing 20% of HealRoad capsules, the ITS was reduced by 55%. There are two possible reasons for this:

1. The increased amounts of circular aggregates in the mix reduces the quantity of large aggregates, thereby reducing the interlocking strength of the material,
2. The release of the rejuvenator during the mixing stage results in the softening of the bitumen.

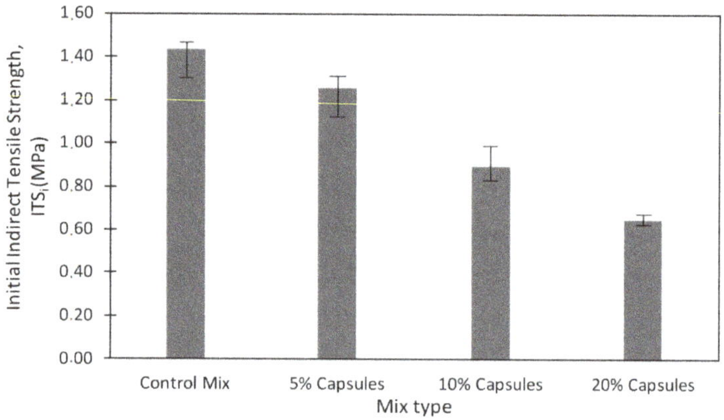

Figure 12. Initial ITS results.

Figure 13 illustrates the ITS results after healing. The results demonstrate that mixes containing HealRoad capsules did not achieve the expected healing. Figure 14 shows the healing index of each mix indicating that HealRoad capsules did not improve material strength recovery, whereas the control mix achieved up to 32.5% of strength recovery and the mix containing 20% of HealRoad capsules achieved up to 35.07% of strength recovery. However, looking at the initial material strength (Figure 12) and recovered material strength (Figure 13), it is evident that the control mix has the highest strength level both before and after healing. This indicates that there are no benefits to be accrued from including the HealRoad capsules in the asphalt mix.

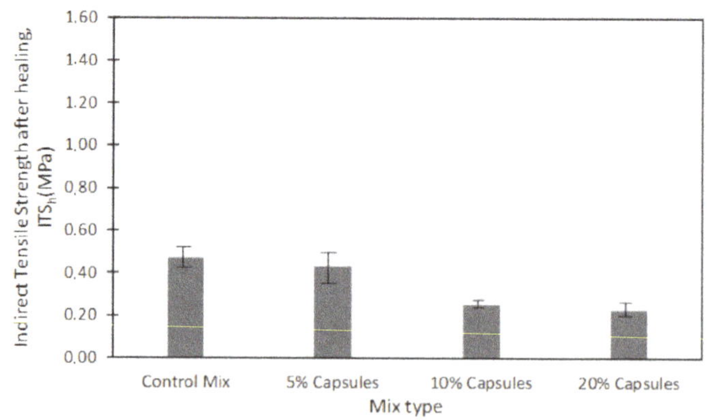

Figure 13. ITS results after healing.

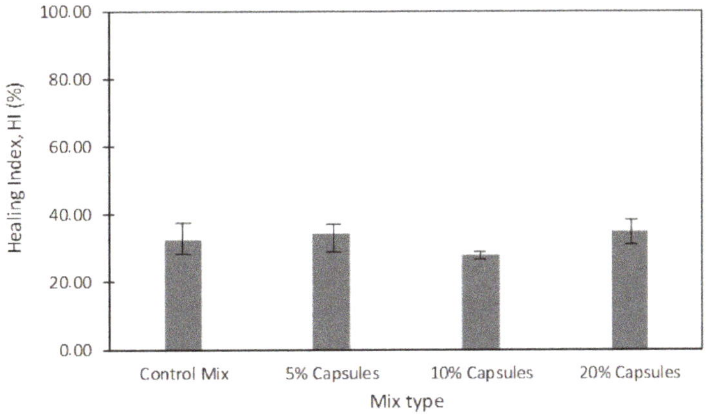

Figure 14. Healing index of the 10 mm SMA mix with and without capsules.

4. Conclusions

The asphalt mix performance test results indicate that mixes containing low amounts of HealRoad capsules (0.44% and 0.64%) have as good and better mix performance as the control mix in terms of strength, water sensitivity and rutting. Unfortunately, mixes containing a low amount of HealRoad capsules, 0–1.45%, do not conduct inductive energy and therefore do not initiate healing. This is due to the poor conductivity properties of iron powder and the high dispersion of the HealRoad capsules in the mix. However, increasing the HealRoad capsule content of the mix to >5% demonstrates increased temperatures when test specimens are subjected to the induction heating. The results show that mixtures containing 20% of HealRoad capsules can reach a temperature >80 °C and can recover up to 35% of its initial ITS. However, mixes without HealRoad capsules recovered 32.5% of their initial strength, without increased temperature. This indicates that although HealRoad capsules are capable of heating the mix and efficiently initiate self-healing, they do not assist the mix in recovering its material strength. Furthermore, increasing the HealRoad capsule content in the mix above 5% gradually decreases the mix strength, whereas the mix containing 20% of HealRoad capsules has a 55% lower initial ITS in comparison to the control mix, i.e., mix without capsules. These results show that HealRoad is not an efficient asphalt damage repair system. The results of our initial investigation [3], where HealRoad capsules were tested in pure bitumen and bitumen mortar mixes, indicated that HealRoad capsules may be suitable for use in asphalt mixtures with high bitumen and low aggregate content. Further studies will therefore focus on testing HealRoad capsules as a healing system in asphalt plug joint mixes. To enhance the conductivity of HealRoad capsules, the pure iron powder (Fe) will be replaced with magnetite (Fe_3O_4) as a conductive material. Wan et al. [4,21] have demonstrated that alginate capsules containing magnetite as a conductive material have good conductive properties. The assumption is that a more conductive material can improve energy conduction and perhaps also the healing efficiency of the system. Despite weak healing performance, HealRoad capsules may still have the potential to improve asphalt mix performance, paving the way for further improvement and development of the self-healing systems for asphalt pavement mixes.

5. Patents

An intellectual property application titled: "A Conductive Alginate Capsule Encapsulating a Healing Agent", was submitted by the TU Dublin and Amir Tabaković to the European Patent Office, international patent application number: PCT/EP2021/075254. The objective of the IP application is to ring-fence specific capsule formulation for potential production and licensees' particular applications.

Author Contributions: Conceptualization, A.T.; methodology, A.T. and D.O.; validation, A.T., C.F. and D.O.; formal analysis, A.T.; investigation, A.T.; resources, A.T.; data curation, A.T. and C.F.; writing—original draft preparation, A.T.; writing—review and editing, A.T., C.F. and D.O.; visualization, A.T.; project administration, A.T.; funding acquisition, A.T. All authors have read and agreed to the published version of the manuscript.

Funding: This research was funded by the Enterprise Ireland, grant number CF 20191063P.

Acknowledgments: The Authors would like to acknowledge Technological University Dublin (TU Dublin) for hosting the project between 2019 and 2021 and TU Dublin CREST research Centre for providing the space and equipment for production of the capsules.

Conflicts of Interest: The authors declare no conflict of interest. The funders had no role in the design of the study; in the collection, analyses, or interpretation of data; in the writing of the manuscript, or in the decision to publish the results.

References

1. Xu, S.; García, A.; Su, J.; Liu, Q.; Tabaković, A.; Schlangen, E. Self-Healing Asphalt Review: From Idea to Practice. *Adv. Mater. Interfaces* **2018**, *5*, 1800536. [CrossRef]
2. Xu, S.; Liu, X.; Tabaković, A.; Schlangen, E. A novel self-healing system: Towards a sustainable porous asphalt. *J. Clean. Prod.* **2020**, *259*, 120815. [CrossRef]
3. Tabaković, A.; Mohan, J.; Karač, A. Conductive Compartmented Capsules Encapsulating a Bitumen Rejuvenator. *Processes* **2021**, *9*, 1361. [CrossRef]
4. Wan, P.; Liu, Q.; Wu, S.; Zhao, Z.; Chen, S.; Zou, Y.; Rao, W.; Yu, X. A novel microwave induced oil release pattern of calcium alginate/ nano-Fe3O4 composite capsules for asphalt self-healing. *J. Clean. Prod.* **2021**, *297*, 126721. [CrossRef]
5. Seneviratne, S.; Tabaković, A. Self-healing Asphalt HealRoad Customer Discovery and Needs Analysis Report. 2020; Unpublished.
6. United Kingdom Government. Road Accidents and Safety Statistics. 2021. Available online: https://www.gov.uk/government/collections/road-accidents-and-safety-statistics (accessed on 18 February 2022).
7. National Work Zone Safety. Work Zone Fatal Crashes and Fatalities. 2021. Available online: https://www.workzonesafety.org/crash-information/work-zone-fatal-crashes-fatalities/#national (accessed on 18 February 2022).
8. Butt, A.A.; Birgisson, B.; Kringos, N. Optimizing the Highway Lifetime by Improving the Self Healing Capacity of Asphalt. *Procedia-Soc. Behav. Sci.* **2012**, *48*, 2190–2200. [CrossRef]
9. García, Á.; Schlangen, E.; van de Ven, M.; Liu, Q. Electrical conductivity of asphalt mortar containing conductive fibers and fillers. *Constr. Build. Mater.* **2009**, *23*, 3175–3181. [CrossRef]
10. García, Á.; Schlangen, E.; van de Ven, M.; Liu, Q. A simple model to define induction heating in asphalt mastic. *Constr. Build. Mater.* **2012**, *31*, 38–46. [CrossRef]
11. Liu, Q. Induction healing of porous asphalt concrete. In *Faculty of Civil Engineering and Geosciences*; TU: Delft, The Netherlands, 2012.
12. Bueno, M.; Arraigada, M.; Partl, M.N. Damage detection and artificial healing of asphalt concrete after trafficking with a load simulator. *Mech. Time-Dependent Mater.* **2016**, *20*, 265–279. [CrossRef]
13. Norambuena-Contreras, J.; Garcia, A. Self-healing of asphalt mixture by microwave and induction heating. *Mater. Des.* **2016**, *106*, 404–414. [CrossRef]
14. Norambuena-Contreras, J.; Serpell, R.; Vidal, G.V.; Gonzalez, A.; Schlangen, E. Effect of fibres addition on the physical and mechanical properties of asphalt mixtures with crack-healing purposes by microwave radiation. *Constr. Build. Mater.* **2016**, *127*, 369–382. [CrossRef]
15. Tabaković, A.; Post, W.; Cantero, D.; Copuroglu, O.; Garcia, S.; Schlangen, E. The reinforcement and healing of asphalt mastic mixtures by rejuvenator encapsulation in alginate compartmented fibres. *Smart Mater. Struct.* **2016**, *25*, 084003. [CrossRef]
16. Xu, S.; Tabaković, A.; Liu, X.; Schlangen, E. Calcium alginate capsules encapsulating rejuvenator as healing system for asphalt mastic. *Constr. Build. Mater.* **2018**, *169*, 379–387. [CrossRef]
17. Gonzalez-Torre, I.; Norambuena-Contreras, J. Recent advances on self-healing of bituminous materials by the action of encapsulated rejuvenators. *Constr. Build. Mater.* **2020**, *258*, 119568. [CrossRef]
18. Tabaković, A.; Schuyffel, L.; Karač, A.; Schlangen, E. An Evaluation of the Efficiency of Compartmented Alginate Fibres Encapsulating a Rejuvenator as an Asphalt Pavement Healing System. *Appl. Sci.* **2017**, *7*, 647. [CrossRef]
19. Tabaković, A.; Schlangen, E. Self-Healing Technology for Asphalt Pavements. In *Self-Healing Materials*; Hager, M.D., van der Zwaag, S., Schubert, U.S., Eds.; Springer International Publishing: Switzerland, Cham, 2016; pp. 285–306.
20. Xu, S.; Liu, X.; Tabaković, A.; Schlangen, E. The influence of asphalt ageing on induction healing effect on porous asphalt concrete. *RILEM Tech. Lett.* **2018**, *3*, 98–103. [CrossRef]
21. Wan, P.; Wu, S.; Liu, Q.; Xu, H.; Wang, H.; Peng, Z.; Rao, W.; Zou, Y.; Zhao, Z.; Chen, S. Self-healing properties of asphalt concrete containing responsive calcium alginate/nano-Fe3O4 composite capsules via microwave irradiation. *Constr. Build. Mater.* **2021**, *310*, 125258. [CrossRef]

22. Zemskov, S.V.; Jonkers, H.M.; Vermolen, F.J. Two analytical models for the probability characteristics of a crack hitting encapsulated particles: Application to self-healing materials. *Comput. Mater. Sci.* **2011**, *50*, 3323–3333. [CrossRef]
23. Tabaković, A.; Braak, D.; van Gerwen, M.; Copuroglu, O.; Post, W.; Garcia, S.J.; Schlangen, E. The compartmented alginate fibres optimisation for bitumen rejuvenator encapsulation. *J. Traffic Transp. Eng.* **2017**, *4*, 347–359. [CrossRef]
24. Tabaković, A.; Karač, A.; Ivanković, A.; Gibney, A.; McNally, C.; Gilchrist, M.D. Modelling the quasi-static behaviour of bituminous material using a cohesive zone model. *Eng. Fract. Mech.* **2010**, *77*, 2403–2418. [CrossRef]
25. Xu, S.; Liu, X.; Tabaković, A.; Lin, P.; Zhang, Y.; Nahar, S.; Lommerts, B.J.; Schlangen, E. The role of rejuvenators in embedded damage healing for asphalt pavement. *Mater. Des.* **2021**, *202*, 109564. [CrossRef]
26. Fladvad, M.; Arnhild, U. Large-size aggregates for road construction—a review of standard specifications and test methods. *Bull. Eng. Geol. Environ.* **2021**, *80*, 8847–8859. [CrossRef]

applied sciences

Article

Characteristics of Warm Mix Asphalt Incorporating Coarse Steel Slag Aggregates

Adham Mohammed Alnadish [1,*], Mohamad Yusri Aman [1], Herda Yati Binti Katman [2,*] and Mohd Rasdan Ibrahim [3]

1 Department of Civil Engineering and Built Environment, University Tun Hussein Onn Malaysia, Parit Raja 86400, Malaysia; mdyusri@uthm.edu.my
2 Department of Civil Engineering, University Tenaga Nasional, Kajang 43000, Malaysia
3 Department of Civil Engineering, University Malaya, Kuala Lumpur 50603, Malaysia; rasdan@um.edu.my
* Correspondence: adhmalnadish@gmail.com (A.M.A.); herda@uniten.edu.my (H.Y.B.K.); Tel.: +60-8-921-2020 (ext. 2252) (H.Y.B.K.)

Abstract: The major goal of sustainable practices is to preserve raw resources through the utilization of waste materials as an alternative to natural resources. Decreasing the temperature required to produce asphalt mixes contributes to environmental sustainability by reducing energy consumption and toxic emissions. In this study, warm mix asphalt incorporating coarse steel slag aggregates was investigated. Warm mix asphalt was produced at different temperatures lower than the control asphalt mixes (hot mix asphalt) by 10, 20, and 30 °C. The performances of the control and warm mix asphalt were assessed through laboratory tests examining stiffness modulus, dynamic creep, and moisture sensitivity. Furthermore, a response surface methodology (RSM) was conducted by means of DESIGN EXPERT 11 to develop prediction models for the performance of warm mix asphalt. The findings of this study illustrate that producing warm mix asphalt at a temperature 10 °C lower than that of hot mix asphalt exhibited the best results, compared to the other mixes. Additionally, the warm mix asphalt produced at 30 °C lower than the hot mix asphalt exhibited comparable performance to the hot mix asphalt. However, as the production temperature increases, the performance of the warm mix asphalt improves.

Keywords: steel slag aggregate; warm mix asphalt; RSM; stiffness modulus; dynamic creep; moisture sensitivity

Citation: Alnadish, A.M.; Aman, M.Y.; Katman, H.Y.B.; Ibrahim, M.R. Characteristics of Warm Mix Asphalt Incorporating Coarse Steel Slag Aggregates. *Appl. Sci.* **2021**, *11*, 3708. https://doi.org/10.3390/app11083708

Academic Editor: Amir Tabakovic

Received: 3 March 2021
Accepted: 31 March 2021
Published: 20 April 2021

1. Introduction

In most cases, hot mix asphalt (HMA) is manufactured at a temperature ranging from 150 °C to 180 °C. A high production temperature is necessary to dry the aggregates, coat them with bitumen, and achieve the desired workability. However, carbon dioxide emissions associated with the production of HMA have become a serious concern due to the negative effects of these emissions on the environment. Accordingly, warm mix asphalt (WMA) has recently acquired salability because of its effective ability to reduce toxic emissions and contribute to the preservation of the environment. The main function of WMA is to produce asphalt mixes at temperatures ranging from 20 °C to 40 °C lower than HMA, at which time the materials are blended, compacted, and placed on roads. The decreased production temperature is attributed to warm-mix asphalt technologies such as chemical additives, organic additives, and foaming that reduces binder viscosity [1,2]. Also, WMA decreases the energy required to produce asphalt mixes and provides a less harmful environment for workers and the areas surrounding asphalt mixing plants because its production process features reduced carcinogenic polycyclic aromatic hydrocarbon (PAH) emissions [3]. The production of WMA may decrease greenhouse emissions and energy consumption up to 40%, compared to the conventional asphalt mix [4]. Young [5] concluded that the fuel consumption is decreased by 2–3% for every 6 °C reduction to the production temperature. Similarly, Olard et al. [6] reported that for every 10 °C decrease in mixing temperature, energy and CO2 are reduced by 5.5%. Furthermore, WMA technologies have minimized the influence of short- and long-term aging on the performance of the asphalt mix in comparison to HMA, due to the reduced oxidation of binders modified by WMA technologies [7,8].

Industrial waste and by-product materials are utilized in pavement applications as an alternative for natural aggregates to contribute toward environmental sustainability. One of the common by-products used in pavement applications is electric arc furnace steel slag aggregate. The common use of steel slag aggregate is attributed to its superior mechanical properties. Many studies have investigated the suitability of utilizing steel slag aggregate in asphalt mixes. Behnood and Ameri [9] assessed the performance of the stone mastic asphalt (SMA) containing steel slag aggregate as a coarse portion. The findings of the study illustrated that utilizing steel slag aggregates in stone mastic asphalt improved the resilient modulus, Marshal stability, and tensile strength. Ziari et al. [10] noted that hot mix asphalt containing steel slag aggregate as a coarse portion showed higher Marshal stability, indirect tensile strength, and higher fatigue life in comparison to the control mix. These enhancements were attributed to the angularity, toughness, and roughness of the steel slag aggregate [11]. Similarly, Ahmedzade and Sengoz [12] conducted a study assessing hot mix asphalt performance incorporated coarse steel slag aggregates. The study concluded that the characteristics of asphalt mixes consisting of coarse steel slag aggregates exhibited higher cracking resistance, stiffness modulus, and moisture sensitivity than the control mixes. Chen and Wei [13] investigated the characteristics of hot mix asphalt incorporating basic oxygen furnace steel slag as a coarse aggregate. The outputs of the laboratory tests indicated that utilizing steel slag aggregate improved the rutting resistance and moisture damage resistance of the asphalt. Alnadish et al. [14,15] carried out several investigations on the performance of hot mix asphalt incorporating coarse steel slag aggregate. The findings of the studies showed that asphalt mixtures composed of steel slag aggregate as a coarse portion enhanced the stiffness modulus, rutting resistance, and cracking resistance, as well as it displayed slightly lower tensile strength ratio as compared to the control mix. Furthermore, limited studies have investigated WMA containing steel slag aggregates. Ameri et al. [16] observed that the WMA produced by Sasobit additive and composed of steel slag aggregates as a coarse portion has improved Marshal stability, resilient modulus, permanent deformation, and moisture sensitivity. Goli et al. [17] investigated the performance of WMA incorporating steel slag aggregates. The incorporation of additives Sasobit and Rediset at a dosage of 2% by weight of the bitumen was used to produce WMA. The findings of the study showed that the WMA composed coarse steel aggregate exhibited better performance than HMA and steel slag aggregate replacements in terms of permanent deformation, fatigue life, and moisture susceptibility. Similarly, Amelian et al. [18] concluded that utilizing coarse basic oxygen furnace steel slag aggregate in warm mix asphalt manifested better performance than asphalt mixtures incorporating conventional aggregates. Ziaee and Behnia [19] assessed WMA modified with 1.5% of Sasobit and coarse steel slag aggregates. The laboratory experiments of the study showed that WMA incorporating coarse steel slag aggregate was significantly improved in stiffness modulus, permanent deformation, and tensile strength, in comparison to HMA.

Permanent deformation and moisture damage are among the main drawbacks of warm mix asphalt, due to the fact that reducing the production temperature may cause inchoate drying of the aggregates, insufficient film thickness, and reduced hardening of the binder [20].

Few studies have evaluated warm mix asphalt at various production temperatures. Furthermore, all the previous studies have assessed the performance of warm mix asphalt incorporating steel slag aggregates using the additive Sasobit. Therefore, the objectives of this study are to investigate the characteristics of warm mix asphalt containing coarse steel slag aggregates, modified with LEADCAP. The warm mix asphalt was investigated at different mixing temperatures of, 150, 140, and 130 °C to introduce a better understanding and evaluation of the behavior of warm mix asphalt at different production temperatures. A response surface methodology by means of DESIGN EXPERT 11 [21] was conducted to develop prediction models for the performance of the warm mix asphalt as a contribution in terms of identifying the performances of the asphalt mixes at various production temperatures. The laboratory performance tests were stiffness modulus at the temperatures of 25 °C and 40 °C, dynamic stability, and moisture sensitivity. This study focused on the tests for permanent deformation and moisture sensitivity because warm mix asphalt is highly susceptible to permanent deformation and moisture damage.

2. Materials and Methods

2.1. Materials

The grade of the bitumen utilized to produce the asphalt mixes was 80/100, which is suitable for the wearing course. The characteristics of the base and modified bitumen are listed in Table 1. Also, an organic additive called LEADCAP was utilized to produce warm mix asphalt. LEADCAP is a wax-based component that consists of a crystal controller and an adhesion promoter to control the crystallization of the wax components at low temperatures and to improve the bonding between the aggregates and the binder [22]. LEADCAP is produced by KUMHO petrochemical (Seoul, South Korea). Figure 1 shows the LEADCAP additive. Electric arc furnace (EAF) steel slag was used as the coarse aggregate, while granite aggregates were utilized as the fine aggregates in the preparation of the asphalt mixes. The EAF steel slag was supplied by NCL chemicals Ltd. chemical products, Singapore. Table 2 displays the characteristics of the electric arc furnace steel slag and granite aggregates. The gradations of the aggregates are listed in Table 3.

Table 1. Characteristics of the base and modified binders.

Properties	Base Binder	Modified Binder	Test Basis
Grade of binder	80/100	80/100	-
Penetration grade (0.1 mm)	93	84	ASTM D5 [23]
Softening Point (°C)	45	47	ASTM D36 [24]
Viscosity @ 135 °C (cP)	487	398	-
Viscosity @ 165 °C (cP)	144	123	ASTM D4402 [25]
Ductility @ 25 °C (cm)	>100	>100	ASTM D113 [26]

Figure 1. LEADCAP additive.

Table 2. Characteristics of the aggregates.

Properties	Granite	EAF Steel Slag	Limits	Test Basis
Loss Angeles Abrasion (%)	22	17.80	≤25%	ASTM C131 [27]
Aggregate Crushing Value (%)	25	22.60	≤25%	IS: 2386 (Part IV) [28]
Density (g/cm^3)	2.63	3.22	N/A	ASTM C127 [29]
Absorption (%)	0.84	2.75	≤3%	ASTM C127
Elongated and Flat Particles (%)	8.40	3.90	≤10%	ASTM D4791 [30]
Coarse angularity (%)	84	95	≥80%	ASTM D5821 [31]
Content of free CaO (%)	-	1.17	≤4%	-

Table 3. Gradation of the EAF steel slag and granite aggregates.

Sieve Size (mm)	Passing (%)	Retained (%)
19	100	-
12.5	95	5
9.5	85	10
4.75	60	25
2.36	43	17
1.18	33	10
0.6	26	7
0.3	20	6
0.075	5	15
Pan	-	5

2.2. Preparation of the Samples

In this study, the base binder (80/100) was modified with the warm mix asphalt additive LEADCAP with the wet process, using a high-shear mixer at 1000 RPM until homogeneity was achieved (i.e., 10 min). LEADCAP was introduced to the base binder at the dosage recommended by the supplier (2% by weight of the bitumen). The hot mix asphalt (HMA) was produced at a mixing temperature of 160 °C and compacted at a temperature of 150 °C. The warm mix asphalt was manufactured at different mixing temperatures (i.e., 150, 140, and 130 °C), and compacted at the compaction temperatures of 140, 130, and 120 °C, respectively. The control asphalt mixes incorporating the granite and steel slag aggregates were labeled as HG160°C and HS160°C, while the produced WMA containing steel slag aggregate at the temperatures of 150, 140, and 130 °C were coded as WS150°C, WS140°C, and WS130°C, respectively. The asphalt mixtures were produced in accordance with the specifications stated in Superpave mix design (SHRP-A-407) [32]. The aggregates were conditioned at the production temperature for at least four hours. Thereafter, the aggregates were blended with a binder using a 20 L auto mixer with a mixing speed of 75 RPM. The loose asphalt mixes were kept in the oven for at least two hours at the compaction temperature to simulate the short-term aging. Then, the conditioned asphalt mixes were compacted using a Superpave gyratory compactor (SGC, Controls Group, Milan, Italy) at 100 revolutions. The diameter and height of manufactured specimens were 100 ± 1 mm and 63 ± 2.5 mm, respectively. The optimum bitumen contents of the hot mix asphalt incorporating the granite and the coarse steel slag aggregate asphalt were 4.78% and 4.9%, respectively, while the optimum bitumen content of the warm mix asphalt was 4.8%. The slight decrease in the optimum bitumen content of the warm mix asphalt was caused by the LEADCAP additive, which decreased the viscosity of the bitumen due to the wax component. The densities of the asphalt mixes containing granite, steel slag aggregate, and LEADCAP were 2.343, 2.56, 2.564 g/cm^3, respectively.

2.3. Laboratory Tests

The stiffness modulus (resilient modulus) test was carried out to investigate the stiffness of the asphalt mixes. The stiffness modulus of the asphalt mixes was used to describe the characteristics of the asphalt mixes. A universal testing machine (UTM-5P)

(Inopave Group, Singapore) was utilized to conduct the stiffness modulus test. The stiffness modulus test was carried out in accordance with ASTM D7369 [33]. The test was conducted at two testing temperatures, i.e., 25 °C and 40 °C. However, a total of 15 samples with air void contents of 4 ± 1% were manufactured for this test. The specimens were conditioned in the environmental chamber for four hours at every testing temperature prior to the test. Thereafter, a load of 1000 N was applied to the specimens. The duration of the applied load with the haversine wave pulse was 0.1 s, while the rest period was 0.9 s.

To assess the permanent deformation of the asphalt mixes, the dynamic creep test was performed. The test was conducted using a universal testing machine (UTM-5P). The procedures of the dynamic stability test were conducted in accordance with BS DD 226 [34]. To accomplish this test, a total of 30 specimens with air void contents of 4 ± 1% were produced. The specimens were kept in the environmental chamber for four hours at the testing temperature of 40 °C. Thereafter, a stress of 10 kPa was applied for 120 s as a pre-load. Then, the specimens were subjected to different compressive stress i.e., 100 kPa and 200 kPa, for 1 h. The stress was applied with a square wave pulse consisting of 1 s for loading and 1 s for the rest period. The total number of load applications was 1800 cycles.

The test of moisture sensitivity was carried out to investigate the resistance to moisture damage of the asphalt mixes. The moisture sensitivity test was conducted according to the procedures stated in AASHTO T 283 [35]. To perform this test, 30 samples were produced. The air void content of the produced specimens was targeted at 7 ± 0.5%. Six specimens were manufactured for every mix. Three represented the dry condition, while the rest corresponded to the wet condition. In the dry condition, the specimens were kept in the environmental chamber for at least two hours at a temperature of 25 °C. Then, the specimens were subjected to an indirect tensile strength. For the wet condition, the samples were exposed to a water saturation of 70–80%. Thereafter, the specimens were conditioned in a water bath for 24 h at a temperature of 60 °C. Then, the specimens were immersed in the water bath for at least 2 h at a temperature of 25 °C. Thereafter, every sample was subjected to the applied indirect tensile strength at a constant rate of 50 mm/minute. The tensile strength of the asphalt mix should be higher than 80%. Figure 2 shows the tests setup of the stiffness modulus, dynamic creep, and moisture sensitivity.

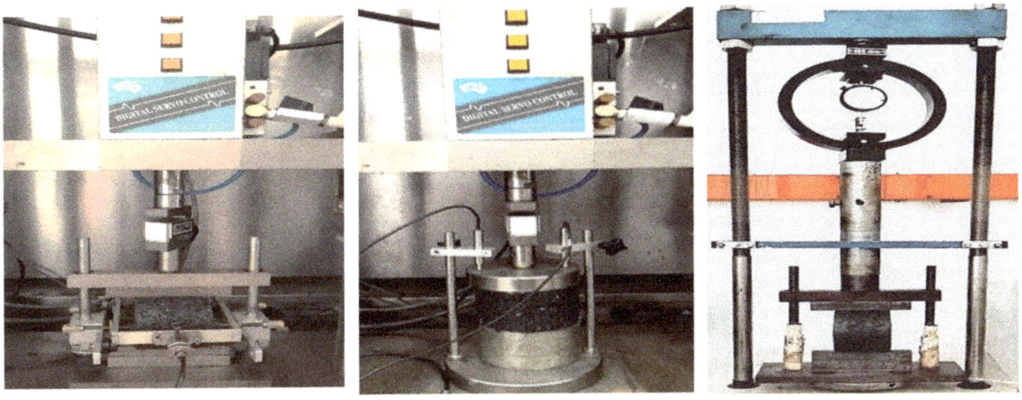

(a) (b) (c)

Figure 2. Laboratory tests: (**a**) stiffness modulus (**b**) dynamic creep; (**c**) Indirect tensile strength.

3. Results and Discussion

3.1. The Stiffness Modulus of the Asphalt Mixes

The stiffness modulus of the asphalt mixes at 25 °C and 40 °C are shown in Figures 3 and 4. As it is seen in the figures, the asphalt mixes composed of steel slag aggregate as the coarse portion exhibited better stiffness modulus results than the reference asphalt mix incorporating conventional aggregates. The higher stiffness modulus of HS160°C was caused by the superior toughness and angularity of the EAF steel slag aggregates. Furthermore, warm mix asphalt produced at 150 °C exhibited the highest stiffness modulus at 25 °C and 40 °C as compared to the other mixtures. The stiffness modulus of WS150°C at 25 °C was higher than that of the HS160°C by 18.6%, while at the 40 °C WS150°C showed a higher stiffness modulus than that of the HS160°C by about 11%. Also, the stiffness modulus of WS140°C at 25 °C and 40 °C were higher than that of the HS160°C by about 10% and 3%, respectively. The warm asphalt mix produced at the mixing temperature of 130 °C, which was lower than the control mix by 30 °C, showed similar stiffness modulus results to the HS160°C at the temperatures of 25 °C and 40 °C. Moreover, WMA produced at 150, 140, and 130 °C revealed better stiffness modulus than HG160°C. The higher stiffness modulus of the warm mix asphalt produced at the temperatures of 150 °C and 140 °C was attributed to the hardening and crystallization of the wax competent in the modified binder during the conditioning of the specimens to simulate short-term aging. As the production temperature of the warm mix asphalt increased, the bitumen hardening increased, and this, in turn, improved the stiffness of the asphalt mixes.

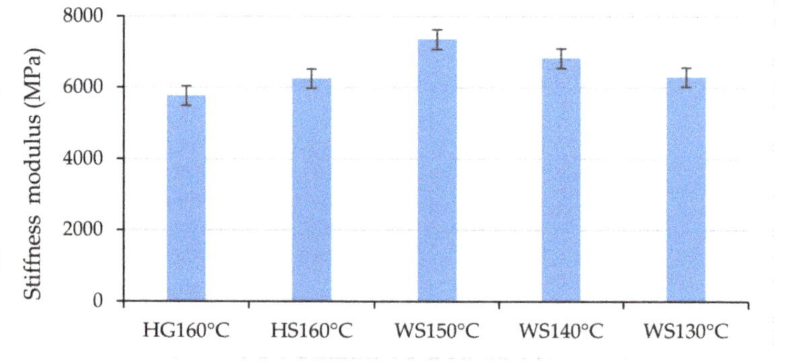

Figure 3. Stiffness modulus and the standard error bars of the mixes at 25 °C.

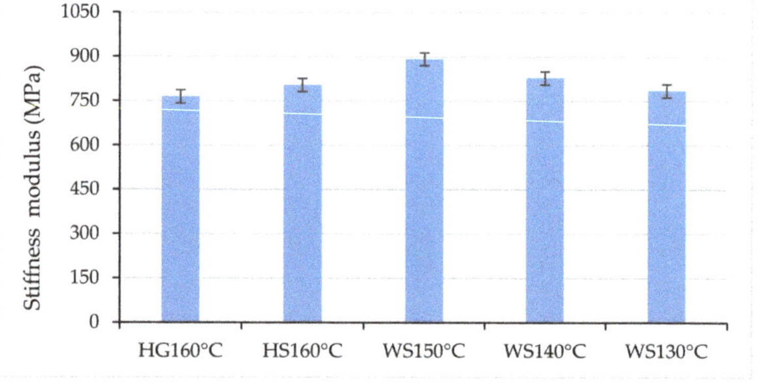

Figure 4. Stiffness modulus and the standard error bars of the mixes at 40 °C.

3.2. *Permanent Deformation of the Asphalt Mixes*

Figure 5 presents the results of the permanent deformation test at the applied stresses of 100 kPa and 200 kPa. It is clearly seen in Figure 5 that the higher accumulated strain indicates the lower resistance to permanent deformation of the asphalt mixes. As shown in Figure 5, HG160°C exhibited a slight decrease in permanent deformation resistance in comparison to HS160°C. On the other hand, the produced asphalt mix at the mixing temperature of 150 °C incorporated 2% of LEADCAP exhibited the lowest accumulated strain in comparison to the other mixes. The accumulated strain was lower than the HS160°C by about 25% and 20% for the applied stresses of 100 kPa and 200 kPa, respectively. Additionally, WS140°C exhibited better permanent deformation resistance than the HS160°C by about 15% and 9% for compressive stresses of 100 kPa and 200 kPa, respectively. The permanent deformation of the mixtures produced at the mixing temperature of 130 °C (WS130°C) was approximately comparable to that of the HS160°C mix. Furthermore, the warm mix asphalt produced at 150, 140, and 130 °C displayed lower accumulated strain than the reference mix incorporating natural aggregates (HG160°C). The better resistance to permanent deformation of the mixes incorporating the LEADCAP additive was caused by the crystallization of the wax component in the additive, which stiffened the binder [22].

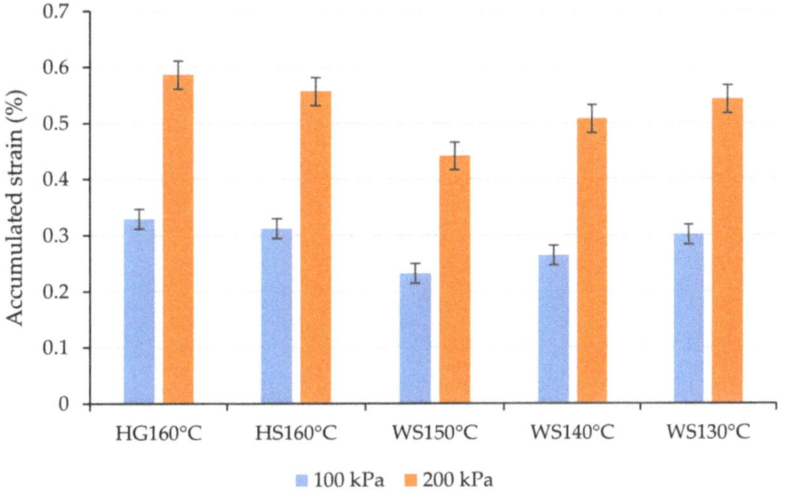

Figure 5. Accumulated strain and the standard error bars of the asphalt mixtures.

3.3. *Resisitance of the Aspahlt Mixes to Moisture Damage*

Moisture susceptibility is the most serious concern associated with producing warm mix asphalt. The damage of moisture sensitivity in WMA is attributed to the incomplete drying of the aggregates and the coating of the aggregate with bitumen due to the lowering of the production temperature. Modifying the binder with WMA additive may decrease the efficacy of the bonding between the binder and the surface of the aggregates. In this study, the efficiency of using LEADCAP in producing WMA was investigated through a test of moisture susceptibility. The conditioned and unconditioned indirect tensile strengths, as well as the ratio of the tensile strength of the asphalt mixes, are demonstrated in Figures 6 and 7. It is shown in Figure 7 that HG160°C demonstrated slightly higher resistance to moisture damage, compared to HS160°C. This was caused by the high porosity of the EAF steel slag aggregates. Additionally, it is seen in the figures that WS150°C displayed a higher tensile strength ratio (TSR) and indirect tensile strength (ITS) than the reference asphalt mixes. Production of the warm mix asphalt at 140 °C exhibited higher ITS and TSR than that of the HS160°C. This is because the LEADCAP additive contained an adhesion promoter. Decreasing the production temperature by 30 °C slightly decreased the indirect tensile strength and tensile strength ratios in comparison to the control mixes produced at 160 °C. As the manufacturing temperature of the warm mix asphalt increased, the tensile strength and ratio of tensile strength improved.

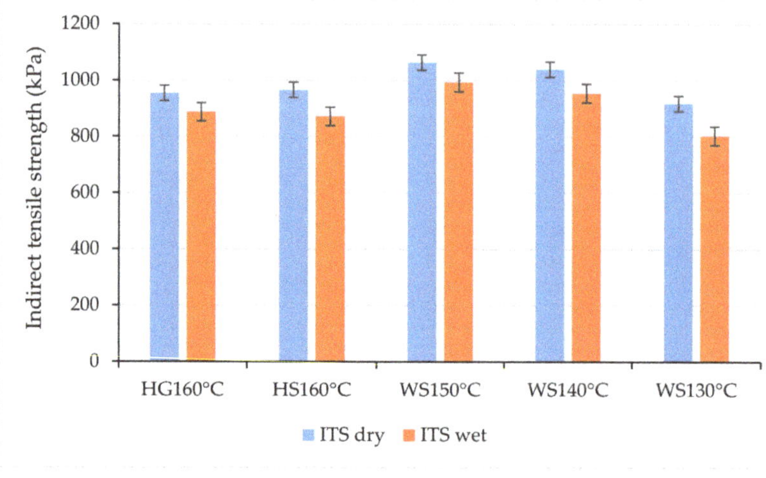

Figure 6. Indirect tensile strength and the standard error bars of the asphalt mixes.

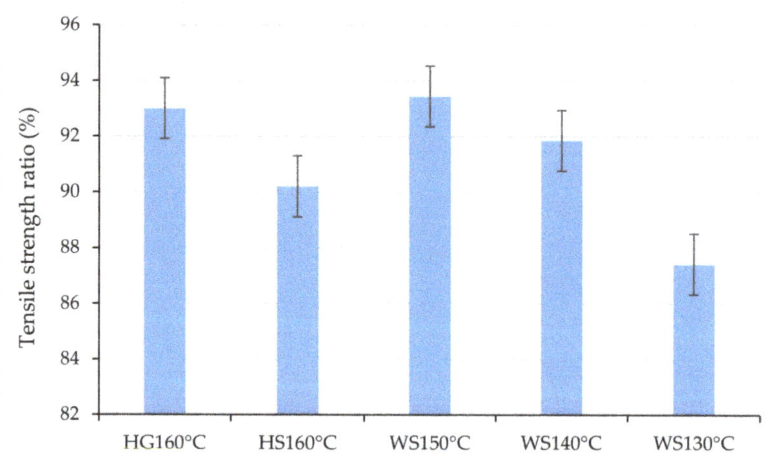

Figure 7. Tensile strength ratio and the standard error bars of the mixes.

4. Developing Prediction Models for the Performance of the Warm Mix Asphalt

Response surface methodology (RSM) was utilized, by means of DESIGN EXPERT 11, to develop prediction models for the stiffness modulus and accumulated strain of the warm mix asphalt incorporating LEADCAP and the coarse steel slag aggregate. The response surface methodology (RSM) approach is used to illustrate the relationship and correlations between different factors (variables) and responses through the development of a mathematical model. Hence, the responses are predicted based on the relationship and correlation between the factors and the responses. The response surface methodology approach offers different mathematical equations i.e., linear, 2FI, quadratic, cubic, fifth-, and sixth-order, to introduce the best equation to most accurately describe the correlations between the factors and the responses.

The prediction models were developed to study the effect of the production temperatures on the performance of WMA. This is because the WMA was produced at different production temperatures i.e., 150, 140, and 130 °C. The prediction models were developed for the stiffness modulus and accumulated strain because the results of these tests are based on the factors of the tests, such as the testing and production temperatures for the stiffness modulus test, while the factors of the dynamic creep test are the applied stress and the production temperature. However, the results of the moisture sensitivity test were not considered in developing the prediction model. This is because the tensile strength ratio (TSR) is determined through the dry indirect tensile strength and the wet indirect tensile strength. Therefore, the factors of the moisture sensitivity are the results of the test, and in this case, the model cannot be considered as a prediction model since the factors are the obtained results. Moreover, developing a prediction model for the moisture susceptibility test based on the factor of production temperature and the response of TSR may introduce an inappropriate model. This is because that the main factors that affect the TSR, the dry and wet indirect tensile strength, are not included.

4.1. Developing a Prediction Model for the Stiffness Modulus of Warm Mix Asphalt

Table 4 summarizes the input data of the response surface methodology to develop a prediction model for the stiffness modulus of the warm mix asphalt. Table 4 consists of two factors: the testing temperatures and the production temperatures, while the response is the results obtained from the stiffness modulus test. Table 5 presents the fit summary of the models. As reported in Table 5, the two-factor interaction model (2FI) adequately fits the data, since the lack of fit is insignificant and the coefficient of determination (R^2) is very strong. Lack of fit is used to describe the adequacy of the model based on the functional relationship between the independent and the dependent variables. The insignificant lack of fit (p-value > 0.05) indicates that the model fits the data well, while the significant lack of fit implies that the model does not accurately fit the data, which in turn requires the addition of new terms or a transformation of the data. Furthermore, the difference between the adjusted and predicted coefficient of determination (R^2) of the 2FI model was less than 0.2, which implies that the model fits the data. Table 6 shows the outputs of the ANOVA for the 2FI model. It can be seen that the model is significant since the p-value is less than 0.05. Also, the significant p-value of the factors indicates that the testing temperatures and production temperatures have a significant influence on the stiffness modulus. This reveals that there is a very strong agreement between the factors and the response. The 2FI model is described in Equation (1).

Table 4. Input data of the stiffness modulus model.

Testing Temperature (°C)	Production Temperature (°C)	Stiffness Modulus (MPa)
25.00	130.00	6079
25.00	130.00	6301
25.00	130.00	6507
40.00	130.00	825
40.00	130.00	811
40.00	130.00	715
25.00	140.00	6570
25.00	140.00	7010
25.00	140.00	6897
40.00	140.00	925
40.00	140.00	725
40.00	140.00	832
25.00	150.00	7105
25.00	150.00	7422
25.00	150.00	7542
40.00	150.00	905
40.00	150.00	815
40.00	150.00	955

Table 5. Fit summary of the models.

Source	Sequential *p*-Value	Lack of Fit *p*-Value	R^2	Adjusted R^2	Predicted R^2	Performance
Linear	<0.0001	0.0032	0.9938	0.9930	0.9906	
2FI	0.0001	0.9962	0.9979	0.9975	0.9966	Suggested
Quadratic	0.9481	0.9533	0.9979	0.9973	0.9961	Aliased

Table 6. ANOVA for the 2FI model.

Source	Sum of Squares	df	Mean Square	F-Value	*p*-Value	Performance
Model	1.633×10^8	3	5.442×10^7	2270.79	<0.0001	Significant
A-Testing temperature (°C)	5.475×10^7	1	5.475×10^7	2284.79	<0.0001	Significant
B-Production temperature (°C)	1.024×10^6	1	1.024×10^6	42.74	<0.0001	Significant
AB	6.807×10^5	1	6.807×10^5	28.40	0.0001	Significant
Residual	3.355×10^5	14	23,964.56			
Lack of Fit	213.78	2	106.89	0.0038	0.9962	Not significant
Pure Error	3.353×10^5	12	27,940.83			
Cor Total	1.636×10^8	17				

Stiffness Modulus (MPa) = −1727.11 + 45.13 × testing temperature (°C) + 132.42 × production temperature (°C) − 3.176 × testing temperature (°C) × production temperature °C (1).

Figure 8 displays the error distributions of the relationships between the testing temperatures, production temperatures, and stiffness modulus at 25 °C and 40 °C. The error distribution refers to normality. As seen in Figure 8, the residuals are close to the straight line, which indicates that the residuals are distributed normally. Also, the Anderson–Darling test was conducted to check the residual distribution. The outputs of the Anderson–Darling displayed test that the *p*-value was insignificant (*p*-value = 0.255 > 0.05), which indicates that the residuals are distributed normally. Figure 9 shows the correlation between the actual and predicted values. The correlation between the actual and predicted values was used to check the ability of the model to predict the dependent variables. It is seen in Figure 9 that the actual and predicted values were close to the 1:1 line, which indicates that the model is capable of predicting the stiffness modulus at 25 °C and 40 °C. Figure 10 illustrates the interaction between the factors and the responses. The function of this figure is to introduce a better understanding of the relationship between the testing temperatures, production temperatures, and stiffness modulus. As can be observed from Figure 10, there was a significant influence of the testing temperatures and production temperatures on the stiffness modulus. The higher the production temperature, the better the stiffness modulus was, while the higher the testing temperature, the lower the stiffness modulus.

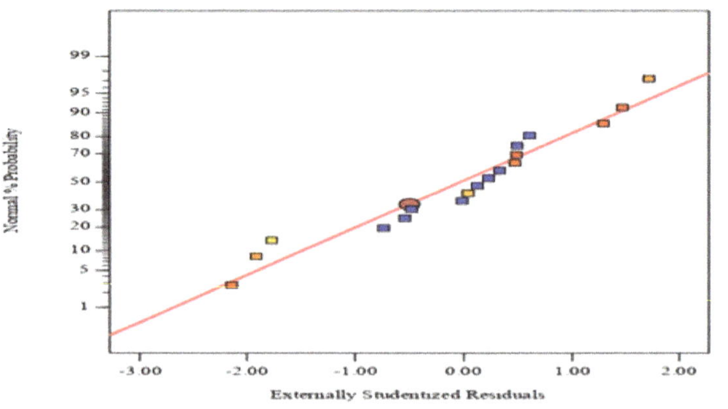

Figure 8. Plot of normality of 2FI model.

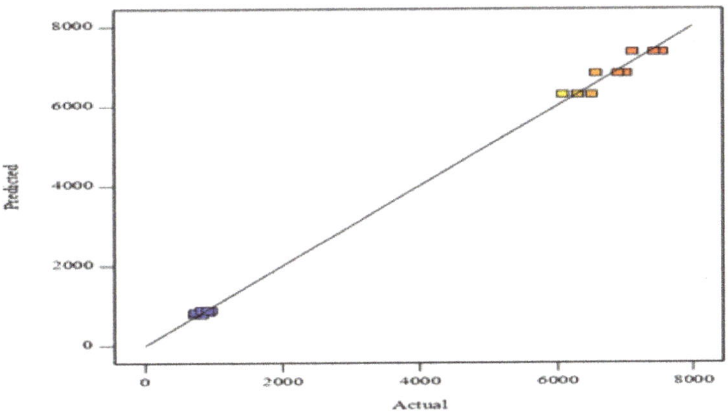

Figure 9. Plot of actual versus predicted values of the 2FI model.

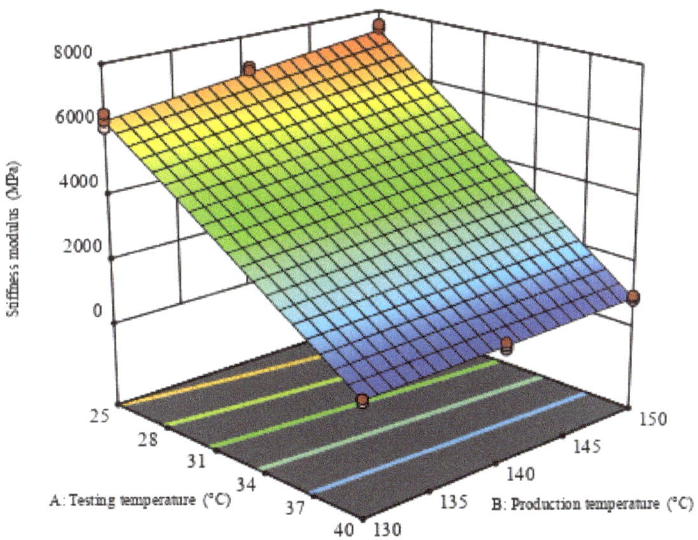

Figure 10. Plot of the interaction between the factors and the response of the 2FI model.

4.2. Developing a Prediction Model for the Accumlated Strain of the Warm Mix Asphalt

Table 7 displays the data input to DESIGN EXPERT. In Table 7, the factors represent the compressive stress and the production temperatures applied, while the response was the observed values of the accumulated strain at the applied loads of 100 kPa and 200 kPa. The fit summary of the models is summarized in Table 8. As reported in Table 8, a linear model is suggested to predict the accumulated strain of the warm mix asphalt. This is because the sequential *p*-value is significant, while the sequential *p*-value of the other models was not significant. Also, the lack of fit of the linear model is insignificant because the *p*-value is greater than 0.05. This, in turn, suggests that the model suits the data. The high coefficient of determination of the linear model implies that there is a very strong relationship between the compressive stress, production temperatures, and accumulated strain applied. In addition, the less difference between the adjusted R^2 and predicted R^2, which was less than 0.2, refers to the high agreement between the factors and the response.

Table 7. Input data of the accumulated strain model.

Compressive Stress (kPa)	Production Temperature (°C)	Accumulated Strain (%)
100.00	130.00	0.284
100.00	130.00	0.315
100.00	130.00	0.305
200.00	130.00	0.559
200.00	130.00	0.542
200.00	130.00	0.525
100.00	140.00	0.270
100.00	140.00	0.241
100.00	140.00	0.281
200.00	140.00	0.520
200.00	140.00	0.512
200.00	140.00	0.490
100.00	150.00	0.239
100.00	150.00	0.245
100.00	150.00	0.212
200.00	150.00	0.459
200.00	150.00	0.427
200.00	150.00	0.438

Table 8. Fit summary of the models.

Source	Sequential *p*-Value	Lack of Fit *p*-Value	R^2	Adjusted R^2	Predicted R^2	Performance
Linear	<0.0001	0.2923	0.9820	0.9796	0.9740	Suggested
2FI	0.1337	0.4513	0.9848	0.9815	0.9749	
Quadratic	0.4662	0.3082	0.9854	0.9809	0.9722	Aliased

Table 9 shows the ANOVA for the linear model. As seen in Table 9, the significant *p*-value of the factors implies that the applied compressive stress and production temperatures had a considerable effect on the accumulated strain of the warm mix asphalt. The linear model is expressed in Equation (2).

Table 9. ANOVA for the linear model.

Source	Sum of Squares	df	Mean Square	F-Value	*p*-Value	Performance
Model	0.2620	2	0.1310	409.13	<0.0001	Significant
A-Compressive stress (kPa)	0.2404	1	0.2404	750.57	<0.0001	Significant
B-Production temperature (°C)	0.0217	1	0.0217	67.69	<0.0001	Significant
Residual	0.0048	15	0.0003			
Lack of Fit	0.0012	3	0.0004	1.39	0.2923	Insignificant
Pure Error	0.0036	12	0.0003			
Cor Total	0.2668	17				

$$\text{Accumulated strain (\%)} = 0.629667 + 0.002311 \times \text{compressive stress (kPa)} - 0.004250 \times \text{production temperature °C} \quad (2).$$

The assumptions of the linear model should be investigated to examine its adequacy. Figure 11 illustrates the assumptions of the normality of the linear regression. The plot of normal probability displays the shape of the error distribution. The straight distribution of the errors indicates that linear regression fits the data well, whereas the other shapes such as right skew, left skew, S, or long and short tails, imply that the error distribution is not normal, and thus that linear regression is not suitable. As seen in Figure 11, the residuals are distributed normally, since the points are close to the line of the theoretical residual. In addition, the Anderson–Darling test was carried out to check the assumption of normality. The *p*-value of the Anderson–Darling test was insignificant (*p*-value = 0.3505 > 0.05), which indicates that the errors were normally distributed.

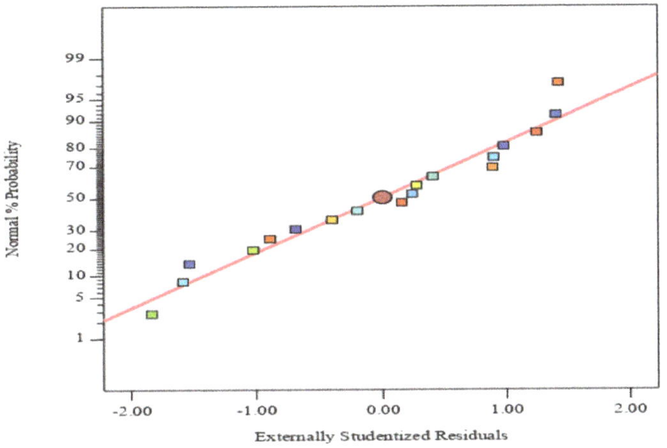

Figure 11. Plot of the normal probability of the linear model.

A plot of the externally studentized residuals versus the predicted values was used to assess the assumption of homoscedasticity (equal variance). In other words, linear regression assumes that the variance in the residual versus predicted value plot is constant, with a symmetric shape. Therefore, a symmetrically shaped distribution of errors around the zero in the scatter plot indicates homoscedasticity, while asymmetric shapes in the scatter plot are heteroscedastic. The asymmetric shapes in the scatter plot indicate that the linear model is not suitable. Figure 12 shows that the predicted values versus residual values are distributed around zero in a symmetric shape. Thus, the linear model was suitable to predict the accumulated strain of the warm mix asphalt.

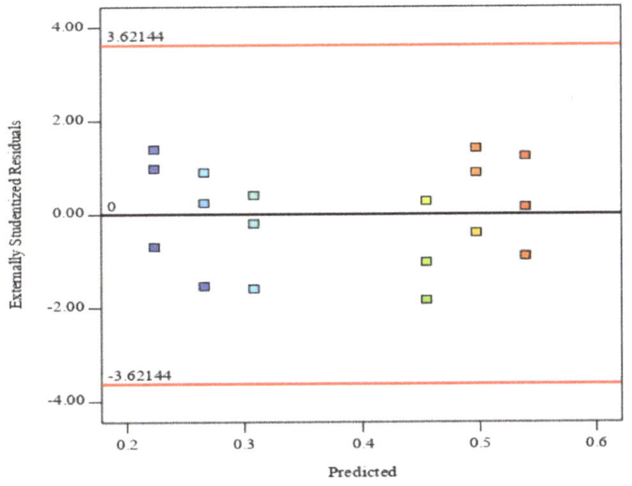

Figure 12. Plot of the externally residuals versus predicted values of the linear model.

To check the assumption of independence in the linear regression, a plot of the externally studentized residuals versus the run number is worthwhile. The residuals versus run number plot is used to detect autocorrelation between disturbances. Autocorrelation occurs when the residuals have a pattern where they remain positive or negative. As shown in Figure 13, there was no autocorrelation among the disturbances, since the residuals randomly transitioned between positive and negative values, which indicates that the dependent variable of accumulated strain was not independent.

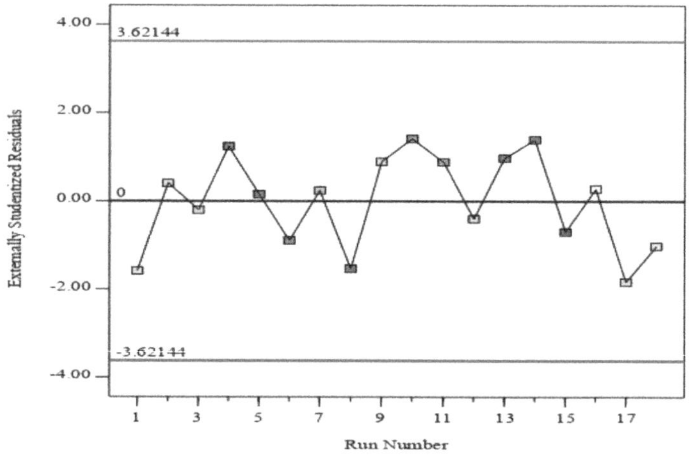

Figure 13. Plot of the externally residuals versus run number of the linear model.

The plot of the actual versus predicted values is influential to understand the linear relationship between the actual and the predicted values. Also, the assumption of linearity was checked through a plot of the relationship between the actual and the predicted values. As observed in Figure 14, there is a strong agreement between the actual and predicted values, since the data of the actual values and predicted values are close to the 1:1 line. The strong agreement indicates the capability of the linear model to predict the accumulated strain. The regression coefficient of the variance inflation factor (VIF) was utilized to check the assumption of multicollinearity in the models. The VIF is determined based on the variance of the predicted model divided by the variance of every factor. Multicollinearity occurs when the independent variables are highly correlated to each other, and thus a large change may occur in the model if one of the factors is removed or changed. To ensure that the multicollinearity assumption is met, the VIF value should not exceed 5. The variance inflation factor for the factors of the linear model was 1, which implies that the multicollinearity assumption is met. To introduce a better understanding of the effect of the applied compressive stress and production temperatures on the accumulated strain of the warm mix asphalt, Figure 15 is worthwhile. As indicated in Figure 15, the applied compressive stress and production temperatures have a great influence on the accumulated strain of the warm mix asphalt. As the applied load increases, the accumulated strain significantly increase. Also, the higher the production temperature, the better the permanent deformation resistance.

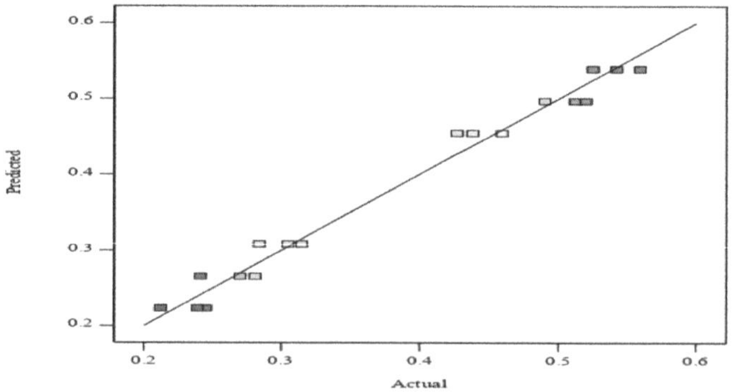

Figure 14. Plot of the actual versus predicted values of the linear model.

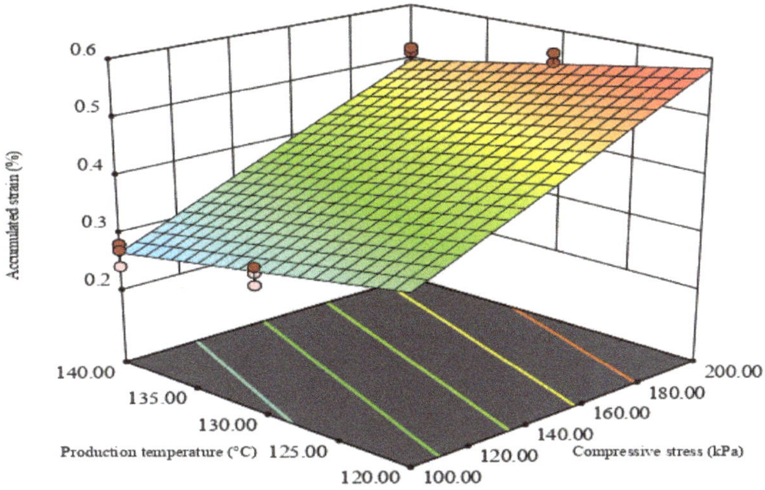

Figure 15. Plot of the interaction between the factors and the response of the linear model.

5. Conclusions

This study aimed to investigate the characteristics of warm mix asphalt incorporating coarse steel slag aggregate. The findings of the study illustrate that asphalt mixes incorporating EAF steel slag exhibited better characteristics than the asphalt mixes made with conventional mixes. The stiffness modulus of the produced warm mix asphalt at 150 °C was the best performing, as compared to the other production temperatures. The warm mix asphalt produced at 30 °C lower than the reference mix (HMA) exhibited a comparable stiffness modulus to the stiffness modulus of the reference mix incorporated steel slag. Also, the warm mix asphalt produced at a mixing temperature of 10 °C lower than the hot mix asphalt exhibited superior permanent deformation resistance than the other mixes. Moreover, decreasing the production temperature up to 30 °C less than the control mix resulted in a similar permanent deformation resistance to that of the control asphalt mix (HMA). The results of the moisture susceptibility experiment demonstrated that the resistance of the produced warm mix asphalt at 130 °C to moisture damage was slightly lower than that of the reference mixes, while the moisture damage resistance of the produced warm mix asphalt at 150 °C was the best. Furthermore, the produced warm mix asphalt at 150, 140, and 130 °C revealed higher stiffness modulus and permanent deformation resistance in comparison to the HMA incorporated natural aggregates. In general, as the mixing temperatures increased, the performance of the warm mix asphalt improved. However, producing the warm mix asphalt mixtures incorporating EAF coarse steel slag aggregate at a temperature of 130 °C is recommended since the results of the performance tests showed performance comparable to that of hot mix asphalt.

Author Contributions: Conceptualization, A.M.A.; methodology, A.M.A. and M.Y.A.; software, A.M.A.; validation, A.M.A., M.Y.A., H.Y.B.K., and M.R.I.; formal analysis, A.M.A.; investigation, A.M.A.; resources, A.M.A.; data curation, A.M.A., M.Y.A., H.Y.B.K., and M.R.I.; writing—original draft preparation, A.M.A., M.Y.A., H.Y.B.K., and M.R.I.; writing—review and editing, A.M.A., M.Y.A., H.Y.B.K., and M.R.I.; visualization, A.M.A.; supervision, M.Y.A.; funding acquisition, H.Y.B.K. All authors have read and agreed to the published version of the manuscript.

Funding: This research is supported by Universiti Tenaga Nasional Malaysia (BOLD 2020 RJO10517844/046).

Institutional Review Board Statement: Not applicable.

Informed Consent Statement: Not applicable.

Data Availability Statement: The presented data are available in the article.

Acknowledgments: The authors would like to express their gratitude for the provided support by Universiti Tun Hussien Onn Malaysia and Universiti Tenaga Nasional.

Conflicts of Interest: The authors declare no conflict of interest.

References

1. D'Angelo, J.A.; Harm, E.E.; Bartoszek, J.C.; Baumgardner, G.L.; Corrigan, M.R.; Cowsert, J.E.; Prowell, B.D. *Warmmix Asphalt: European Practice (No. FHWA-PL-08-007)*; Federal Highway Administration: Washington, DC, USA, 2008.
2. Prowell, B.D.; Hurley, G.C.; Frank, B. *Warm-Mix Asphalt: Best practices*; National Asphalt Pavement Association: Lanham, MD, USA, 2011.
3. Song, M.; Lee, K.; Oh, S.H.; Bae, M.S. Impact of Polycyclic Aromatic Hydrocarbons (PAHs) from an Asphalt Mix Plant in a Suburban Residential Area. *Appl. Sci.* **2020**, *10*, 4632. [CrossRef]
4. Cheraghian, G.; Falchetto, A.C.; You, Z.; Chen, S.; Kim, Y.S.; Westerhoff, J.; Wistuba, M.P. Warm mix asphalt technology: An up to date review. *J. Clean. Prod.* **2020**, *268*, 122128. [CrossRef]
5. Young, T.J. *Energy Conservation in Hot-Mix Asphalt Production*; National Asphalt Pavement Association: Lanham, MD, USA, 2007.
6. Olard, F.; Le Noan, C.; Beduneau, E.; Romier, A. Low energy asphalts for sustainable road construction. In Proceedings of the 4th Eurasphalt and Eurobitume Congress, Copenhagen, Denmark, 21–23 May 2008.
7. Li, B.; Wang, Y.; Ren, X.; Teng, X.; Su, X. Influence of ultraviolet aging on adhesion performance of warm mix asphalt based on the surface free energy theory. *Appl. Sci.* **2019**, *9*, 2046. [CrossRef]
8. Cheraghian, G.; Wistuba, M.P. Ultraviolet aging study on bitumen modified by a composite of clay and fumed silica nanoparticles. *Sci. Rep.* **2020**, *10*, 1–17. [CrossRef] [PubMed]
9. Behnood, A.; Ameri, M. Experimental investigation of stone matrix asphalt mixtures containing steel slag. *Sci. Iranica* **2012**, *19*, 1214–1219. [CrossRef]
10. Ziari, H.; Nowbakht, S.; Rezaei, S.; Mahboob, A. Laboratory investigation of fatigue characteristics of asphalt mixtures with steel slag aggregates. *Adv. Mater. Sci. Eng.* **2015**, *2015*. [CrossRef]
11. Alnadish, A.; Aman, Y. A study on the economic using of steel slag aggregate in asphalt mixtures reinforced by aramid fiber. *ARPN J. Eng. Appl. Sci.* **2018**, *13*, 276–292.
12. Ahmedzade, P.; Sengoz, B. Evaluation of steel slag coarse aggregate in hot mix asphalt concrete. *J. Hazard. Mater.* **2009**, *165*, 300–305. [CrossRef] [PubMed]
13. Chen, J.S.; Wei, S.H. Engineering properties and performance of asphalt mixtures incorporating steel slag. *Constr. Build. Mater.* **2016**, *128*, 148–153. [CrossRef]
14. Alnadish, A.M.; Aman, M.Y.; Katman, H.Y.B.; Ibrahim, M.R. Influence of the Long-Term Oven Aging on the Performance of the Reinforced Asphalt Mixtures. *Coatings* **2020**, *10*, 953. [CrossRef]
15. Alnadish, A.M.; Aman, M.Y.; Katman, H.Y.B.; Ibrahim, M.R. Laboratory assessment of the performance and elastic behavior of asphalt mixtures containing steel slag aggregate and synthetic fibers. *Int. J. Pavement Res. Technol.* **2021**, *14*, 473–481. [CrossRef]
16. Ameri, M.; Hesami, S.; Goli, H. Laboratory evaluation of warm mix asphalt mixtures containing electric arc furnace (EAF) steel slag. *Constr. Build. Mater.* **2013**, *49*, 611–617. [CrossRef]
17. Goli, H.; Hesami, S.; Ameri, M. Laboratory evaluation of damage behavior of warm mix asphalt containing steel slag aggregates. *J. Mater. Civil Eng.* **2017**, *29*, 04017009. [CrossRef]
18. Amelian, S.; Manian, M.; Abtahi, S.M.; Goli, A. Moisture sensitivity and mechanical performance assessment of warm mix asphalt containing by-product steel slag. *J. Cleaner Prod.* **2018**, *176*, 329–337. [CrossRef]
19. Ziaee, S.A.; Behnia, K. Evaluating the effect of electric arc furnace steel slag on dynamic and static mechanical behavior of warm mix asphalt mixtures. *J. Cleaner Prod.* **2020**, *274*, 123092. [CrossRef]
20. Rubio, M.C.; Martínez, G.; Baena, L.; Moreno, F. Warm mix asphalt: An overview. *J. Cleaner Prod.* **2012**, *24*, 76–84. [CrossRef]
21. Design-Expert. Available online: https://www.statease.com/docs/v11/ (accessed on 31 March 2021).
22. Kim, Y.; Lee, J.; Baek, C.; Yang, S.; Kwon, S.; Suh, Y. Performance evaluation of warm-and hot-mix asphalt mixtures based on laboratory and accelerated pavement tests. *Adv. Mater. Sci. Eng.* **2012**, *2012*. [CrossRef]
23. ASTM. *Standard Test Method for Penetration of Bituminous Materials*; American Society for Testing and Materials: West Conshohocken, PA, USA, 2013.
24. ASTM. *Standard Test Method for Softening Point of Bitumen (Ring-and-Ball Apparatus)*; American Society for Testing and Materials: West Conshohocken, PA, USA, 2014.
25. ASTM. *Standard Test Method for Viscosity Determination of Asphalt at Elevated Temperatures Using a Rotational Viscometer*; American Society for Testing and Materials: West Conshohocken, PA, USA, 2015.
26. ASTM. *Standard Test Method for Standard Test Method for Ductility of Bituminous Materials*; American Society for Testing and Materials: West Conshohocken, PA, USA, 2007.
27. ASTM. *Resistance to Abrasion of Small-Size Coarse Aggregate by Use of the Los Angeles Machine*; American Society for Testing and Materials: West Conshohocken, PA, USA, 1989.
28. IS. *Methods of Test for Aggregates*; Bureau of Indian Standards: Manek Bhawan, India, 1963.
29. ASTM. *Standard Test Method for Relative Density (Specific Gravity) and Absorption of Coarse Aggregate*; American Society for Testing and Materials: West Conshohocken, PA, USA, 2015.

30. ASTM. *Standard Test Method for Flat Particles, Elongated Particles, or Flat and Elongated Particles in Coarse Aggregate*; American Society for Testing and Materials: West Conshohocken, PA, USA, 2010.
31. ASTM. *Standard Test Method for Determining the Percentage of Fractured Particles in Coarse Aggregate*; American Society for Testing and Materials: West Conshohocken, PA, USA, 2013.
32. Cominsky, R.J.; Huber, G.A.; Kennedy, T.W.; Anderson, M. *The Superpave Mix Design Manual for New Construction and Overlays*; No. SHRP-A-407; Strategic Highway Research Program: Washington, DC, USA, 1994.
33. ASTM. *Standard Test Method for Determining the Resilient Modulus of Bituminous Mixtures by Indirect Tension Test*; American Society for Testing and Materials: West Conshohocken, PA, USA, 2011.
34. Standard, B. *Method for Determining Resistance to Permanent Deformation of Bituminous Mixtures Subject to Unconfined Dynamic Loading*; Draft for Development; British Standards Institution: London, UK, 1996; Volume 226.
35. Aashto, T. *Standard Method of Test for Resistance of Compacted Asphalt Mixtures to Moisture-Induced Damage*; American Association of State and Highway Transportation Officials: Washington, DC, USA, 2007.

applied sciences

MDPI

Article

Recycling Aged Asphalt Using Hard Asphalt Binder for Hot-Mixing Recycled Asphalt Mixture

Jian Zhou, Jing Li, Guoqiang Liu, Tao Yang and Yongli Zhao *

School of Transportation, Southeast University, Nanjing 210096, China; 230139228@seu.edu.cn (J.Z.); 230149195@seu.edu.cn (J.L.); guoqiangliu2016@seu.edu.cn (G.L.); ytaochd@126.com (T.Y.)
* Correspondence: yonglizhao2016@126.com

Abstract: Increasing the content of reclaimed asphalt pavement material (RAP) in hot-mix recycled asphalt mixture (RHMA) with a satisfactory performance has been a hot topic in recent years. In this study, the performances of Trinidad lake asphalt (TLA), virgin asphalt binder, and aged asphalt binder were first compared, and then the modification mechanism of TLA on virgin asphalt and aged asphalt was explored. Furthermore, the RHMA was designed in accordance with the French norm NF P 98-140 containing 50% and 100% RAP, and their high-temperature stability, low-temperature cracking resistance, and fatigue performances were tested to be compared with the conventional dense gradation AC-20 asphalt mixture. The results show that the addition of TLA changes the component proportion of virgin asphalt binder, but no new functional groups are produced. The hard asphalt binder modified by TLA has a better rutting resistance, while the fatigue and cracking resistance is lower, compared to both aged and virgin asphalt. The high-modulus design concept of RHMA is a promising way to increase the RAP content in RHMA with acceptable performance. Generally, the RHMA with 50% RAP has similar properties to AC-20. And, when the RAP content reaches 100%, the high- and low-temperature performance and anti-fatigue performance of RHMA are better than AC-20 mixture. Thus, recycling aged asphalt using hard asphalt binder for hot-mixing recycled asphalt mixture to increase the RAP content is feasible.

Keywords: Trinidad lake asphalt; aged asphalt; hot-mix recycled asphalt mixture; reclaimed asphalt pavement; high-modulus asphalt mixture

Citation: Zhou, J.; Li, J.; Liu, G.; Yang, T.; Zhao, Y. Recycling Aged Asphalt Using Hard Asphalt Binder for Hot-Mixing Recycled Asphalt Mixture. *Appl. Sci.* **2021**, *11*, 5698. https://doi.org/10.3390/app11125698

Academic Editor: Amir Tabakovic

Received: 24 April 2021
Accepted: 17 June 2021
Published: 19 June 2021

1. Introduction

Recycling the old asphalt pavement is a common operation used worldwide for saving natural materials, investment, and energy in highway maintenance projects [1–4]. In practice, the hot-mix recycled asphalt mixture (RHMA) has similar durability and performance to the conventional dense asphalt mixture if the RAP content is limited at 10–40%, depending on the design of mixture [5–7]. According to recent official data, China produces approximately 790 million tons of reclaimed asphalt pavement (RAP) per year through asphalt pavement rehabilitation projects [8]. This figure is expected to significantly increase in the next five years as the demand for road maintenance is increasing. Following the issuance of the most stringent environmental protection law in China's history, the exploitation of natural stone has been significantly restricted, leading to an estimated capacity reduction of over 60%. Consequently, in some parts of the country, such as the Jiangxi Province, the price of natural aggregate has nearly tripled in the past five years [9]. Thus, the Chinese transportation infrastructure sector faces years of ineffective reuse of accumulated RAP and increasing natural aggregate costs in future asphalt pavement rehabilitation projects. Therefore, increasing the RAP content in RHMA has attracted increasing attention over the past few years [10,11].

Although the aged asphalt binder in RAP is a considerable asset for construction, the incorporation of RAP could potentially reduce the durability in terms of fatigue resistance, low-temperature cracking resistance, and moisture stability due to its poor rheological properties after volatilization and oxidation in the field [12]. To this end, the most widely and traditional solution is adding rejuvenators or using soft asphalt binder as the modifier to recover the performance of aged asphalt binder [13,14]. This solution was shown to be effective in previous publications and engineering projects. However, when RAP content reaches a high percentage (for example, over 50% by the weight of RHMA), the extra softening additives could reduce the rutting resistance of RHMA [15,16]. Thus, the use of a high percentage of RAP in RHMA has not been properly addressed in current recycling technology.

To overcome the negative effect of softening additives on RHMA, previous studies reported that the high-modulus asphalt concrete (HMAC) that originated in France in the early 1980s could be a potential solution [17]. The basic design concept of high-modulus asphalt mixtures is that through the use of hard-grade asphalt binder (with a penetration of about 10–20), fine aggregate gradation, high asphalt binder content (about 6%), and low porosity (generally 3%), the high-modulus asphalt mixture is obtained through the high modulus properties of the hard-grade asphalt binder itself [18]. For example, Ma et al. indicated that the HMAC has a higher binder/aggregate ratio, lower void content, and finer aggregate gradation compared to the conventional dense asphalt concrete [19]. The results show that the addition of RAP can be incorporated into HMAC at relatively higher contents, due to the similar stiffness between the aged asphalt binder in RAP and the high-modulus asphalt binder. To achieve good rutting resistance, hard asphalt binder is a suitable modifier of HMAC. In most cases, the hard asphalt is produced in refineries, and it also can be obtained by blending the virgin asphalt and natural asphaltite, for example, Trinidad lake or Gilsonite-like materials, whose penetration is in range of 8–28 (0.1 mm) [20]. Thus, the design concept of HMAC has the potential of incorporating more RAP contents in the asphalt mixture. However, using hard asphalt binder as the rejuvenator to recover the aged binder is rarely reported in previous studies. The traditional asphalt mixture regeneration technology is to add rejuvenator or soft asphalt to reduce the viscosity of the aged asphalt, thereby restoring its performance. Therefore, this topic may go against the concept of the traditional solutions. The objectives of this work can be summarized as follows.

- Compare the performances of Trinidad lake asphalt, virgin asphalt binder, hard asphalt binder, aged asphalt binder, and recovered asphalt binder.
- Reveal the modification mechanism of Trinidad lake asphalt on aged asphalt binder.
- Verify the feasibility of increasing RAP content in RHMA by using the concept of HMAC.

2. Experiments

2.1. Materials

2.1.1. Binder

The TH-70# virgin asphalt binder produced by Xinjiang Tarim Petrochemical Co., Ltd, Urumqi, China. was used. Its performance was tested according to a Chinese specification entitled Standard Test Methods of Bitumen and Bituminous Mixtures for Highway Engineering (JTG E20-2011), and the results are shown in Table 1.

Aged binder was obtained using the thin film oven test (TFOT), and each virgin asphalt binder sample with 3.2 mm thickness of asphalt film (50 g) was placed in the oven at 163.5 ± 0.5 °C for a designated duration. To create samples with different aging levels, the heating durations of virgin asphalt binder were 0 h, 5 h, 12 h, 19 h, and 26 h, while the penetration of samples after heating was 64.2, 48.6, 42.4, 34.2, 27.5, respectively.

The hard asphalt binder was used as newly-added virgin asphalt binder in the recycled mixture for recovering the performances of aged asphalt binder. The hard asphalt binder was blended with virgin asphalt binder and Trinidad lake asphalt (TLA). TLA is a stiffer binder with high modulus and poor workability. It is a common modifier for producing hard asphalt binder. The basic performance of TLA was tested according to JTG E20-2011, and the results are shown in Table 2. The TLA was cut into the 1 cm diameter pieces. Virgin asphalt binder and TLA were respectively heated to 150 °C, and then they were mixed using a high-speed shear mixer for 15 min at 165 °C to reach a uniform state.

Table 1. Typical characteristics of TH-70# asphalt binder.

Item	TH-70#	Test Methods
Penetration at 25 °C (0.1 mm)	64.2	ASTM D5
Softening point (°C)	49.0	ASTM D36
Ductility at 15 °C (cm)	>100	ASTM D113
After thin film oven test (TFOT for short, 163 °C, 5 h)		
Mass loss (%)	0.21	ASTM D2872
Penetration at 25 °C (0.1 mm)	48.6	ASTM D5
Softening point (°C)	60.5	ASTM D36
Ductility at 15 °C (cm)	40.1	ASTM D113

Table 2. Typical characteristics of Trinidad lake asphalt (TLA).

Item	TLA	Specification
Penetration at 25 °C (0.1 mm)	2.5	0–5
Softening point (°C)	95.0	≥90
Ductility at 15 °C (cm)	-	-
Ash content (%)	19.2	-
Density (g/m^3)	1.383	1.3–1.5
After TFOT (163 °C, 5 h)		
Mass loss (%)	0.24	<2.0
Penetration ratio (%)	81	≥50

To achieve the objectives of this study, 11 kinds of asphalt binder samples were prepared or used for the laboratory tests. To better organize and introduce this study's work, the specific component, mark, and description of the asphalt binder samples are listed in Table 3.

Table 3. Details of the prepared asphalt binder samples.

Number	Description	Mark	Component
1	Virgin asphalt (VA) binder, TH-70#	VA	100% VA
2	Trinidad lake asphalt (TLA), the modifier for virgin asphalt binder to produce the hard asphalt (HA) binder	TLA	100% TLA
3	Hard asphalt binder, produced by TLA and VA, used as newly-added virgin asphalt binder for recycled mixture	HA	40% TLA + 60% VA
4	Aged asphalt binder, obtained by heating VA, numbered by their aging hours	A5h A12h A19h A26h	100% aged asphalt binder
5	Recycled asphalt binder, produced by mixing the different aging level of asphalt binder with HA	A5h + HA A12h + HA A19h + HA A26h + HA	50% aged asphalt binder +50% HA

2.1.2. Mixture

Research shows that the asphalt mixture with a finer gradation and a higher asphalt binder content has better low-temperature crack resistance and fatigue performance [21]. Therefore, for different RAP contents, different asphalt mixture gradations were designed in this paper. The greater the RAP content, the finer the asphalt mixture gradation. Three mixtures were designed and tested in this work: (i) the conventional hot-mix dense asphalt mixture using natural aggregates, AC-20; hot-mix recycled asphalt mixture designed with EME-20 (a high-modulus asphalt mixture originally recommended by the French industry) containing (ii) 50% reclaimed asphalt pavement (RAP), and (iii) 100% RAP. The designed aggregate gradation is shown in Figure 1.

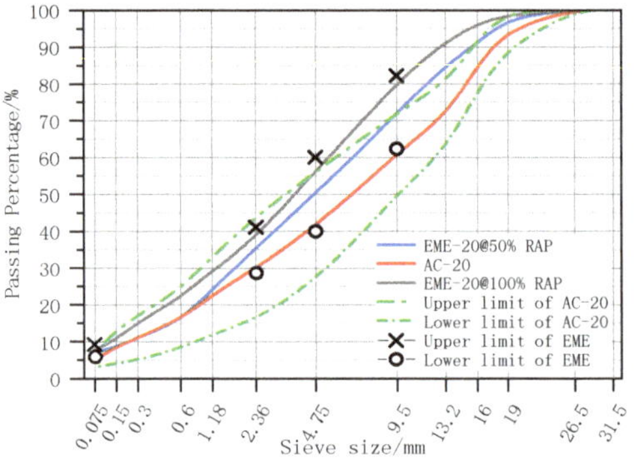

Figure 1. The aggregate gradation of the three designed mixtures.

The RAP was collected from a pavement maintenance project; it was then sieved into three classifications: 0–8, 8–12, and 12–20 mm. The aged binder contained in the RAP was extracted from the aggregates according to ASTM D2172-05, and its properties are listed in Table 4. The gradation of three classifications was determined according to ASTM C136, and the results are shown in Figure 2. The gradation was then used in the recycled mixture design by adjusting the ratio of the three classifications. In addition, the asphalt content of the three RAP classifications was tested, and the results were 6.42% (0–8 mm), 3.21% (8–12 mm), and 2.05% (12–20 mm). The physical properties of the virgin and old aggregates are shown in Table 5. The produced hard asphalt (HA), made with 40% TLA and 60% virgin asphalt (VA) was used as the newly-added binder for EME-20@50%RAP and EME-20@100%RAP. The performance of HA is listed in Table 6. Based on the Marshall test, the optimal asphalt contents for AC-20, EME-20@50%RAP, and EME-20@100%RAP were determined. Details are shown in Table 7.

Table 4. The performances of aged asphalt binder extracted from reclaimed asphalt pavement (RAP).

Penetration at 25 °C (0.1 mm)	Softening Point (°C)	Ductility at 15 °C (cm)	Density (g/m^3)
27	57	21	1.44

Table 5. Physical properties of the aggregates tested according to JTG E42—2005.

Properties	Limestone		RAP Aggregates		Mineral Filler	Test Method
	Coarse	Fine	Coarse	Fine		
Bulk-specific gravity (kg/m^3)	2751	2715	2720	2622	2648	T0308
Flat-elongated Particles (%)	11.4	-	13.5	-	-	T0311
Water absorption (%)	0.5	-	0.4	-	-	T0308
Crush value (%)	12.6	-	14.2	-	-	T0316
LA abrasion (%)	22	-	24	-	-	T0317

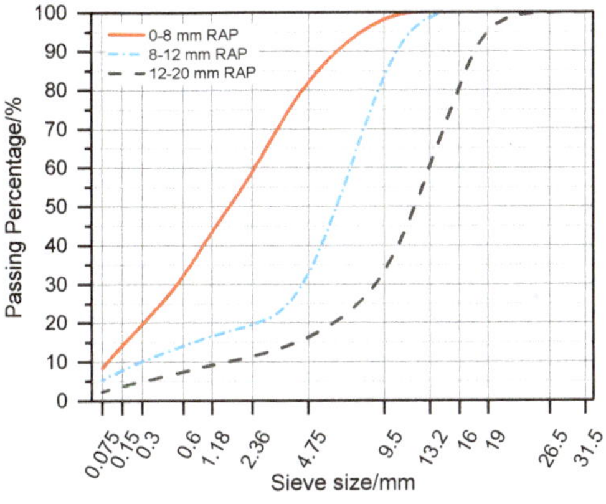

Figure 2. Gradation of the aggregates in the three RAP classifications.

Table 6. Typical characteristics of used hard asphalt binder.

Item		HA	Specification
Penetration at 25 °C (0.1 mm)		31	20–40
Softening point (°C)		59	≥54
Ductility at 15 °C (cm)		45	-
Density (g/m^3)		1.42	1.3–1.5
After TFOT (163 °C, 5 h)	Mass loss (%)	0.22	<2.0

Table 7. Asphalt contents of the three mixtures.

Mixtures	Newly-Added Binder	Optimal Binder/Aggregate Ratio	Newly-Added Binder/Aggregate Ratio (in Mass of Recycled Mixture)
AC-20	VA	4.20%	0
EME-20@50%RAP	HA	5.50%	2.79%
EME-20@100%RAP	HA	6.10%	1.36%

2.2. Test Methods

2.2.1. Dynamic Modulus Test

The objective of the dynamic modulus test is to determine the optimal TLA content in hard asphalt binder considering the improvement of hard asphalt binder on the dynamic modulus of recycled asphalt mixture (RHMA). The LPC Bituminous Mixture Design Guide requires that the dynamic modulus of the high-modulus asphalt mixture is over 14,000 MPa under the conditions of a temperature of 15 °C and a load frequency of 10 Hz [7]. On the other hand, considering that TLA is more expensive than ordinary asphalt, the minimum TLA content that meets the above requirements is the optimal TLA content in hard asphalt. Therefore, the optimal TLA can be determined when the modulus of RHMA reaches 14,000 MPa under 15 °C, 10Hz. A UTM-35 test machine was used, and the investigated frequency was 0–25 Hz. The specimens were prepared with EME-20@50%RAP as it was previously introduced in Table 7, wherein the TLA content in hard asphalt binder was 30%, 40%, 50%, and 70%. The RHMA specimens were compacted using a gyration compactor. The diameter was 150 mm, and the height was 170 mm. The specimens were cored to a diameter of 100 mm and a height of 150 mm, as shown in Figure 3.

Figure 3. The testing of the dynamic modulus of the recycled asphalt mixture.

2.2.2. Rheology Test

A dynamic shear rheometer (DSR) was used to evaluate the high-temperature performance of asphalt binder samples according to AASHTO T315. The temperature sweep mode was used to investigate the performance of samples under 58–82 °C. The complex modulus G* and phase angle δ were recorded to calculate the fatigue factor (G*sinδ) and rutting-resistance factor (G*/sinδ). Furthermore, a bending beam rheometer (BBR) was used to determine the low-temperature performance of asphalt binder samples wherein the creep stiffness (S) and creep rate (M) were recorded to evaluate the low-temperature crack resistance of prepared samples. The investigated temperatures were −12 °C and −18 °C.

2.2.3. FTIR Spectra Analysis

The FTIR spectra analysis was conducted to characterize and compare the difference in the functional group of asphalt binder samples. A Nicolet iS10 Fourier transform infrared spectrometer produced by Thermo Scientific Co., Ltd., Waltham, MA USA was used. The spectral resolution was 0.4 cm^{-1} and the spectral range was 4000–400 cm^{-1}. Lake asphalt is a solid at room temperature; thus, a small amount of TLA powder and potassium bromide crystal were ground and mixed in an agate mortar, and then the samples were pressed by a YP-2 tablet pressing machine to obtain the samples for the FTIR spectra test.

2.2.4. Atomic Force Microscope Test

The atomic force microscope (AFM) observation of asphalt binder samples aims to reveal the difference in the microstructure of samples. A Bruker's cloth Dimension ICON AFM was used with a tap model. The scan area was 90 μm × 90 μm × 10 μm, the data sampling rate was 50 MHz, and the imaging temperature was 27 °C. The asphalt binder sample was heated to 150 °C and was dropped on a glass slide to obtain the specimen. The bee-like structures in the AFM images were identified by color deference using a commercial software, and then the features of bee-like structure were extracted and calculated. Three parameters, including the maximum bee-like area, the average area of all bee-like structures, and the area ratio of bee-like structures to image area, can be obtained.

2.2.5. Performances Test of Recycled Asphalt Mixture

To investigate the effect of hard asphalt binder on the performance of produced asphalt mixtures, the high-temperature rutting test, low-temperature bending test, and fatigue test were conducted. The rutting test was performed under 60 °C ambient temperature. The size of the specimen was 30 cm × 30 cm × 5 cm; each test required three specimens. The high-temperature rutting resistance is characterized by dynamic stability and can be calculated by Equation (1).

$$\mathrm{DS} = \frac{(t_2 - t_1) \times N}{d_2 - d_1} \quad (1)$$

where DS is the dynamic stability, load/mm; t_1 and t_2 are the 45th and 60th min, respectively; d_1 and d_2 are the deformations at the 45th and 60th min, respectively; N is the wheeling speed, 42 load/min.

The three-point bending test was performed to evaluate the low-temperature crack resistance of the mixture, as shown in Figure 4. The specimens were 250 mm × 30 mm × 35 mm cuboids, and the loading rate was 50 mm/min. Bending stiffness modulus, bending strength, and maximum bending strain were obtained by the test. Generally, greater maximum bending strain and higher bending strength lead to a better low-temperature performance of the mixture. Furthermore, the fracture energy can be calculated from the area of the stress-strain curve before the material reaches the maximum stress. According to research experience, the evaluation of low-temperature performance by fracture energy is more reasonable than the traditional index of maximum bending strain [22]. Therefore, the fracture energy was selected as the indicator to describe the low-temperature performance of the mixtures.

Figure 4. Three-point bending test of recycled asphalt mixture.

A UTM-35 test machine was used to conduct the three-point fatigue test of the asphalt mixture, and a stress control model under 15 °C with 10 Hz was used. Three specimens were tested under each stress level. The specimens were 250 mm × 30 mm × 35 mm cuboids. The fatigue characteristic is given by Equation (2), and after taking logarithms on both sides of the equation, it can be written as a linear equation, Equation (3). In Equation (3), the intercept k denotes the average fatigue life of the specimens, and the absolute value of slope σ stands for the decay rate of the fatigue life with an increasing stress ratio.

$$N_f = K\left(\frac{1}{\sigma}\right)^n \tag{2}$$

$$\lg N_f = -n\lg\sigma + k \tag{3}$$

where N_f is the loading time when specimens break; σ is the stress of repeat loading; 20%, 30%, and 40% of maximum bending stress were used, respectively.

Maximum bending stress of specimens of different mixtures was first tested, and the results are shown in Table 8. A maximum bending stress of 20%, 30%, and 40% was calculated for each type of mixture. We performed the three-point fatigue test of different asphalt mixtures according to the calculation results in Table 8.

Table 8. Bending stress of specimens of different mixtures.

Mixture Type	Maximum Bending Stress (KN)	Stress Ratio		
		0.2	0.3	0.4
AC-20	8.2	1.6	2.5	3.3
EME-20@50%RAP	10.1	2.0	3.0	4.1
EME-20@100%RAP	9.7	1.9	2.9	3.9

3. Results and Discussion

3.1. Performance of Hard Asphalt Modified Aged Asphalt

3.1.1. Determining the Optimal Content of Lake Asphalt in Hard Asphalt

The asphalt mixture with a higher modulus tends to have better rutting resistance to traffic. Generally, its modulus can be improved by using hard asphalt binder and additives. This study investigated the feasibility of using TLA as the modifier to increase the modulus of the recycled asphalt mixture. To this end, the dynamic modulus of the EME-20 mixture containing 50% RAP (EME-20@50%RAP) was determined. Its newly-added asphalt binder was hard asphalt made with 30% to 70% proportion of TLA, and the results are shown in Figure 5.

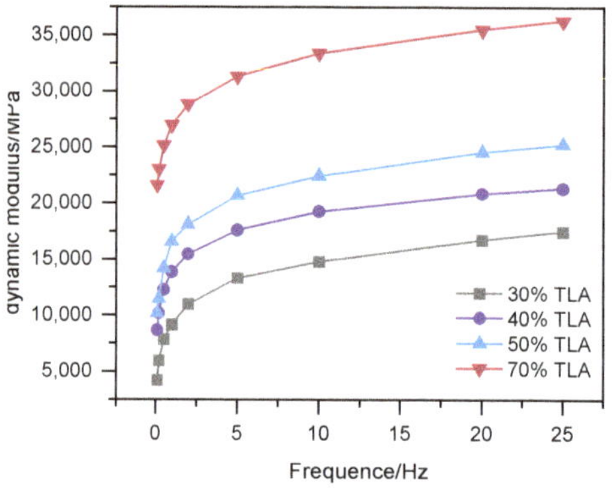

Figure 5. Test results of the EME-20 asphalt mixture containing 50% of RAP under 0–25 Hz.

According to Figure 5, the first finding is that the increasing TLA content in hard asphalt binder can significantly increase the modulus of the recycled mixture at every frequency. The optimal TLA content can be obtained when the modulus of RHMA reaches 14,000 MPa under 15 °C, 10 Hz. Therefore, the TLA content in hard asphalt binder was determined as 40%.

3.1.2. FTIR Spectra Analysis of Recycled Asphalt Binder

Figure 6a presents the FTIR spectra of virgin asphalt (VA) and aged virgin asphalt. In Figure 6, a frequency band present at 3800–3500 cm^{-1} is caused by the vibration of bound water or free water and CO_2 in ambient air; the strong peaks around 2922–2852 cm^{-1} are associated with the stretching vibrations in the aliphatic chain of the asphalt binder [23]; the peak at 1600 cm^{-1} denotes to the vibration of C=C in the benzene skeleton; the peak at 1462 and 1376 cm^{-1} relates to the deformation vibrations of the aliphatic chain; the peak at 1031 cm^{-1} is the stretching vibrations of the sulfoxide group (S=O); the peaks at 863, 811, and 747 cm^{-1} are resulting from the bending vibration of C-H in the aromatic ring. The aging level of asphalt binder samples can be determined by comparing the absorbance of the functional group. The significant difference between three samples is that the transmittance of the functional group of C=C, C=O, and S=O increased after the virgin asphalt binders were aged, and a greater aging level leads to a higher absorbance of those functional groups. The increase in C=O indicates that the oxygen-containing functional groups such as aldehydes, ketones, esters, and carboxylic acids are formed during the aging process of the asphalt binder. Furthermore, the increase in S=O demonstrates that the sulfur element reacts with oxygen to form a polar functional sulfoxide group.

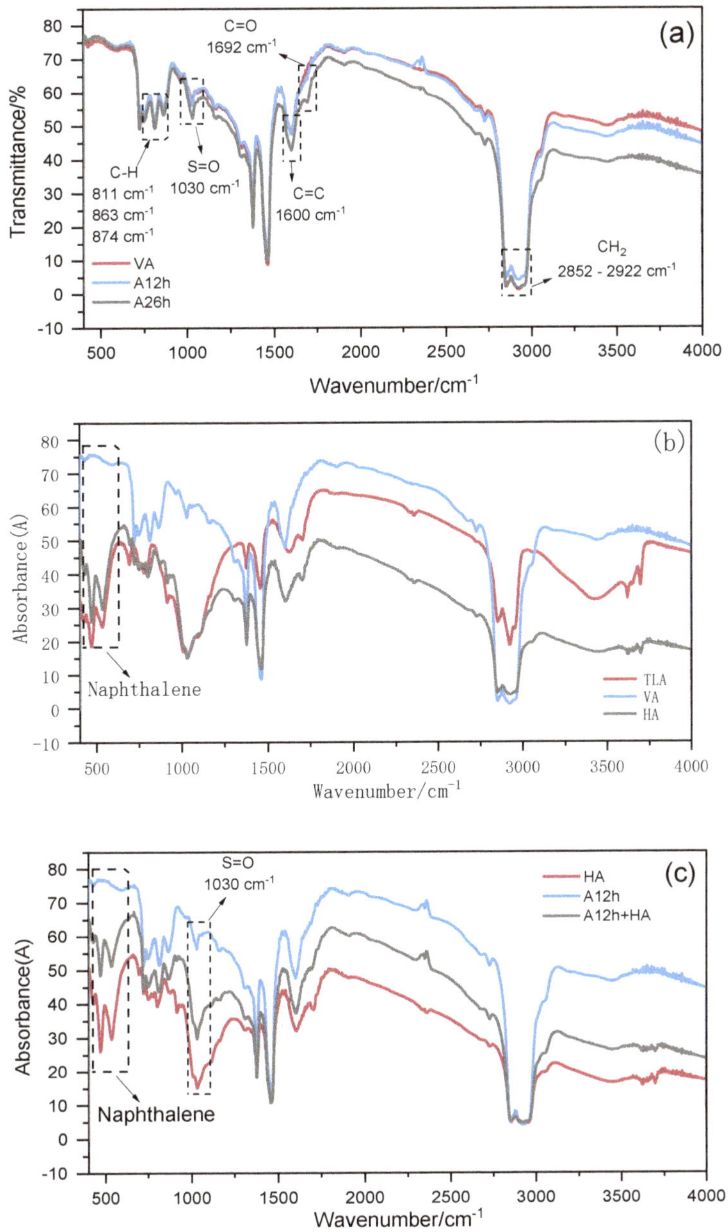

Figure 6. *Cont.*

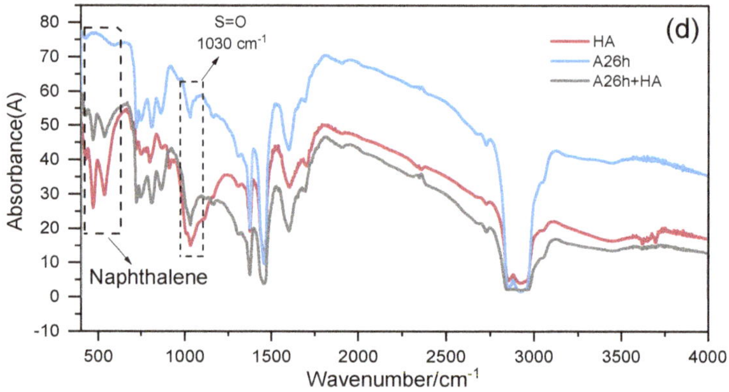

Figure 6. Fourier transform infrared (FTIR) spectra of virgin and aged asphalt binder samples: (**a**) VA, A12h, and A26h; (**b**) TLA, VA, and HA; (c) HA, A12h, and A12h + HA; (**d**) HA, A26h, and A26h + HA.

Figure 6b illustrates the FTIR spectra of virgin asphalt (VA), Trinidad lake asphalt (TLA), and hard asphalt produced with 60% of VA and 40% of TLA. By comparing the FTIR spectra of VA and TLA, most of the transmittance peaks of TLA are near the absorption peaks of the matrix asphalt. The only new absorption peaks appear at 535, 467, and 428 cm^{-1}, which correspond to naphthalene functional groups. The addition of a naphthalene functional group is helpful for enhancing the structural strength of petroleum asphalt molecules, which will increase the viscosity of asphalt and enhance the cohesion of asphalt binder and the adhesion with aggregate. In addition, the transmittance peak area of sulfoxide group S=O at 1031cm^{-1} of TLA is significantly larger than that of ordinary asphalt binder. The TLA was intensively exposed in an open-air environment, which accelerated the oxidation of TLA. The sulfur and oxygen have sufficient chemical reaction to form more sulfoxide groups. Furthermore, HA and TLA have the same characteristic peaks, in which the transmittance of each functional group in HA is smaller than that of TLA. The above results show that the addition of TLA only changes the proportion of each component in VA, and there is no chemical reaction in this process.

One of the objectives of this work is to rejuvenate the aged asphalt binder using HA containing TLA for producing a high-modulus recycled mixture with high RAP content. Thus, blended asphalt binder samples containing 50% aged asphalt (A12h, A26h) and 50% HA were prepared. Their FTIR spectra are shown in Figure 6c,d. From the effect of HA on aged asphalt, it can be concluded that the sulfoxide group S=O peaks at 1031 cm^{-1} in both A12h + HA and A26h + HA are sharper compared to that in aged asphalt. This is because the sulfoxide group tends to disintegrate under high temperatures. Furthermore, the appearance of peaks at 535, 467, and 428 cm^{-1} in both A12h+HA and A26h+HA indicates that the naphthalene functional group in TLA was introduced in the rejuvenated asphalt binder.

3.1.3. Analysis of Microstructure Characteristics of Recycled Asphalt Binder

Based on the colloid structure theory of the asphalt binder, asphaltene is considered as the colloid center with a high molecular weight, and the micelles are formed by absorbing the colloidal particles that have greater polarity. The micelles are dispersed in oil [24]. To date, asphaltene is widely believed to be the main component of the convex part of the bee-like structure [25]. By comparing the characteristics of the bee-like structure, the differences in colloid structure of the asphalt binder can be determined. In this section, Figure 7a–c shows virgin asphalt and the aged virgin asphalt after 12 h and 26 h of thermal aging, respectively. The surfaces of aged asphalt samples are rougher than the virgin asphalt when aging time is increased, as shown in Figure 8a. Furthermore, the characteristics including maximum area, average area, and total area ratio of bee-like structures in A12h and A26h are significantly higher than those in VA. This means thermal aging contributes to a higher content of asphaltene and changes the microstructure of asphalt binder.

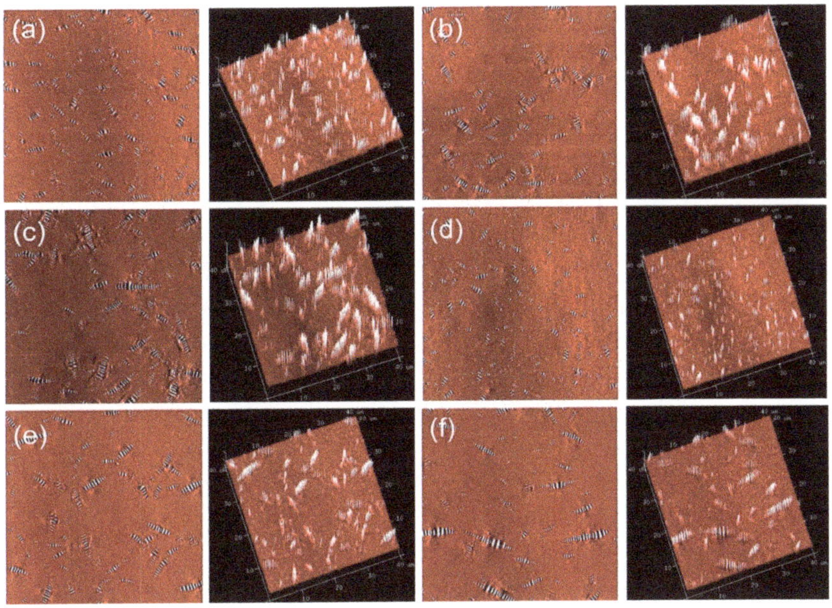

Figure 7. Microstructure and 3D structural side view of the asphalt binder imaged using atomic force microscopy (AFM): (**a**) virgin asphalt, VA; virgin asphalt after (**b**) 12 h and (**c**) 26 h thermal aging, denoted by A12h and A26h, respectively; (**d**) hard asphalt mixed with (**e**) 12 h and (**f**) 26 h aged virgin asphalt, denoted by A12h + HA and A26h + HA, respectively.

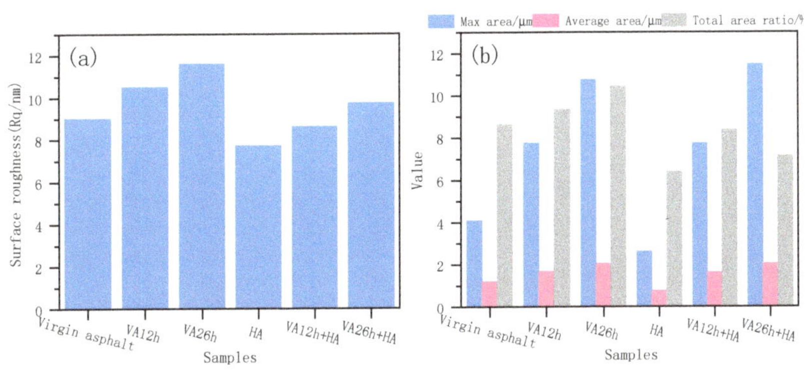

Figure 8. Results of the microstructure characteristics of asphalt binder samples: (**a**) roughness of AFM surface; (**b**) statistics of bee-like structure.

Furthermore, there is no obvious difference in surface morphology between VA and HA, and they have a similar bee-like structure, as shown in Figure 7a,d. This means the incorporation of TLA in VA did not significantly change the microstructure of VA. To evaluate the rejuvenation effect of HA on aged asphalt with different aging levels, AFM images of HA, A12h + HA, and A26h + HA were obtained, as shown in Figure 7d–f, respectively. By comparing A12h and A12h + HA, as shown in Figure 8b, the characteristic values of the bee-like structure of A12h + HA were determined as slightly lower than those of A12h, indicating that there was a limited rejuvenation effect on aged asphalt in view of the microstructure. On the other hand, there was a greater improvement in the microstructure of the A26h sample because the characteristic values in A12h + HA were higher than those in A26h. Finally, when HA was added into the asphalt binder with a higher aging degree, the characteristics of the bee-like structure changed dramatically, showing a significant improvement in the maximum area, average area, and total area ratio of bee-like structures.

3.1.4. Rheological Properties of Recycled Asphalt Binder

The creep stiffness (S) and creep rate (m) of asphalt binder samples were measured using the BBR test at −12 °C. Generally, the larger the S and the lower the m are, the worse the resistance of the asphalt binder to brittle fracturing in low-temperature environments, which is not conducive to the durability of asphalt pavement. It can be seen from Figure 9a,b that the low-temperature crack resistance of HA was worse than that of VA, which was caused by the addition of TLA. In addition, the low-temperature crack resistance of the virgin asphalt significantly deteriorated after aging, and the S value of a 26 h was 3.2 times that of the VA. In addition, when 50% HA was added to the aged asphalt, the low-temperature stability of the recycled asphalt binder was significantly worse than that of the VA after the corresponding aging time. Therefore, the use of HA as a rejuvenator for aged asphalt leads to worse low-temperature stability, and this conclusion suggests that the HA modifier is not recommended for use in cold regions.

Figure 9c shows that the high-temperature rutting factor of HA was significantly higher than that of VA, especially at 58 °C and 64 °C. In addition, with the increase of the VA aging degree, the rutting factor of A26h asphalt binder increased significantly. The rutting resistance factor of A26h asphalt binder was about seven times higher than that of VA. Furthermore, the rutting factor of asphalt binder with different aging degrees was greatly improved after mixing with 50% HA. For example, at 58 °C, the rutting factor of A26h + HA was about 20% higher than that of A26. The results of Figure 9a show that HA had a significant effect on improving the rutting resistance of aged asphalt binder. Compared to the conventional asphalt binder, TLA has a high asphaltene and ash content, which contribute to a high stiffness of hard asphalt binder produced with TLA, increasing the high-temperature performance of the hard asphalt binder.

The fatigue factor $G^{*}\sin\delta$ was used to compare and evaluate the anti-fatigue characteristics of asphalt binder samples; $G^{*}\cdot\sin\delta$ refers to the energy lost by the asphalt binder under repeated loads. The smaller the $G^{*}\cdot\sin\delta$, the less energy is lost and the better the fatigue resistance. The fatigue factor test results are shown in Figure 9. Figure 9 shows that the fatigue factors of HA and VA were very close, indicating that they have similar fatigue resistance. In addition, with the increase of aging time, the anti-fatigue performance of virgin asphalt decreased significantly. For example, the fatigue factor of A26h was about 30 times higher than that of VA. The reason is that as the aging time increased, the asphaltene and gum content in the asphalt binder increased, and the saturated and aromatic components decrease, resulting in an increase in the complex modulus and the fatigue factor. In addition, when HA was mixed with aged asphalt, its fatigue factor was slightly higher than that of aged asphalt. For instance, the fatigue factor of A26h + HA was about 10% higher than that of A26h. The above results show that the fatigue performance of aged asphalt decreases with the addition of HA. This is due to the fact that TLA has a relatively large ash content. After the ash particles precipitate, the elasticity and toughness of HA are reduced, which leads to a decrease in fatigue performance. From this point of view, rejuvenating the aging asphalt binder using HA is not preferred for high-traffic pavements.

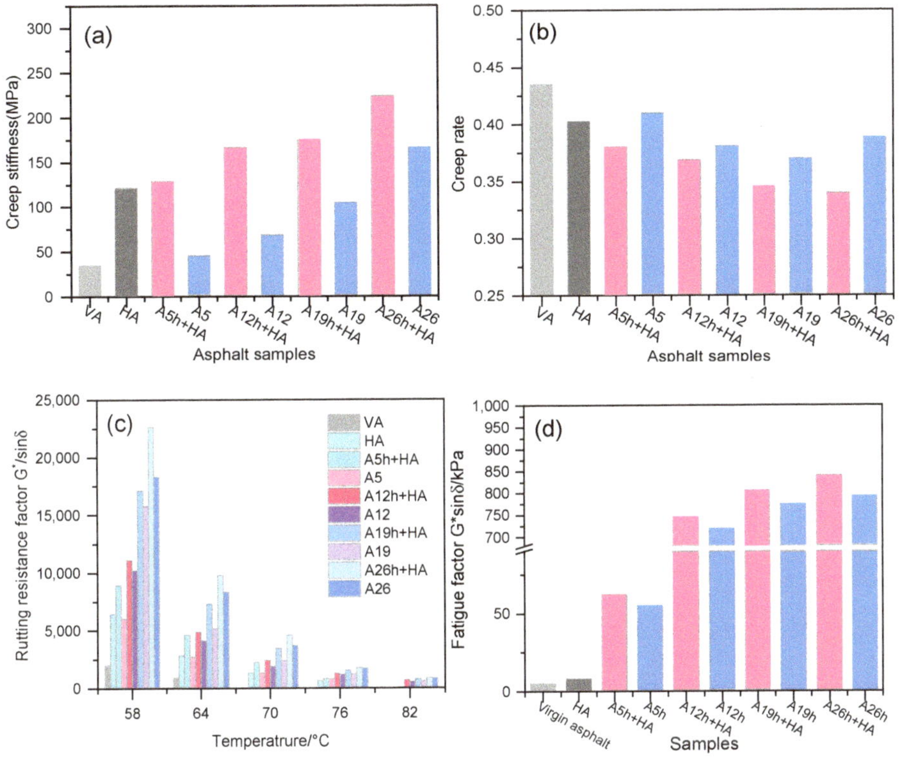

Figure 9. Rheological properties of recycled asphalt binder: (**a**) creep stiffness; (**b**) creep rate; (**c**) rutting factor; (**d**) fatigue factor.

3.2. Performance of Hot-Mix Recycled Asphalt Mixture Containing Hard Asphalt

Three kinds of asphalt mixtures were prepared to reveal the difference in performance between the conventional and the hot-mix recycled asphalt mixture (RHMA). The RHMA was designed based on the principle of the high-modulus mixture using the hard asphalt as a binder, containing 50% and 100% RAP, denoted by EME-20@50% and EME-20@100%, respectively. Figure 10a indicates that EME-20@100% and EME-20@50% had similar dynamic stability (DS). Compared with the ordinary asphalt mixture containing hard asphalt binder, the high-temperature performance was lower.

As shown in Figure 10b, the AC-20 specimens had the highest bending strain compared to others, indicating the conventional AC-20 could bear a higher deformation than EME-20 specimens. Furthermore, there was no significant difference between EME-20@50% and EME-20@100%, which means that they have a similar deformation resistance. Figure 10c shows the bending strength of tested samples; the results indicate that AC-20 has the lowest bending strength. In summary, the specimens designed using the concept of high-modulus asphalt mixture were stiffer than the conventional one, which it led to a greater bending strength and lower bending strain. To compare the low-temperature performance of samples, fracture energy was used, which can reveal the performance differences in terms of energy; the results are shown in Figure 10d. Figure 10d shows that EME-20@50% had a similar fracture energy to AC-20, indicating that the incorporation of hard asphalt and 50% RAP had no negative effect on the low-temperature cracking resistance of the asphalt mixture. Moreover, when the RAP content was increased to 100%, the fracture energy of EME-20@100% increased by 23% compared to that of AC-20 due to the fine gradation and high asphalt binder content.

Moreover, Figure 10e suggests that EME-20@100% had the highest fatigue life under different stress ratios. The fatigue lives of EME-20@100% and EME-20@50% were both higher than that of AC-20, but the slope of the fatigue curve was greater, indicating that they were more sensitive to stress. This paper used a stress-controlled mode to carry out the fatigue experiment. The experimental results obtained under strain-controlled mode may be different, which should be further studied. In summary, fine gradation and a high asphalt binder content improved low-temperature crack resistance and fatigue life while maintaining a good high-temperature performance of RHMA. Therefore, the high modulus design concept can be adopted to increase RAP content in recycling.

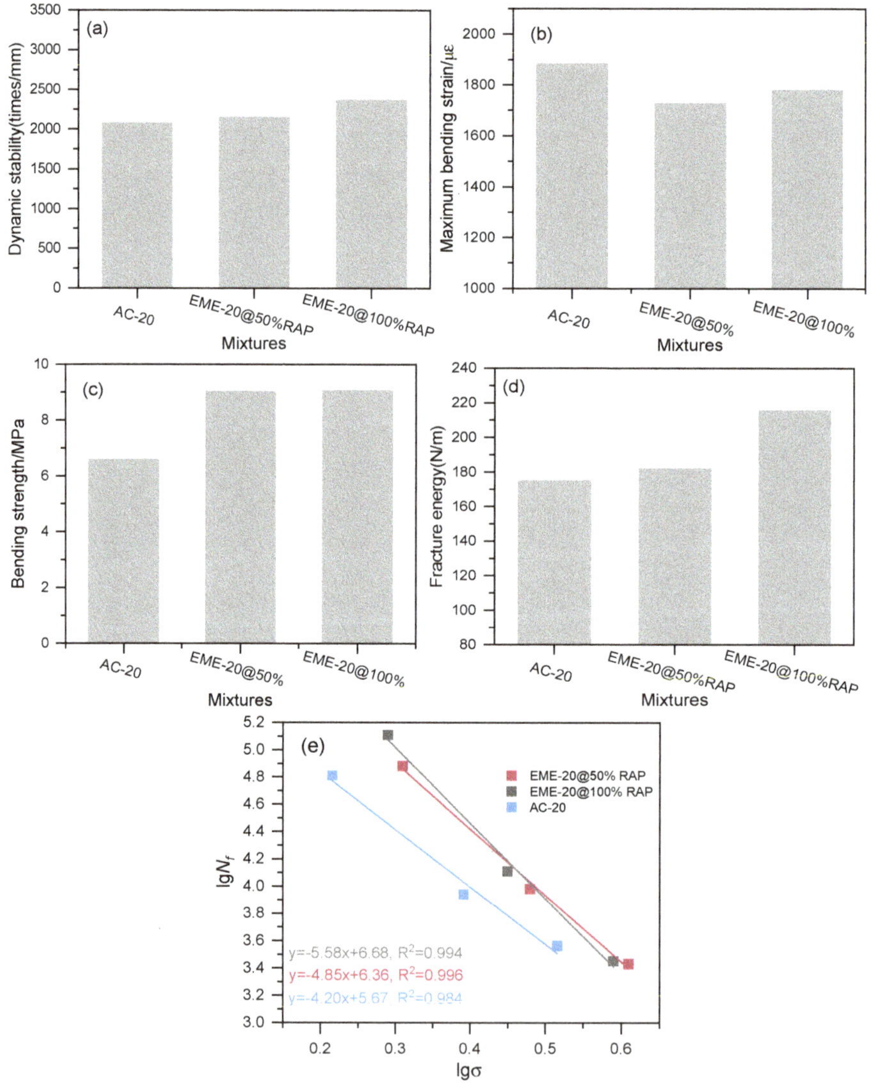

Figure 10. Performance of hot-mix recycled asphalt mixture containing hard asphalt: (**a**) High-temperature rutting resistance; (**b**) Maximum bending strain; (**c**) Bending strength; (**d**) Fracture energy; (**e**) Fatigue performance.

4. Conclusions

In this paper, the modification mechanism, microstructure, and rheological properties of hard asphalt binder produced by mixing TLA and VA were analyzed using infrared spectroscopy, atomic force microscopy, and DSR and BBR experiments. The performance of RHMA produced with hard asphalt was tested, and the following conclusions can be drawn.

(1) The addition of TLA leads to a change in the component proportion of asphalt binder, but no new functional groups are produced; the addition of TLA leads to a decrease in the number and total area of bee structures, which indicates that lake asphalt changes the interaction between asphaltene and other asphalt molecules. This is conducive to the dissolution of asphaltene and forms a more stable system.

(2) The low-temperature performance of recycled asphalt binder and aged asphalt binder has the same change rule. With the extension of aging time, the low-temperature performance of recycled asphalt binder gradually decreases, the high-temperature performance increases, and fatigue damage develops rapidly.

(3) As high-modulus asphalt mixture and RAP have similarity in terms of asphalt binder performance and fine gradation, adopting the high modulus design concept for recycling can increase the RAP content. Through the optimization of gradation and the use of hard asphalt, the high- and low-temperature performance and anti-fatigue performance of 100% recycled asphalt mixture are better than when using the AC-20 mixture. Thus, recycling with a high-modulus asphalt mixture design concept to increase the RAP content is feasible.

Author Contributions: Conceptualization, J.Z. and Y.Z.; methodology, J.Z., J.L. and G.L.; formal analysis, J.Z., J.L., G.L. and T.Y.; data curation, J.Z. and J.L.; writing—original draft preparation, J.Z., G.L., and T.Y.; writing—review and editing, G.L., T.Y. and Y.Z. All authors have read and agreed to the published version of the manuscript.

Funding: This research was funded by the National Natural Science Foundation of China through project 51378123 and the Graduate Research and Innovation Projects of Jiangsu Province (KYLX_0167).

Institutional Review Board Statement: Not applicable.

Informed Consent Statement: Not applicable.

Data Availability Statement: No new data were created or analyzed in this study. Data sharing is not applicable to this article.

Acknowledgments: The authors would like to acknowledge the financial support provided by the National Natural Science Foundation of China through project 51378123. The present article is also supported by the Fundamental Research Funds for the Graduate Research and Innovation Projects of Jiangsu Province (KYLX_0167). The funding supported the FTIR spectra analysis and atomic force microscope test. The authors thank them for their support in the development of the project.

Conflicts of Interest: The authors declare no conflict of interest.

References

1. Debbarma, S.; Ransinchung, G.D. Achieving sustainability in roller compacted concrete pavement mixes using reclaimed asphalt pavement aggregates—State of the art review. *J. Clean. Prod.* **2021**, *287*, 125078. [CrossRef]
2. Georgiou, P.; Loizos, A. Environmental assessment of warm mix asphalt incorporating steel slag and high reclaimed asphalt for wearing courses: A case study. *Road Mater. Pavement Des.* **2021**, *22*, S662–S671. [CrossRef]
3. Wang, H.; Thakkar, C.; Chen, X.; Murrel, S. Life-cycle assessment of airport pavement design alternatives for energy and environmental impacts. *J. Clean. Prod.* **2016**, *133*, 163–171. [CrossRef]
4. Wang, F.; Xie, J.; Wu, S.; Li, J.; Barbieri, D.M.; Zhang, L. Life cycle energy consumption by roads and associated interpretative analysis of sustainable policies. *Renew. Sustain. Energy Rev.* **2021**, *141*, 110823. [CrossRef]
5. Pradyumna, T.A.; Mittal, A.; Jain, P.K. Characterization of Reclaimed Asphalt Pavement (RAP) for Use in Bituminous Road Construction. *Procedia Soc. Behav. Sci.* **2013**, *104*, 1149–1157. [CrossRef]
6. Hussain, A.; Yanjun, Q. Effect of Reclaimed Asphalt Pavement on the Properties of Asphalt Binders. *Procedia Eng.* **2013**, *54*, 840–850. [CrossRef]

7. Gao, J.; Yang, J.; Yu, D.; Jiang, Y.; Ruan, K.; Tao, W.; Sun, C.; Luo, L. Reducing the variability of multi-source reclaimed asphalt pavement materials: A practice in China. *Constr. Build. Mater.* **2021**, *278*, 122389. [CrossRef]
8. Delorme, J.L.; De la Roche, C.; Wendling, L. *LPC Bituminous Mixtures Design Guide*; Laboratoire Central des Ponts et Chaussées: Paris, France, 2007.
9. Yan, J.; Zhang, Z.; Zhu, H.; Li, F.; Liu, Q. Experimental Study of Hot Recycled Asphalt Mixtures with High Percentages of Reclaimed Asphalt Pavement and Different Recycling Agents. *J. Test. Eval.* **2014**, *42*, 1183–1190. [CrossRef]
10. Gao, L.; Li, H.; Xie, J.; Yu, Z.; Charmot, S. Evaluation of pavement performance for reclaimed asphalt materials in different layers. *Constr. Build. Mater.* **2018**, *159*, 561–566. [CrossRef]
11. Hu, X.; Nie, Y.; Feng, Y.; Zheng, Q. Pavement Performance of Asphalt Surface Course Containing Reclaimed Asphalt Pavement (RAP). *J. Test. Eval.* **2012**, *40*, 1162–1168. [CrossRef]
12. Wang, H.; Dang, Z.; Li, L.; You, Z. Analysis on fatigue crack growth laws for crumb rubber modified (CRM) asphalt mixture. *Constr. Build. Mater.* **2013**, *47*, 1342–1349. [CrossRef]
13. Al-Saffar, Z.H.; Yaacob, H.; Satar, M.K.I.M.; Saleem, M.K.; Lai, J.C.; Putra Jaya, R. A review on rejuvenating materials used with reclaimed hot mix asphalt. *Can. J. Civ. Eng.* **2021**, *48*, 233–249. [CrossRef]
14. Yu, X.; Zaumanis, M.; dos Santos, S.; Poulikakos, L.D. Rheological, microscopic, and chemical characterization of the rejuvenating effect on asphalt binders. *Fuel* **2014**, *135*, 162–171. [CrossRef]
15. Yousefi, A.; Behnood, A.; Nowruzi, A.; Haghshenas, H. Performance evaluation of asphalt mixtures containing warm mix asphalt (WMA) additives and reclaimed asphalt pavement (RAP). *Constr. Build. Mater.* **2021**, *268*, 121200. [CrossRef]
16. Wróbel, M.; Woszuk, A.; Ratajczak, M.; Franus, W. Properties of reclaimed asphalt pavement mixture with organic rejuvenator. *Constr. Build. Mater.* **2021**, *271*, 121514. [CrossRef]
17. Yan, J.; Leng, Z.; Ling, C.; Zhu, J.; Zhou, L. Characterization and comparison of high-modulus asphalt mixtures produced with different methods. *Constr. Build. Mater.* **2020**, *237*, 117594. [CrossRef]
18. Zhu, J.; Ma, T.; Fan, J.; Fang, Z.; Chen, T.; Zhou, Y. Experimental study of high modulus asphalt mixture containing reclaimed asphalt pavement. *J. Clean. Prod.* **2020**, *263*, 121447. [CrossRef]
19. Ma, T.; Wang, H.; Huang, X.; Wang, Z.; Xiao, F. Laboratory performance characteristics of high modulus asphalt mixture with high-content RAP. *Constr. Build. Mater.* **2015**, *101*, 975–982. [CrossRef]
20. Liu, J.; Yan, K.; Liu, J.; Guo, D. Evaluation of the characteristics of Trinidad Lake Asphalt and Styrene–Butadiene–Rubber compound modified binder. *Constr. Build. Mater.* **2019**, *202*, 614–621. [CrossRef]
21. Fang, M.; Park, D.; Singuranayo, J.L.; Chen, H.; Li, Y. Aggregate gradation theory, design and its impact on asphalt pavement performance: A review. *Int. J. Pavement Eng.* **2018**, *20*, 1408–1424. [CrossRef]
22. Mull, M.A.; Stuart, K.; Yehia, A. Fracture resistance characterization of chemically modified crumb rubber asphalt pavement. *J. Mater. Sci.* **2002**, *37*, 557–566. [CrossRef]
23. Wang, J.; Yuan, J.; Kim, K.W.; Xiao, F. Chemical, thermal and rheological characteristics of composite polymerized asphalts. *Fuel* **2018**, *227*, 289–299. [CrossRef]
24. Zhang, H.; Wang, Y.; Yu, T.; Liu, Z. Microstructural characteristics of differently aged asphalt samples based on atomic force microscopy (AFM). *Constr. Build. Mater.* **2020**, *255*, 119388. [CrossRef]
25. Wang, Y.; Zhang, H. Influence of asphalt microstructure to its high and low temperature performance based on atomic force microscope (AFM). *Constr. Build. Mater.* **2021**, *267*, 120998. [CrossRef]

applied sciences

MDPI

Article

A Preliminary Laboratory Evaluation on the Use of Shredded Cigarette Filters as Stabilizing Fibers for Stone Mastic Asphalts

Piergiorgio Tataranni * and Cesare Sangiorgi

Department of Civil, Chemical, Environmental and Materials Engineering, University of Bologna, 40131 Bologna, Italy; cesare.sangiorgi4@unibo.it
* Correspondence: piergiorg.tataranni2@unibo.it

Abstract: Cigarette butts can be considered as one of the most common contemporary sources of waste, considering the large consumption of cigarettes all over the world. Despite the fact that different solutions have been developed and tested in the recent years aiming to recycle them, cigarette butts are currently landfilled and incinerated. Following the circular economy principles, the experimental application proposed in this paper is an exploratory investigation on the use of shredded cigarette filters as sustainable alternative to the addition of fibers into Stone Mastic Asphalts (SMAs). This represents the preliminary step for a wider research project, aiming to find a possible recycling solution for cigarette butts as fibers in bituminous materials. The use of fibers is a common and well-established solution for the production of high bitumen content mixtures. The fibers have a double function: acting, generally, as a stabilizing agent and, where possible, improving the mechanical performance of the bituminous mixtures. In the present research, two different SMAs were produced and tested aiming to analyze the effects given by the addition of the shredded cigarette filters. The first asphalt concrete, produced with traditional cellulose fibers was taken as a reference mixture, while the experimental mixture was produced with the shredded cigarette filters. The data highlight interesting and promising results for future development, making the use of waste cigarette filters a potential eco-friendly alternative to common cellulose fibers for SMAs.

Keywords: cigarette filters; recycled cigarette butts; asphalt concrete; SMA; fibers

Citation: Tataranni, P.; Sangiorgi, C. A Preliminary Laboratory Evaluation on the Use of Shredded Cigarette Filters as Stabilizing Fibers for Stone Mastic Asphalts. *Appl. Sci.* **2021**, *11*, 5674. https://doi.org/10.3390/app11125674

Academic Editors: Amir Tabakovic, Jan Valentin and Liang He

Received: 1 June 2021
Accepted: 16 June 2021
Published: 18 June 2021

1. Introduction

Sustainable and responsible development is probably the most important task for a modern society finally aware of the damage caused to the environment by the uncontrolled growth of the last 30 years. As stated in the 7th Environment Action Programme (EAP) produced by the EU, the recycling and reuse of waste can be considered as one of the principal tools for addressing this new challenge [1].

Cigarette butts (CBs) can be classified as one on the most common sources of waste, considering their large consumption all over the world [2]. According to the most recent estimates, around 1.2 million tons of cigarette butts are produced every year [3]. From an environmental point of view, the CBs represent a real problem considering the lack of disposal control and their slow degradation rate. Several researches based on interviews to cigarette smokers, highlighted that around the 73% of the users do not have specific practices for the CBs disposal [4,5]. Furthermore, as stated in the report "Tobacco and its Environmental Impact: an overview" from the World Health Organization, around 70% of CBs are released into the environment, generally discarded directly on the ground [6]. The vast majority of cigarette filters are mainly made of cellulose acetate with a very limited degradation rate [7]. Brodof (1996) assessed that the breakdown of a common cellulose acetate generally takes around 18 months under normal conditions [8]. Recent studies verified that the slow process of degradation of CBs discarded on the ground could last up to 13 years [9].

Furthermore, the presence of toxic chemicals within CBs is now a well-known fact [10,11]. A cigarette filter is mainly designed to reduce the inhalation of several substances during tobacco combustion; as a result, it absorbs and retains toxic substances that can be released into the environment when discarded. Several detailed studies and researches highlighted that around 7000 chemicals (USDHHS, 2010) are released by CBs into the environment, with potential effects on humans and ecosystems [12–16].

Still, the lack of standardized protocols or best practices for CB disposal makes the management of this kind of waste a real issue. Thus, the majority or CBs are currently landfilled and incinerated. In the light of the above, it is clear that these disposal solutions will no longer be feasible, and it is evident that finding recycling alternatives to disposing of CBs in landfill is now crucial.

Different recycling methods and applications have been tested and are currently under investigation in order to find possible solutions in line with the principles of circular economy [17]. Several alternative solutions are available in the literature for the recycling and re-use of CBs. Good results have been obtained for the recycling of CBs in the production of new construction materials (i.e., fired bricks, lightweight bricks, precast concrete blocks) or acoustic insulation systems [18–20]. A few studies have also been developed on the use of CBs as bituminous materials for road pavements, and the proposed experimental applications were mostly focused on the use of waste CBs as bitumen modifiers/additives [21,22].

The research proposed in this paper is a preliminary study on the direct application of brand new cigarette filters (CF) for the production of asphalt concretes (AC). The work is the preliminary and exploratory step of a wider research project aiming to study the re-use of waste cigarette butts as a potentially sustainable alternative to fibers (both traditional and innovative ones) for ACs. The use of fibers, in fact, is a common practice for the production of high bitumen content mixtures for pavements such as stone mastic asphalt (SMA) and porous asphalt [23]. The fibers can have the double function of working as a stabilizing agent, avoiding the separation between the aggregates and the bitumen during the storage and transport operations, and are also used to improve the mechanical performance of the final AC [24,25].

Considering the wide use of fibers, a new market to recycle waste in the production process of new cellulose-based fibers has been generated, taking into account the good properties of the waste in terms of binder absorption, when compared to mineral fibers [26–31]. In order to evaluate the effects given by the addition of the grinded brand new CFs in the bituminous mixture, two different SMAs were produced in the laboratory and tested: one (labelled SMA0) with common cellulose stabilizing fibers, and another (labelled SMAF) containing CFs. The aim of the present research is to explore the possible use of this waste as a pavement material, and therefore to develop a new recycling application to limit the landfilling of waste cigarette butts.

2. Materials and Methods

As aforementioned, the experimental application of CFs in substitution of common stabilizing fibers has been carried out, adding common shredded CFs during the mixing operation of a traditional SMA.

2.1. Cigarette Filters and Stabilizing Fibers

In the case being studied, brand new cigarettes were collected and filters were cut off. In order to represent what could happen on a real scale, cigarette from different producers were used in this experimental application. However, it is worth mentioning that all the collected CFs were made of cellulose acetate-based material.

The CFs were grinded using a mechanical shredder, set up in order to have particles up to 10 mm. From a visual analysis of the obtained shredded CFs (Figure 1), the material is mostly composed of cellulose acetate fibers, with traces of plug wrap paper and tipping paper.

Figure 1. Shredded cigarette filters and a EUR 1 coin as size reference.

Some traditional cellulose-based stabilizing fibers were used as reference fibers. The physical and mechanical properties of the reference fibers are reported in Table 1.

Table 1. Properties of cellulose stabilizing fibers.

Fibers Technical Data	
Avg. fibers length	min 200 μm, max 1100 μm
Avg. fibers diameter	min 25 μm, max 45 μm
Melting point	>230 °C
Water solubility (%)	0.450–0.500 kg/m^3

2.2. SMA Mix Design

Stone mastic asphalt was first developed in Germany in the 1960s and it was soon used in the rest of Europe, the United States, Canada and Australia due to its good performance. The combination of selected aggregates according to a gap-graded gradation and the adoption of modified bitumen allows for a high resistance to plastic deformation, making SMA suitable for high- and heavy-traffic roads. Furthermore, the use of high quality aggregates ensures remarkable skid-resistance properties.

In the present research, the adopted mix design for the SMA followed the District of Bologna technical specification [32]. The grading distribution is presented in Figure 2 and Table 2, while the rheological properties of the used bitumen are reported in Table 3.

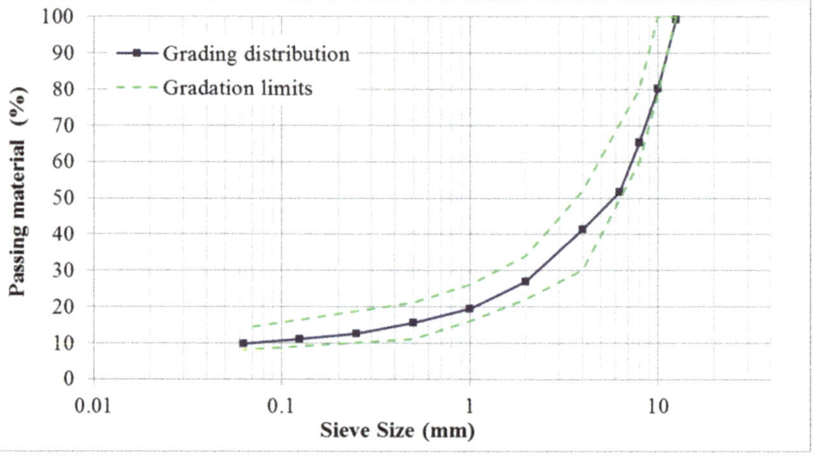

Figure 2. SMA grading distribution.

Table 2. SMA gradation.

Sieve (mm)	Retained Material (%)	Passing Material (%)
14	0.00	100.00
12	0.68	99.32
10	19.10	80.22
8	15.01	65.21
6.3	13.65	51.57
4	10.14	41.42
2	14.62	26.80
1	7.36	19.45
0.5	3.96	15.49
0.25	2.88	12.61
0.125	1.64	10.97
0.063	1.16	9.81
<0.063	9.81	-

Table 3. Conventional properties of the adopted bitumen.

Property	Unit	Characteristic Value	Standard
Penetration @ 25 °C	dmm	50–70	EN 1426 [33]
Softening Point	°C	46–54	EN 1427 [34]
Penetration Index	-	1.05–0.70	EN 12591 [35]
Dynamic Viscosity @ 60 °C	Pa·s	145	EN 12596 [36]
Ductility	%	80	ASTM D 113 [37]

Based on previous studies and laboratory tests on the same SMA mix design, the optimum binder content was fixed at 6% on the weight of aggregates, while the amount of filler was 10% on the weight of aggregates [38].

Following the technical specification of the commercial fibers and the optimum bitumen content, their ratio was 0.3% of the weight of the aggregates for the reference mix (SMA0). As for the SMAF, as shown in the following paragraph, specific drain-down tests (ASTM D6390-11 [39]) were carried out in order to evaluate the absorption properties of the shredded CFs and to define their optimum dosage within the bituminous mixture.

2.3. Experimental Program

The experimental program was based on the physical and mechanical characterization of two different SMAs produced with and without CF as stabilizing fibers. Once the mix design and the absorption properties of the shredded CFs were defined, tests were carried out on SGC (Superpave Gyratory Compactor, EN 12697-31 [40]) specimens. The mixing and compaction operations and temperatures (170 °C) were fixed univocally in order to obtain comparable results.

The volumetric properties of the two different SMAs were assessed with the air voids (AV) content analysis (EN 12697-8 [41]) on three samples for each mixture according to three different compaction levels (10, 100 and 180 gyrations). The evaluation of the workability and compactability properties of the mixtures was corroborated also by the analysis of the gyratory compaction curves.

For each mixture, nine SGS samples (100 gyrations [40]) were prepared for the mechanical characterization. The cohesion properties were assessed in terms of indirect tensile strength (ITS) at 25 °C, according to the EN 12697-23 standard [42]. The mechanical analysis was supported with indirect tensile stiffness modulus (ITSM) tests following the EN 12697-26 standard [43]. The thermal sensitivity of the mixtures was evaluated, repeating the ITSM test at three reference temperatures: 10, 20 and 30 °C. Furthermore, the water susceptibility of the SMAs was investigated in terms of ITS reduction (ITSR, EN 12697-12 [44]), carrying out ITS tests on specimens after saturation in a water bath at 40 °C for 72 h.

The durability of the experimental bituminous mixture was also evaluated in terms of resistance to permanent deformation, in compliance with the EN 12697-25 standard [45], according to a uniaxial test configuration. The repeated load axial test (RLAT) is a laboratory-based, quick method for determining the creep characteristics of bituminous mixtures under cyclic loads at a high temperature (40 °C).

3. Results

3.1. Evaluation of the Absorption Properties of CFs

As mentioned before, drain-down tests were performed in order to assess the bitumen absorption properties of the CFs. The test was carried out in compliance with the [39] standard. This test allows the amount of drain-down in a loose bituminous mixture to be determined, when this is held at high temperature, compared to the amount obtained during the mixture in-plant production, storage and transportation. It is very useful for high bitumen content mixtures, where the addition of fibers is required precisely to prevent the separation of bitumen from aggregates.

Four mixtures were tested, containing shredded CFs at a ratio of 0, 0.2, 0.3 and 0.4% of the weight of the aggregates, and two samples were produced for each mixture. The average results are presented in Table 4.

Table 4. Average drain-down results for the shredded CFs.

Fibers Amount	Drain-Down (%)
0% Fibers	0.6 (±0.2)
0.2% Fibers	0.4 (±0.2)
0.3% Fibers	0.4 (±0.1)
0.4% Fibers	0.3 (±0.1)

Based on these results, the addition of experiment fibers equal to 0.4% of the weight of the aggregates seemed to be suitable, showing average drain-down values below the recommended 0.3% [46]. All in all, the shredded CFs exhibit similar absorption properties to the reference cellulose fibers, even if an increase by 0.1% in their dosage is required to achieve the same drain-down limit effect.

3.2. Physical Characterization

The physical characterization was based on the evaluation of the air voids content (AV, [41]) in SGC samples according three different compaction levels equal to 10, 100 and 180 gyrations. This method represents a rapid control of the compactability and workability properties of bituminous mixtures. The average results are shown in Table 5, while the gyratory compaction curves are plotted in Figure 3.

Table 5. Average AV content for SGC samples.

AV Content	SMA0	SMAF
AV (%) at 10 gyrations	15.3 (±0.6)	15.5 (±0.4)
AV (%) at 100 gyrations	6.0 (±0.2)	6.4 (±0.3)
AV (%) at 180 gyrations	4.5 (±0.2)	4.9 (±0.1)

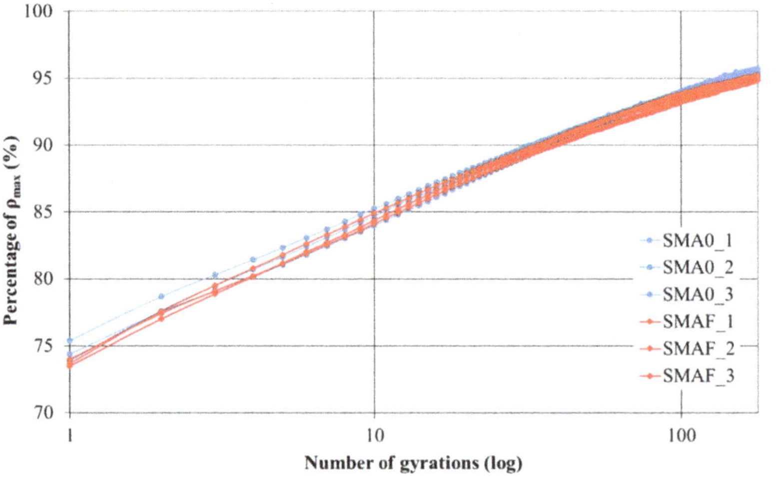

Figure 3. Bituminous mixtures gyratory compaction curves.

All in all, the results are in line with the adopted technical specification requirements (AV < 5% at N_{max}). However, a slight reduction in AV content was recorded for the reference SMA mixture at every compaction level considered.

Overall, there is not a relevant difference between the two mixtures: the presence of CFs does not negatively affect the volumetric properties of the bituminous mixture. Still, the addition of the experimental fibers did not modify or influence the laboratory mixing and compaction operations, confirming the potential application of this material for a real asphalt concrete production. The analysis of the gyratory compaction curves verified the absence of difference in terms of compactability properties due to the presence of CFs. Both SMAs showed comparable densification properties.

3.3. Mechanical Characterization

The mechanical characterization was based on the evaluation of the cohesion properties (ITS, [42]) and the mixtures' stiffness (ITSM, [42]). Furthermore, the stiffness depends on temperature variation in terms of ITSM. Furthermore, the durability of the SMAs was analyzed considering their water susceptibility and the rutting resistance properties.

3.3.1. Indirect Tensile Strength Analysis

The mechanical properties were analyzed in terms of cohesion between aggregates and bituminous mastic with the evaluation of indirect tensile strength (ITS, [42]) at 25 °C. Three SGC samples (100 gyrations [40]) were tested, and the results are presented in Table 6.

Table 6. Indirect tensile strength at 25 °C results.

Mixture	Thickness (mm)	Max Load (N)	Displacement (mm)	ITS (MPa)
SMA0_1	60.6	11400	2.44	1.20
SMA0_1	60.6	12860	1.94	1.35
SMA0_1	61.1	12720	2.28	1.33
avg. SMA0	-	-	-	1.29
SMAF_1	60.8	9250	2.96	0.97
SMAF_1	61.8	9990	2.60	1.03
SMAF_1	61.3	9250	2.60	0.96
avg. SMAF	-	-	-	0.99

Overall, both SMAs show ITS values above the threshold limit suggested by the adopted technical specification (ITS > 0.90 MPa). However, the reference mixture has the best performance. This phenomenon might be attributed to presence of the commercial stabilizing fibers, able to create a stronger cohesion between particles. It is worth noting that, as the SMA mixtures are rich in bitumen, the presence of fibers is fundamental in order to fix the bituminous mastic with the aggregates. The better performance of the reference fibers could be further confirmed by analyzing the displacement data from the ITS tests. Thus, the lower displacement registered for the SMA0 can be attributed to its stiffer structure. Furthermore, the higher ITS results for the reference mixture could also be related to the lower AV content.

3.3.2. Indirect Tensile Stiffness Modulus Analysis

The evaluation of the stiffness modulus was carried out according to the [43] with an indirect tensile test configuration. Three SGC samples (100 gyrations [40]) for each mixture were tested at 10, 20 and 30 °C in order to investigate the thermal sensitivity of the bituminous mixtures. The relation between the stiffness modulus and temperature can be described by the following equation:

$$logS = -\alpha \cdot T + \beta \tag{1}$$

where S is the stiffness modulus at the specific temperature T, while α and β are experimental parameters depending on the material properties. The former parameter is directly related to the temperature sensitivity, as it is generally higher for very temperature-sensitive materials.

The ITSM results are presented in Table 7. In Figure 4, the average results are depicted, but the equations, as well as the coefficients of determination (R^2), are related to the individual values.

Table 7. Indirect tensile stiffness modulus results.

Mixture	ITSM @ 10 °C (MPa)	ITSM @ 20 °C (MPa)	ITSM @ 30 °C (MPa)
SMA0_1	11,114	4518	1493
SMA0_1	12,608	5809	2262
SMA0_1	11,873	5577	2372
avg. SMA0	11,865	5301	2014
SMAF_1	11,041	4486	1683
SMAF_1	11,422	4665	1721
SMAF_1	11,592	4649	1653
avg. SMAF	11,352	4600	1686

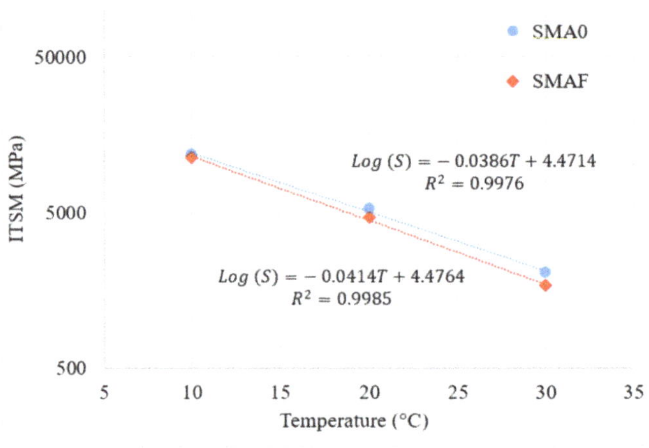

Figure 4. ITSM average results and temperature sensitivity.

In terms of ITSM, the only indication from the technical specification taken as reference is the lower limit for the stiffness modulus at 20 °C (ITSM > 3500 MPa). In this case, both mixtures exceed the threshold limit. The reference SMA shows higher values for each test temperature. Nevertheless, there is not a remarkable difference in terms of stiffness between the two mixtures. The data are in line with the previous physical characterization, with the SMA0 being stiffer than the experimental mixture. In terms of thermal sensitivity, both SMAs show the same trend as confirmed by the comparison of the α parameter in the equations. Thus, the presence of shredded CFs does not modify the mechanical behavior of the SMA mixture in terms of a response to dynamic loads at different test temperatures.

3.3.3. Water Susceptibility

The durability of both mixtures has been assessed in terms of water susceptibility. Following the [44] standard, three SGS samples (100 gyrations, [40]) were submerged in a water bath at 40 °C for 72 h. The water susceptibility is addressed as indirect tensile strength reduction (ITSR).

The ITS results after saturation in water are reported in Table 8, while the ITSR average results are reported in Table 9.

Table 8. Indirect tensile strength results after saturation in water at 40 °C for 72 h.

Mixture	Thickness (mm)	Max Load (N)	Displacement (mm)	ITS_{wet} (MPa)
SMA0_1	60.4	10,540	2.40	1.11
SMA0_1	60.0	10,420	2.32	1.11
SMA0_1	60.2	11,250	2.40	1.19
avg. SMA0	-	-	-	1.13
SMAF_1	61.8	7280	4.06	0.75
SMAF_1	60.9	10,290	2.76	1.08
SMAF_1	61.2	9020	3.50	0.94
avg. SMAF	-	-	-	0.92

Table 9. Average ITSR results.

Mixture	ITS (MPa)	ITS_{wet} (MPa)	ITSR (%)
SMA0	1.29 (±0.04)	1.13 (±0.02)	88
SMAF	0.99 (±0.03)	0.92 (±0.03)	93

All in all, both SMAs are above the ITSR limit imposed by the reference technical specification (ITSR > 75%). The experimental mixture shows a reduced ITS both in normal and wet conditions if compared to the reference SMA. However, it is worth underlining that, in absolute terms, the ITSR is higher for the SMAF. Thus, it can be stated that the presence of CFs does not negatively affect the water susceptibility of the bituminous mixtures.

3.3.4. Repeated Load Axial Test

The durability of the bituminous mixtures was also evaluated in terms of rutting resistance. In fact, this represents a common pavement disease for high bitumen content mixtures for surface layers. According to the [45] standard, method A, a bituminous sample is placed between two parallel plates. The upper plate applies a cyclic compression load and has a dimension smaller than the bituminous sample in order to achieve a certain confinement. The standard specifies a load frequency of 0.5 Hz and a pressure of 100 ± 2 kPa. The specimen's accumulated axial deformation is measured after 3600 loading cycles at 40 °C. Three SGS samples (100 gyrations, [40]) were tested for each mixture and the average results are presented in Table 10 and Figure 5.

Table 10. Average RLAT results.

Mixture	Accumulated Strain (%)	Creep Modulus (MPa)
SMA0	0.36 (±0.1)	29.6
SMAF	0.38 (±0.1)	26.5

Overall, in terms of accumulated strain, there is not a remarkable difference between the two mixtures. The reference SMA, confirming the higher performance highlighted in the previous mechanical characterization, recorded the lowest accumulated strain and, therefore, the highest creep modulus. In this case, the reference technical specifications do not specify any limit for the RLAT but the obtained results were in line with values recorded in previous studies on SMAs with polymer-modified bitumen and modified fibers [38].

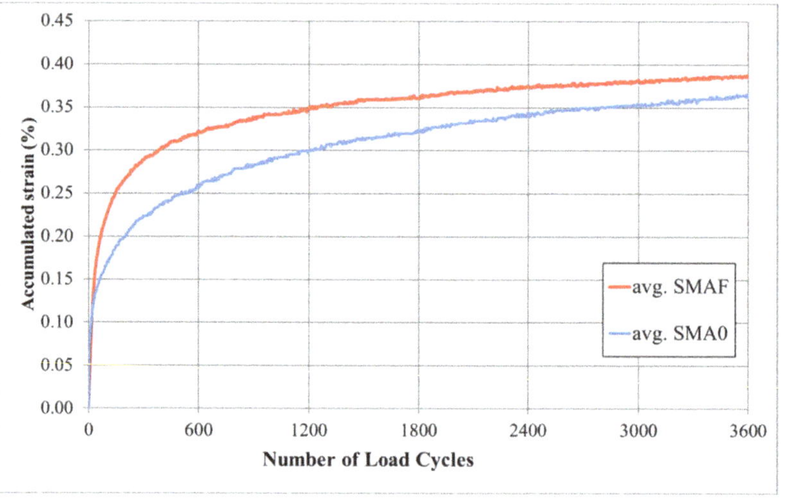

Figure 5. RLAT average results at 40 °C.

The lower performance of the experimental SMA might be related to the higher porosity of the mixture and therefore to the lower stiffness recorded for the SMAF mixture during the ITSM characterization at high temperatures.

4. Discussion

In the present work, an exploratory experimental application of CFs as stabilizing fibers for the production of a bituminous mixture has been proposed. The final aim of this research project is to develop a new recycling solution in order to promote the recycling of waste cigarette butts. Of course, a life cycle analysis (LCA) will assess the sustainability of the proposed recycling process and final application. However, it should be highlighted that the lack of disposal control for this waste is a significant problem, and only with a specific collecting and disposal protocol would the recycling of cigarette butts be feasible.

As for this first experimental application, based on the results:

- The production of the experimental fibers was based on the simple shredding of brand new CFs. The entire outcome of this process was added to the bituminous mixture as a common commercial stabilizing fiber. From the drain-down test results, the shredded CFs show properties similar to the reference fibers. Thus, from an absorption point-of-view, the behavior of the experimental fibers can be compared to the normal cellulose-based stabilizing fibers.
- The addition of shredded CFs into the experimental SMA does not affect the workability and compactability properties of the mixture. The air voids content is in line with the reference mixture and with the technical specifications for this type of asphalt concrete.
- In terms of mechanical properties, the presence of the experimental fibers leads to a slight reduction in the ITS, despite the fact that the obtained values are above the requirements imposed by the reference technical specifications. This reduction could be an indication of a lower cohesion, probably due to the limited effect of the experimental fibers in stabilizing the high bitumen content, and in generating a strong mastic with the finer particles of the mixture. However, the stiffness of the experimental SMA, as well as its thermal sensitivity, was not negatively affected by the addition of CFs.

- As for the water susceptibility, a minimum ITS reduction was registered for the experimental mixture. Furthermore, this reduction was lower even if compared to the reference SMA.
- The presence of shredded CFs does not alter the rutting resistance properties of the bituminous mixture, which were in line with common values for stone mastic asphalt mixtures.

5. Conclusions

All in all, these preliminary results confirm that the addition of shredded CFs as stabilizing fibers for bituminous mixture could be a feasible application.

However, it is worth noting that in this exploratory study, the CFs were simply shredded and directly added to the mixture in a total substitution of commercial fibers. This could represent a simple way to re-use this waste, which can be easily reproduced on an industrial scale.

Nevertheless, the obtained promising results highlighted some downsides that might be exceeded with a proper CFs treatment that could make this material much more similar to the common stabilizing fibers or could mean that it gives an even better performance. In fact, the presence of limited quantities of plug wrap paper and tipping paper after the CF shredding are probably responsible for the lower cohesion properties of the experimental SMA, if compared to the reference one.

Following the positive outcomes of this preliminary experimental application, in the development of this research project, the use of waste cigarette butts will be investigated.

Author Contributions: Conceptualization, P.T.; methodology, P.T. and C.S.; validation, C.S.; formal analysis, P.T.; investigation, P.T.; data curation, P.T.; writing—original draft preparation, P.T.; writing—review and editing, P.T. and C.S.; supervision, C.S. All authors have read and agreed to the published version of the manuscript.

Funding: This research received no external funding.

Conflicts of Interest: The authors declare no conflict of interest.

References

1. DECISION No 1386/2013/EU OF THE EUROPEAN PARLIAMENT AND OF THE COUNCIL of 20 November 2013 on a General Union Environment Action Programme to 2020 'Living Well, within the Limits of Our Planet'. Available online: http://data.europa.eu/eli/dec/2013/1386/oj (accessed on 20 November 2013).
2. Torkashvand, J.; Farzadkia, M.; Sobhi, H.R.; Esrafili, A. Littered cigarette butt as a well-known hazardous waste: A comprehensive systematic review. *J. Hazard. Mater.* **2020**, *383*, 121242. [CrossRef]
3. Mohajerani, A.; Kadir, A.A.; Larobina, L. A practical proposal for solving the world's cigarette butt problem: Recycling in fired clay bricks. *Waste Manag.* **2016**, *52*, 228–244. [CrossRef]
4. Novotny, T.E.; Zhao, F. Consumption and production waste: Another externality of tobacco use. *Tob. Control* **1999**, *8*, 75–80. [CrossRef]
5. Patel, V.; Thomson, G.W.; Wilson, N. Cigarette butt littering in city streets: A new methodology for studying and results. *Tob. Control* **2013**, *22*, 59–62. [CrossRef] [PubMed]
6. World Health Organization. *Tobacco and Its Environmental Impact: An Overview*; WHO: Geneva, Switzerland, 2017.
7. Haske-Cornelius, O.; Pellis, A.; Tegl, G.; Wurz, S.; Saake, B.; Ludwing, R.; Sebastian, A.; Nyanhogo, G.S.; Guebitz, G.M. Enzymatic Systems for Cellulose Acetate Degradation. *Catalysts* **2017**, *7*, 287. [CrossRef]
8. Brodof, T.A. The mechanisms of cellulose acetate degradation and their relationships to environmental weathering. In Proceedings of the 50th Tobacco Chemists' Research Conference, Richmond, VA, USA, October 1996. Paper 19.
9. Joly, F.X.; Coulis, M. Comparison of cellulose vs. plastic cigarette filter decomposition under distinct disposal environments. *Waste Manag.* **2018**, *72*, 349–353. [CrossRef] [PubMed]
10. Micevska, T.; Warne, M.S.; Pablo, F.; Patra, R. Variation in, and causes of, toxicity of cigarette butts to a cladoceran and microtox. *Arch. Environ. Contam. Toxicol.* **2006**, *50*, 205–212. [CrossRef]
11. Dobaradaran, S.; Schmidt, T.C.; Lorenzo-Parodi, N.; Jochmann, M.A.; Nabipour, I.; Raeisi, A.; Stojanovic, N.; Mahmoodi, M. Cigarette butts: An overlooked source of PAHs in the environment? *Environ. Pollut.* **2019**, *249*, 932–939. [CrossRef]
12. U.S. Department of Health and Human Services (USDHHS). *A Report of the Surgeon General: How Tobacco Smoke Causes Disease: What It Means to You*; U.S. Department of Health and Human Services, Centers for Disease Control and Prevention, National Center for Chronic Disease Prevention and Health Promotion, Office on Smoking and Health: Atlanta, GA, USA, 2010.

13. Huerta-Fontela, M.; Galceran, M.T.; Ventura, F. Stimulatory drugs of abuse in surface waters and their removal in a conventional drinking water treatment plant. *Environ. Sci. Technol.* **2008**, *42*, 6809–6816. [CrossRef]
14. Ariza, E.; Jiménez, J.A.; Sardá, R. Seasonal evolution of beach waste and litter during the bathing season on the Catalan coast. *Waste Manag.* **2008**, *28*, 2604–2613. [CrossRef] [PubMed]
15. Slaughter, E.; Gersberg, R.M.; Watanabe, K.; Rudolph, J.; Stransky, C.; Novotny, T.E. Toxicity of cigarette butts, and their chemical components, to marine and freshwater fish. *Tob. Control* **2011**, *20*, 25–29. [CrossRef]
16. Roder Green, A.L.; Putschew, A.; Nehls, T. Littered cigarette butts as a source of nicotine in urban waters. *J. Hydrol.* **2014**, *519*, 3466–3474. [CrossRef]
17. Marinello, S.; Lolli, F.; Gamberini, R.; Rimini, B. A second life for cigarette butts? A review of recycling solutions. *J. Hazard. Mater.* **2020**, *384*, 121245. [CrossRef]
18. Kadir, A.A.; Mohajerani, A. Recycling cigarette butts in lightweight fired clay bricks. *Proc. Inst. Civ. Eng. Constr. Mater.* **2011**, *164*, 219–229. [CrossRef]
19. Wadalkar, S.; Kulkarni, R.; Sawant, N.; Kashyap, A.; Pathan, Z.; Kale, A. Design of precast concrete blocks for paving with the use of cigarette butts (Cellulose acetate). *Int. J. Res. Appl. Sci. Eng. Technol.* **2018**, *6*, 2958–2965. [CrossRef]
20. Gomez Escobar, V.; Maderuelo-Sanz, R. Acoustical performance of samples prepared with cigarette butts. *Appl. Acoust.* **2017**, *125*, 166–172. [CrossRef]
21. Mohajerani, A.; Tanriverdi, Y.; Nguyen, B.T.; Wong, K.K.; Dissanayake, H.N.; Johnson, L.; Whitfield, D.; Thomson, G.; Alqattan, E.; Rezaei, A. Physico-mechanical properties of asphalt concrete incorporated with encapsulated cigarette butts. *Constr. Build. Mater.* **2017**, *153*, 69–80. [CrossRef]
22. Rahman, M.T.; Mohajerani, A.; Giustozzi, F. Possible Recycling of Cigarette Butts as Fiber Modifier in Bitumen for Asphalt Concrete. *Materials* **2020**, *13*, 734. [CrossRef]
23. Wu, B.; Wu, X.; Xiao, P.; Chen, C.; Xia, J.; Lou, K. Evaluation of the Long-Term Performances of SMA-13 Containing Different Fibers. *Appl. Sci.* **2021**, *11*, 5145. [CrossRef]
24. Chen, H.; Xu, Q. Experimental study of fibers in stabilizing and reinforcing asphalt binder. *Fuel* **2010**, *89*, 1616–1622. [CrossRef]
25. Slebi-Acevedo, C.J.; Lastra-Gonzalez, P.; Pascual-Muñoz, P.; Castro-Fresno, D. Mechanical performance of fibers in hot mix asphalt: A review. *Constr. Build. Mater.* **2019**, *200*, 756–769. [CrossRef]
26. Chen, J.; Lin, K. Mechanism and behavior of bitumen strength reinforcement using fibers. *J. Mater. Sci.* **2005**, *40*, 87–95. [CrossRef]
27. Eskandarsefat, S.; Hofko, B.; Oliviero Rossi, C.; Sangiorgi, C. Fundamental properties of bitumen binders containing novel cellulose-based poly-functional fibres. *Compos. Part B* **2019**, *163*, 339–350. [CrossRef]
28. Salas, M.A.; Pérez-Acebo, H. Introduction of recycled polyurethane foam in mastic asphalt. *Građevinar* **2018**, *70*, 5.
29. Wu, S.; Montalvo, L. Repurposing waste plastics into cleaner asphalt pavement materials: A critical literature review. *J. Clean. Prod.* **2021**, *280*(Pt. 2), 124355. [CrossRef]
30. Moghaddam, T.B.; Karim, M.R.; Syammaun, T. Dynamic properties of stone mastic asphalt mixtures containing waste plastic bottles. *Constr. Build. Mater.* **2012**, *34*, 236–242. [CrossRef]
31. Polaczyk, P.; Ma, Y.; Xiao, R.; Hu, W.; Jiang, X.; Huang, B. Characterization of aggregate interlocking in hot mix asphalt by mechanistic performance tests. *Road Mater. Pavement Des.* **2021**, *22* (Suppl. 1), S498–S513. [CrossRef]
32. Capitolato Speciale d'Appalto. Comune di Bologna. Capo III—Opere Stradali. 2018.
33. EN 1426: 2015. Bitumen and bituminous binders. Determination of needle penetration.
34. EN 1427: 2015. Bitumen and bituminous binders. Determination of the softening point. Ring and Ball method.
35. EN 12591: 2009. Bitumen and bituminous binders. Specifications for paving grade bitumens.
36. EN 12596: 2014. Bitumen and bituminous binders. Determination of dynamic viscosity by vacuum capillary.
37. ASTM D113—17. Standard Test Method for Ductility of Asphalt Materials.
38. Eskandarsefat, S.; Dondi, G.; Sangiorgi, C. Recycled and rubberized SMA modified mixtures: A comparison between polymer modified bitumen and modified fibres. *Constr. Build. Mater.* **2019**, *202*, 681–691. [CrossRef]
39. ASTM D6390—11. Standard Test Method for Determination of Draindown Characteristics in Uncompacted Asphalt Mixtures.
40. EN 12697-31:2019. Bituminous mixtures. Test methods. Specimen preparation by gyratory compactor.
41. EN 12697-8: 2018. Bituminous mixtures. Test methods. Determination of void characteristics of bituminous specimens.
42. EN 12697-23: 2017. Bituminous mixtures. Test methods. Determination of the indirect tensile strength of bituminous specimens.
43. EN 12697-26: 2018. Bituminous mixtures. Test methods. Stiffness.
44. EN 12697-12: 2018. Bituminous mixtures. Test methods. Determination of the water sensitivity of bituminous specimens.
45. EN 12697-25: 2020. Bituminous mixtures. Test methods. Cyclic compression test.
46. Blazejowski, K. *Stone Matrix Asphalt: Theory and Practice*; CRC Press: Boca Raton, FL, USA, 2016.

applied sciences

MDPI

Article

A Rapid Method for the Determination of SBS Content Based on the Principle of Orthogonal Testing

Gang Xu, Xiaojing Gong, Yunhong Yu and Xianhua Chen *

National Demonstration Center for Experimental Road and Traffic Engineering Education, School of Transportation, Southeast University, 2 Sipailou, Nanjing 210096, China; 230198682@seu.edu.cn (G.X.); 220194492@seu.edu.cn (X.G.); iiyyhong@163.com (Y.Y.)

* Correspondence: chenxh@seu.edu.cn

Abstract: The performance of Styrene-butadiene-styrene (SBS) modified asphalt is closely related to the content of SBS modifier. In the production process of modified asphalt, a certain amount of additive such as sulfur and rubber oil may be added to reduce the segregation and promote the swelling of the polymer, but the effect of these additives on determining SBS content in asphalt is not yet clear. This paper presents the calibration curves of SBS content based on rutting factor and creep slope and points out its defects according to the temperature scanning test and the bending beam rheometer test. Subsequently, using Fourier transform infrared spectroscopy (FTIR) for rapid determination of polymer content in SBS modified asphalt based on orthogonal test and then the effects of additives such as asphalt type, SBS content, rubber oil and sulfur on the accuracy of polymer content determination by FTIR were investigated. Moreover, in the orthogonal tests of adding sulfur and rubber oil, the multivariate analysis of variance (MANOVA) was firstly used to analyze the additives influence on the test accuracy of FITR. Results indicated that the influencing degree of different additives is different. The influence of sulfur on the determination accuracy is greater than that of rubber oil. Therefore, the rapid determination method needs further improvement.

Keywords: SBS modified asphalt; high and low temperature performance; infrared spectroscopy; additives; orthogonal test

Citation: Xu, G.; Gong, X.; Yu, Y.; Chen, X. A Rapid Method for the Determination of SBS Content Based on the Principle of Orthogonal Testing. *Appl. Sci.* **2021**, *11*, 10911. https://doi.org/10.3390/app112210911

Academic Editors: Amir Tabakovic, Jan Valentin, Liang He and Luís Picado Santos

Received: 15 August 2021
Accepted: 11 November 2021
Published: 18 November 2021

1. Introduction

By virtue of its excellent performance and environmental protection characteristics, SBS-modified asphalt has become the most commonly used modified asphalt around the world by blending SBS modifier and base asphalt [1–3]. SBS polymer is used as a dispersed phase and physically dispersed into the continuous phase of base asphalt. Due to the large differences between SBS polymer and matrix asphalt in chemical structure and physical parameters such as density, polarity, solubility, and molecular weight, SBS is distributed only in the form of particles in asphalt medium. It has a two-phase structure, which is an unstable thermodynamic incompatible system [4]. Therefore, in order to reduce the segregation and promote the swelling of the polymer in the production process of the modified asphalt, a certain amount of additive such as solubilizer and stabilizer may be added. The solubilizer usually uses petroleum distillate with high aromatic hydrocarbon content and low asphaltene content to improve the ratio of four components in the base asphalt [5–7]. The aromatic hydrocarbon can promote the dissolution of polystyrene blocks in the SBS polymer, making the SBS modifier more compatible with the matrix asphalt. Among all the solubilizers used in the market, rubber oil is widely accepted because of its low cost and good solubilizing performance. In terms of stabilizer, it is often used to improve the storage stability of SBS-modified asphalt and reduce the problem of segregation. The stabilizers currently used for SBS-modified asphalt mainly include the following four categories [8]: sulfur containing stabilizer, such as elemental sulfur and sulfur compounds, polyolefin stabilizer, like polyisobutene, inorganic acid and inorganic metal oxide stabilizers, such as phosphoric acid and zinc oxide, and inorganic clay stabilizer like montmorillonites and kaolin clay. Among them, sulfur-containing stabilizers, especially sulfur, stand out and are widely used in the industrial production of SBS-modified asphalt because of construction convenience, low requirements for equipment and low cost. It can be used directly after mixing with modified asphalt, and the vulcanization crosslinking reaction is mild and easy to control [5,9,10].

SBS is a block polymer of polystyrene-polybutadiene-polystyrene. After being added to asphalt, it is sheared into fine particles and swelled, gradually forming a network-like crosslinked structure. The modification effect of asphalt will be affected by its content. Studies have shown that the microstructure of SBS modified asphalt is largely determined by the content of SBS [11]. The morphology and interaction between modified asphalt molecules will be different with a different SBS content, thus affecting the rheological properties and mechanical properties of modified asphalt. At present, the main methods to determine the modifier content of SBS-modified asphalt include fluorescence microscopy, gel permeation chromatography, chemical analysis, and infrared spectroscopy [12–15]. Fourier-Transform Infrared (FTIR) spectroscopy stands out among various methods, because of the simple test process, quick and easy operation, short data processing time and little error. Being able to quickly determine the content of SBS modifier with higher accuracy, it has become the main method for rapid determination of SBS content in engineering.

However, existing studies on the rapid determination of SBS content by FTIR had only considered a single factor of SBS content and had not considered the influence of admixtures on the test results [16,17]. Therefore, for the SBS modified asphalt with additives such as sulfur and rubber oil, the results of determining SBS content by FTIR may have a certain bias, and the accuracy is not known. It is necessary to design tests to explore the influence of additives on the accuracy of using FTIR to determine the polymer content in modified asphalt. For the case with multiple influencing factors, the commonly used test methods include comprehensive test and orthogonal test. Compared with comprehensive test, the orthogonal test can reduce test times and improve efficiency, which is essential for the cases with multiple influencing factors and levels.

Therefore, in view of this problem, this paper firstly analyzes the effect of SBS content on the high and low temperature performance of modified asphalt, and then establishes the calibration curve of SBS content and characteristic absorption peak absorbance in modified asphalt according to the infrared spectrum. Finally, the orthogonal test is designed, and the results are analyzed by the multivariate analysis of variance to explore the influence of additives on the determination results of SBS in modified asphalt.

2. Experimental

2.1. Temperature Scanning Test by Dynamic Shear Rheometer

The DSR (Dynamic Shear Rheometer) can be used to measure the complex shear modulus (G^*) and phase angle (δ) of the asphalt binder at multiple temperatures to characterize the viscoelastic characteristics of the asphalt materials. At the same time, the rutting factor ($G^*/sin\delta$) is calculated and used as the main index to evaluate the high temperature performance of asphalt. The complex shear modulus (G^*) can be considered to measure the total resistance to deformation of the asphalt when it is repeatedly sheared [18]. The higher the temperature, the smaller the complex shear modulus (G^*) and the worse the anti-deforming capability of the asphalt. The phase angle (δ) is defined as the ratio of the time of strain hysteresis stress to the corresponding stress cycle in a loading cycle. It can be used to characterize the sensitivity of asphalt binders under certain temperature as environmental conditions change. The smaller the phase angle (δ), the more elastic the material. When $\delta = 0^{\circ}$, the material is purely elastic; and When $\delta = 90^{\circ}$, the material is purely viscous [18].

2.2. Bending Beam Rheometer Test

A BBR test was employed to characterize the low-temperature performance of SBS modified asphalt in all combinations according to ASTM D6648. The test temperature was -12, -18 °C and the average results of three replicates were used as the testing results. Creep stiffness modulus (S) and creep rate (m-value) was employed to characterize the low temperature performance. The greater the creep stiffness modulus S value, the more elastic, less viscous and more brittle the asphalt, and the worse the low temperature resistance to deformation. The larger the creep slope m value, the stronger the anti-deformation ability of the asphalt and the better the low temperature performance [19,20].

2.3. Infrared Spectroscopic Analysis

The content of SBS in modified asphalt is quantitatively detected by infrared spectroscopy. As literature indicates, when continuous infrared light is irradiated on the material, the energy level transition occurs in the material molecules, and the substance absorbs infrared light of a specific wavelength to obtain a graph showing changes in absorbance at different wavelengths, which is an infrared spectrum. Its molecular structure is reflected by the infrared spectrum, thereby identifying heteroatom compounds in the asphalt and functional groups of SBS [17,21].

Generally, in the SBS modified asphalt, the base asphalt and the SBS modifier are physically mixed without chemical reaction, so the functional groups of the two do not disappear or add. In the modified asphalt, the vibration frequency of the SBS polymer molecules and the matrix asphalt molecules after infrared light irradiation is different, and the wavelengths of the infrared light absorption are also different, so the positions of the absorption peaks displayed on the infrared spectrum differ. For the modified asphalt with different SBS content, although the absorption peaks positions of the SBS polymer molecules and the matrix asphalt molecules are roughly the same in the infrared spectrum, their absorbance is not the same. Therefore, by comparing the absorption peaks of matrix asphalt and modified asphalt in the infrared spectrum, the existence of SBS modifier can be qualitatively identified and the content of modifier can be quantitatively analyzed [22].

Fourier infrared spectroscopy is based on Lambert-Beer law to achieve quantitative analysis of SBS modifier in modified asphalt as shown in Equation (1). Lambert-Beer law is that when a beam of light penetrates a sample of material, the absorbance at a certain wavenumber (v) is related to the concentration and the optical path length of the material sample, i.e.,

$$A(v) = \lg\left(\frac{1}{T(v)}\right) = a(v)bc \tag{1}$$

where: $A(v)$ is the absorption intensity (absorbance) at wavenumber v, $T(v)$ is the transmittance at wavenumber v, $a(v)$ is the absorbance coefficient at wavenumber v, is the optical path length (the thickness of the sample, mm), is the concentration of the material sample (%).

Therefore, under the same test conditions, the characteristic absorption peak area ratio of SBS has linear relation with its content. The absorption peak area is selected as the characterization of absorption intensity in the infrared test. Infrared spectroscopy tests are carried out on a series of modified asphalt samples with known SBS content to establish a calibration curve of characteristic absorption peak area ratio and SBS content. By measuring the ratio of characteristic peak areas of SBS modified asphalt under the same test conditions the SBS content can be determined according to this calibration curve.

In order to establish the calibration curve of different SBS content, the matrix asphalt, SBS modifier and different modified asphalt samples with the SBS content of 2%, 4%, 6% are completely dissolved in tetrahydrofuran and titrated to KBr tablets. These tablets are placed in an oven at 60 °C for 20 min, and the infrared spectrum samples are prepared after the solvent is completely evaporated. By scanning each sample with an infrared spectrometer, the infrared spectrum is obtained. The spectral acquisition interval is 366~4000 cm^{-1} and the resolution is 2 cm^{-1}.

2.4. Orthogonal Experimental Design

Three factors including SBS, rubber oil and sulfur, which could influence the testing index of infrared spectroscopy, are considered in this paper. Each factor includes three levels of content, which are listed in Table 1. The properties of SBS-modified asphalt and base asphalt are listed in Table 2. Orthogonal tests can be used for multi-factor and multi-level tests. The orthogonal test is based on the orthogonality to select some representative points from the comprehensive test. These representative points have the characteristics of uniformity, dispersion and comparability, which can analyze the comprehensive influence of multiple factors with a small amount of test. It can reduce the number of tests, shorten the test period, and then improve the test efficiency. Therefore, for the case where there are many influencing factors and levels in this subject, the orthogonal test can not only reduce the number of tests, but also make the testing results intuitive and easy to analyze, which is of high rationality.

Table 1. Levels of different factors.

Level	B (SBS Content/%)	C (Sulfur Content/%)	D (Rubber Oil Content/%)
I	2%	0.0%	0%
II	4%	0.1%	2%
III	6%	0.2%	4%

Table 2. Properties of Asphalt.

Test Parameters	SBS Modified Asphalt	Base Asphalt
Penetration(25 °C, 100 g, 5 s)/0.1 mm	56	67.4
Softening point/°C	85.8	46.5
Ductility(5 °C, 5 cm/min)/cm	32	–
Ductility(15 °C, 5 cm/min)/cm		115
Mass change/%	0.11	0.04

Three factors and three levels are involved in this study. Due to the repeated tests, the L_9 (3^3) orthogonality table without blank columns can be used for the sake of simplicity. If the three-factor and three-level tests are carried out in accordance with the requirements of comprehensive test, $3^3 = 27$ combinations of tests must be carried out, and the parallel tests of each combination have not been considered. However, if the test is arranged according to the L_9 (3^3) orthogonality table, only 9 rounds of tests need to be carried out, obviously reducing the workload. The factor level table is listed in Table 1.

2.5. MANOVA

The results of the orthogonal test are usually analyzed in two ways: direct-viewing analysis and analysis of variance (ANOVA). The direct-viewing analysis method is also called the range analysis method, which determines the primary and secondary relationships among the influencing factors by calculating the range of data and comparing its size. However, when the test results corresponding to different levels of a certain factor are different, the direct-viewing analysis method cannot distinguish whether this difference is caused by different levels of the factor or experimental errors. ANOVA can make up for the limitations of direct-viewing analysis and provide reliable analysis results [23]. ANOVA is usually used in the significance test for differences between two or more sample data, which is also known as *F* test or analysis of variability [24]. Generally, due to the influence of various factors, the data obtained from the test will fluctuate. ANOVA can be used to evaluate the volatility of the test data and the influence of various factors on the test results to find certain regularity. ANOVA for orthogonal test can be carried out by SPSS (Statistical Program for Social Sciences) data analysis tool.

Therefore, this paper uses ANOVA to analyze the results of orthogonal tests. The orthogonal test is a multi-factor and single-index test. The influencing factors are SBS content, sulfur content and rubber oil content. It is assumed that the effects of the three are independent of each other and there is no correlation. Thus, the multi-factor univariate ANOVA model with no interaction in SPSS is used to compare and analyze the test results of three factors and three levels. Multivariate analysis of variance (MANOVA) uses the *F* test, whose null hypothesis is H_0: the observed variable values at different levels of each factor have no significant difference. SPSS will automatically calculate the F value and provide the corresponding probability *p* value according to the *F* distribution table. It can be judged whether the different levels of each factor have a significant influence on the observed variables by comparing the *p* value with the magnitude of the significance level α.

3. Results and Discussion

3.1. Rheological Test Results

3.1.1. Temperature Scanning Test Results

The temperature sweep test is carried out on different virgin SBS modified asphalt samples by DSR. The complex shear modulus G^* and phase angle δ of three modified asphalt at multiple temperatures are measured to calculate the rutting factor $G^*/sin\delta$. According to the test results, the curves of rutting factor $G^*/sin\delta$ with temperature under different SBS content are plotted, as shown in Figure 1.

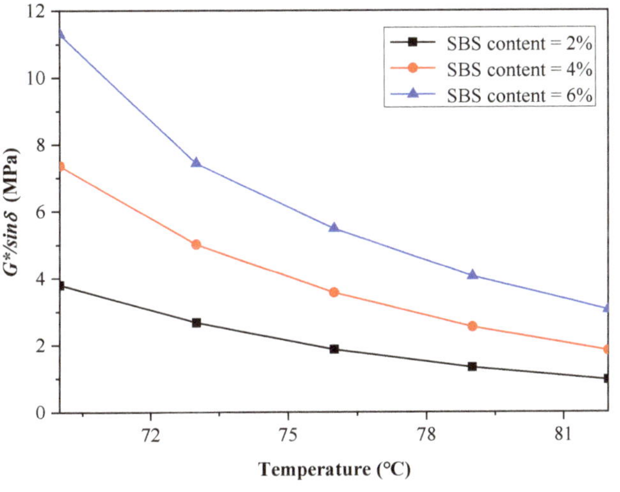

Figure 1. Curves of rutting factor with temperature under different SBS content.

Figure 2 reflects the temperature curves of rutting factor $G^*/sin\delta$ of the three modified asphalt. With the increase of the modifier content, the $G^*/sin\delta$ value of the modified asphalt increases; and as the test temperature increases, the $G^*/sin\delta$ value tends to decrease. This means that the higher the SBS content, the lower the temperature, and the larger the value of $G^*/sin\delta$, indicating that the asphalt is more elastic, the high-temperature flow deformation is smaller, the high-temperature performance is better, and the anti-rutting ability is stronger.

The high temperature performance index of modified asphalt increases with the increase of SBS content, indicating that the higher the SBS content, the better the high temperature performance of the modified asphalt. This is because SBS polymer and matrix asphalt are two completely different materials. SBS is dispersed as particulates in the matrix asphalt under mechanical force. With a different SBS content, SBS has a different distribution in the asphalt. When the SBS content is small, SBS is scattered as a dispersed phase in the continuous asphalt phase. When the SBS content is high, the SBS dispersed phase forms a network structure in the asphalt. If the SBS content is continuously increased, the asphalt will change phase. Therefore, the larger the content of SBS, the more uniform it is distributed in the asphalt. The formation of a three-dimensional network structure can well limit the movement of the entire molecular system of the modified asphalt at high temperature, thereby improving the high temperature performance of the asphalt.

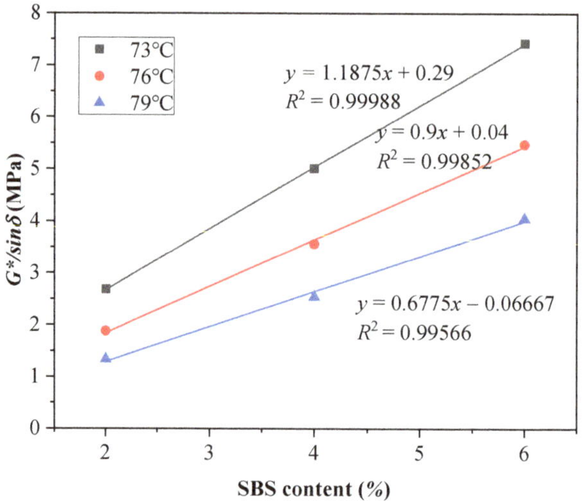

Figure 2. Calibration curves of SBS content based on rutting factor.

According to the SBS content and the calculation results of rutting factor, the fitting curves of $G^*/sin\delta$ value and SBS content at different test temperatures can be drawn, taking 73 °C, 76 °C and 79 °C as examples, as shown in Figure 2. It is obvious that the rutting factor has a good linear relation with the SBS content at different temperatures, which can be used for the calibration of SBS content in modified asphalt. However, the good linear relation is related to temperature. The rutting factor increases linearly with the increase of SBS content at different temperatures, but its growth slope is significantly different. When using this result to determine the SBS content in the modified asphalt, the specific calibration curve at the actual temperature of the asphalt needs to be found. So, this method is not simple. In addition, there is another problem with the determination of SBS content in modified asphalt using a calibration curve based on the rutting factor: this method is susceptible to other modifiers in the asphalt. Due to the high market price of SBS, some modified asphalt manufacturers privately reduce the SBS content and add some cheap modifiers to reduce the cost for personal gain. This makes the rutting factor value of the produced SBS modified asphalt can also meet the requirements of the specification, but its road performance is unknown. In this case, the SBS content measured by the calibration curve based on the rutting factor is qualified with a quite different truth.

3.1.2. Bend Beam Rheometer Test Results

The creep stiffness modulus S, creep rate m of the matrix asphalt and the three modified asphalts in long-term aging were determined by bending beam rheometer at different test temperatures. The test results are recorded in Table 3.

According to the above test results, the linear fitting curves of the stiffness modulus S and the creep slope m with the SBS content at different test temperatures are plotted respectively, as shown in Figure 3. It is obvious that at −18 °C, the creep stiffness S of modified asphalt decreases with the increase of the SBS content, which indicates that as the SBS content increases, the asphalt becomes more elastic, more viscous and less brittle, and the resistance to deformation increases. At −12 °C, although the S value of modified asphalt is smaller than that of matrix asphalt, its variation with the SBS content is not stable.

At the same time, the creep stiffness *m* value of modified asphalt is lower than that of matrix asphalt at both test temperatures, both decreasing with the increase of SBS content. Moreover, when the SBS content is the same, the lower the test temperature, the lower the *m* value, which means that with the decrease of temperature, the anti-deformation ability of the asphalt decreases, and the low temperature performance becomes worse.

Table 3. BBR test results.

Temperature (°C)	SBS Content	S (MPa)	*m*
−12	0%	125	0.420
	2%	102	0.417
	4%	109	0.409
	6%	103	0.387
−18	0%	387	0.287
	2%	256	0.329
	4%	232	0.314
	6%	209	0.298

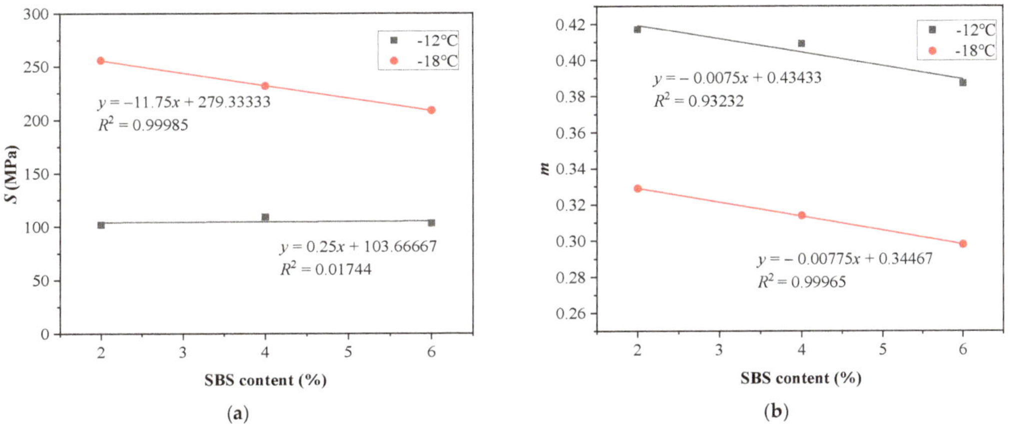

Figure 3. Calibration curves of SBS content based on (**a**) stiffness modulus *S* and (**b**) creep slope *m*.

It can be seen from the figures that the linear fitting results of creep slope *m* with SBS content at both test temperatures are more precise. The stiffness modulus *S* and SBS content also have a good linear relation at −18 °C, but the linear fitting results at −12 °C are not accurate. This is because, as can be seen from Table 3 the stiffness modulus of the matrix asphalt has reached 125 MPa at −12 °C, which is not much different from that of the modified asphalt after the addition of 2% modifier. Therefore, the change of *S* value with the increase of the amount at −12 °C is not obvious. The fitting result of creep slope *m* with SBS content can be used to calibrate the SBS content in modified asphalt, but the good linear relation of the result is also related to temperature, which means the calibration curves differ at different temperatures. The problem with this method is also the same as the calibration method based on the rutting factor.

3.2. FTIR Test Results

3.2.1. Analysis of FTIR Test Results

The infrared spectrum of each sample is obtained by scanning it with an infrared spectrometer. Figure 4 shows the infrared spectrum of matrix asphalt, SBS, rubber oil and SBS-modified asphalt in the wave number region between 500–1400 cm^{-1}. Comparing the infrared spectrum of matrix asphalt and SBS modifier, it can be found that the methyl and methylene groups in matrix asphalt produce a characteristic absorption peak near 1377 cm^{-1}, which is not affected by SBS modifier because SBS does not have this peak. The infrared spectrum of SBS has two strong absorption peaks, one is formed by the out-of-plane rocking vibration of C-H (polystyrene) in benzene ring at 699 cm^{-1}, and the other is formed by the out-of-plane rocking vibration of trans-butadiene = CHC_2 (polybutadiene) at 966 cm^{-1} [25].

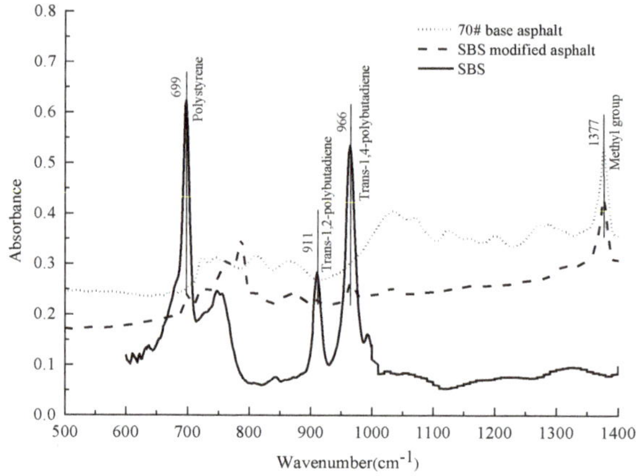

Figure 4. Infrared spectrum of SBS, base asphalt, rubber oil and SBS modified asphalt.

Since the SBS modifier is physically compatible with matrix asphalt, the infrared spectrum of SBS-modified asphalt is a simple superposition of that of SBS and base asphalt, where no new peaks appear, or existing peaks disappear [13]. It can be seen from Figure 4 that the modified asphalt retains the SBS characteristic absorption peaks at 699 cm^{-1} and 966 cm^{-1}, as well as a characteristic absorption peak of matrix asphalt at 1377 cm^{-1}. The absorption peaks of modified asphalt at 699 cm^{-1} and 966 cm^{-1} can be used to judge the existence of SBS modifier, and the SBS content can be quantitatively analyzed according to the intensity of the absorption peak.

Figure 5 is the infrared spectrum of the modified asphalt with different SBS content. The peak area of the absorption peak at 699 cm^{-1} and 966 cm^{-1} has a certain relationship with SBS modifier content, and the intensity of the three absorption peaks does not affect each other. Therefore, the content of SBS modifier can be quantitatively analyzed based on the ratio of the absorption peak area at 699 cm^{-1}, 966 cm^{-1} and 1377 cm^{-1} in the infrared spectrum of the modified asphalt.

The absorbance A is measured by spectral peak area. In this paper, the ratio of characteristic peak area at 699 cm^{-1} to that at 1377 cm^{-1} is A_1, and the ratio of the peak area at 966 cm^{-1} to that at 1377 cm^{-1} is A_2, i.e.,

$$A_1 = \frac{peak\ area\ (intensity)\ at\ 699\ \text{cm}^{-1}}{peak\ area\ (intensity)\ at\ 1377\ \text{cm}^{-1}}$$

$$A_2 = \frac{peak\ area\ (intensity)\ at\ 966\ \text{cm}^{-1}}{peak\ area\ (intensity)\ at\ 1377\ \text{cm}^{-1}}$$

Without any other additives, two kinds of base asphalt (SK70# and Zhongshiyou asphalt) are used to respectively prepare SBS-modified asphalt with the SBS content of 2%, 4% and 6%. Five sets of parallel tests were conducted for each content. By using the OMNIC software reading tool to read the absorbance value the infrared spectrum of these two modified asphalt groups and the two kinds of matrix asphalt is determined. The standard curves of characteristic absorption peak area ratio A_1, A_2 and SBS content are drawn, as shown in Figure 6.

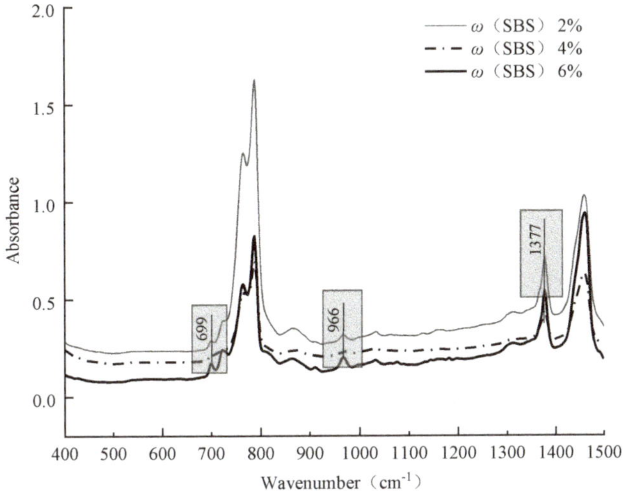

Figure 5. Infrared spectrum of SBS modified asphalt with different SBS content.

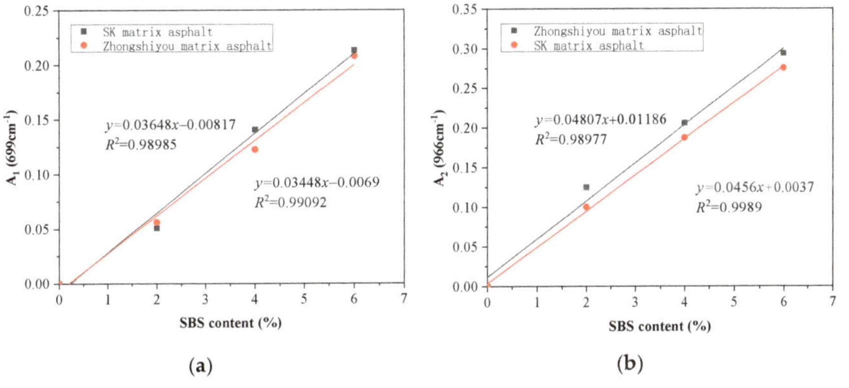

Figure 6. Standard curves of A value with SBS content at different absorbance (**a**) 699 cm^{-1} and (**b**) 966 cm^{-1}.

Regardless of the type of matrix asphalt, the A_1 and A_2 values at 699 cm^{-1} and 966 cm^{-1} also increase with the increase of SBS content. The A value has a good linear relation with the SBS content, and the correlation coefficient is very close to 1, it is consistent with the results of existing studies [16,17]. Both the A_1 and A_2 values can be used as the basis for determining the SBS content in modified asphalt with the correlation not being affected by the type of matrix asphalt. However, the standard equations of the two kinds of base asphalt are quite different, so the original matrix asphalt must be obtained when using this method to determine SBS content, and it cannot be replaced by matrix asphalt of different sources.

3.2.2. Analysis Impact of Additives Based on Orthogonal Test

According to the orthogonality table, nine kinds of modified asphalt are prepared by 0.0%, 0.1%, 0.2% sulfur combined with 0%, 2%, 4% rubber oil. Five parallel samples are prepared for each kind of modified asphalt. FTIR test is carried out on each sample, reading the corresponding absorption peak areas and calculating the infrared spectrum A value as a test index. The test plan is listed in Table 4.

Table 4. Orthogonal test plan.

Test Number	B (SBS Content/%)	C (Sulfur Content/‰)	D (Rubber Oil Content/%)	Test Plan
1	2	0	0	$B_1C_1D_1$
2	2	1	2	$B_1C_2D_2$
3	2	2	4	$B_1C_3D_3$
4	4	0	2	$B_2C_1D_2$
5	4	1	4	$B_2C_2D_3$
6	4	2	0	$B_2C_3D_1$
7	6	0	4	$B_3C_1D_3$
8	6	1	0	$B_3C_2D_1$
9	6	2	2	$B_3C_3D_2$

Taking A_1 and A_2 values as the test indexes, MANOVA is conducted on the orthogonal test results respectively using SPSS21.0, which adopts the system default significance level $\alpha = 0.05$. The results of MANOVA are listed in Tables 5 and 6, which show the calibration model test in ANOVA. The original hypothesis is that the SBS content, sulfur content and rubber oil content in the model have no effect on A value that is the test index. If the probability p value is less than the significance level α, it means that the variance model is statistically significant, that is, at least one of the three influencing factors has a significant influence on the A value.

Table 5. Inter-subject effect test of three factors.

Source	Type III Sum of Squares	df	Mean Square	F	p
Corrected Model	0.152 [a]	6	0.025	131.704	1.000×10^{-13}
Intercept	0.474	1	0.474	2462.990	1.000×10^{-13}
SBS content	0.145	2	0.072	376.333	1.000×10^{-13}
Sulfur content	0.004	2	0.002	10.373	2.771×10^{-4}
Rubber oil content	0.002	2	0.001	6.094	5.255×10^{-13}
Error	0.007	36	0.000	—	—
Total	0.642	43	—	—	—
Corrected Total	0.159	42	—	—	—

[a]. $R^2 = 0.956$ (Adjusted $R^2 = 0.949$), Dependent Variable: A_1 Value (699 cm^{-1}).

Table 6. Inter-subject effect test of three factors.

Source	Type III Sum of Squares	df	Mean Square	F	p
Corrected Model	0.246 [a]	6	0.041	162.999	1.000×10^{-13}
Intercept	1.533	1	1.533	6093.735	1.000×10^{-13}
SBS content	0.222	2	0.111	441.598	1.000×10^{-13}
Sulfur content	0.017	2	0.008	33.542	1.000×10^{-13}
Rubber oil content	0.005	2	0.002	9.867	1.000×10^{-13}
Error	0.009	36	0.000	—	—
Total	1.824	43	—	—	—
Corrected Total	0.255	42	—	—	—

[a]. $R^2 = 0.964$ (Adjusted $R^2 = 0.959$), Dependent Variable: A_2 Value (966 cm^{-1}).

1. A_1 value as the test index

Comparing the p value with the significance level $\alpha = 0.05$, it can be known whether each factor has a significant influence on the observation results, while the comparison of F value shows the influencing degree of each factor. It can be seen from Table 5 that the p values of SBS content, sulfur content and rubber oil content are much smaller than the significance level α, indicating that these three factors have significant influence on the infrared spectrum absorbance A_1 value. However, it's quite different from the influencing degree of the three factors. Table 5 showed $F_B > F_C > F_D$, the F value of SBS content is 376.333, which is much larger than that of sulfur and rubber oil content, and the F value of sulfur content is also larger than that of rubber oil content. It indicates that the most influential factor for the absorbance A_1 value is the SBS content, followed by sulfur content and rubber oil content.

2. A_2 value as the test index

Similarly, as can be seen from the results of MANOVA in Table 6, for the absorbance A_2 value, the probability p values of SBS content, sulfur content and rubber oil content are all close to 0, all significantly less than the significance level $\alpha = 0.05$, which means all these three factors have significant influence on the A_2 value. According to the critical value of the 3-factor significance level $\alpha = 0.05$ for the 2 degrees of freedom is $F_{(2,3)0.05} = 19.164$, Table showed $F_{sbs} = 441.598 > F_{(2,3)0.05}$, $F_{sulfur} = 33.542 > F_{(2,3)0.05}$, $F_{rubber\ oil} = 9.867 < F_{(2,3)0.05}$. Moreover, comparing the F values of the three factors, it is found that the F value of SBS content is much larger than that of sulfur content and rubber oil content, and the F value of sulfur content is also larger than that of rubber oil content. This means that the influence of these three factors on the absorbance A_2 value is different, which is the same as the case where the A_1 value is the test index. In order of their effect, SBS content is the largest, sulfur content followed, and rubber oil content is the smallest. Furthermore, Using SPSS to remove the effects of other variables, the marginal mean was estimated at different levels of each factor and was shown in Figure 7. Figure 7 illustrated that the A_2 increased when the content of SBS and rubber oil increased, but reduced when the content of sulfur increased. Therefore, at the significance level $\alpha = 0.05$, the three factors of SBS content, sulfur content and rubber oil content have significant influence on the A_1 and A_2 values, which are the characteristic peak area ratios of SBS-modified asphalt at 669 cm^{-1} and 966 cm^{-1} in the infrared spectrum. The influencing degree is both: SBS content > sulfur content > rubber oil content. This means under different SBS content, the addition of sulfur and rubber oil will have a significant effect on the accuracy of FTIR test results, and the influencing degree of sulfur is greater than that of rubber oil. The reason may be that the rubber oil added to the modified asphalt promotes the cross-linking of SBS polymer without changing the characteristic functional groups in SBS. However, after the addition of sulfur, the double bond of polybutadiene block in the SBS polymer reacts with the hetero atom in asphalt to form SBS-asphalt grafts, which changes the characteristic functional groups in SBS, thereby affecting the magnitude of A value. Therefore, the method of rapidly determining the SBS content in modified asphalt by infrared spectroscopy needs to be further improved.

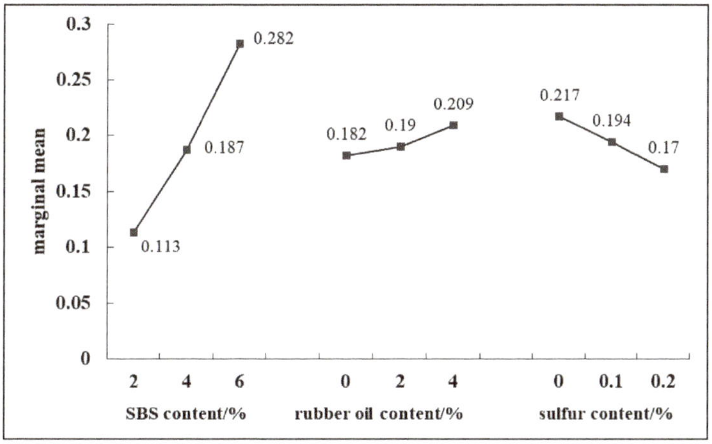

Figure 7. Marginal mean of all three factors.

4. Conclusions

In this paper, based on temperature shear test and bending beam rheological test, the influence of SBS content on the high and low temperature performance of modified asphalt is preliminarily analyzed. The calibration curves of SBS content based on rutting factor and creep slope are presented and its defects are pointed out. Meanwhile, the calibration curve of determining SBS content by FTIR is established, and the orthogonal test is designed to explore the influence of additives sulfur and rubber oil on the determination results. The main conclusions are as follows:

(1) The rutting factor $G^*/sin\delta$ and the creep slope m both have a good linear relation with the SBS content. The calibration curves of the two indexes can be used to determine the SBS content in modified asphalt. However, the linear fitting of these two factors is related to temperature, and the determination results are also susceptible to other modifiers in the asphalt, so it is not worth promoting.

(2) It can be seen from the infrared spectrum that the SBS modified asphalt retains the characteristic absorption peaks of SBS modifier at 699 cm^{-1} and 966 cm^{-1}, as well as the characteristic peak of matrix asphalt at 1377 cm^{-1}. The area of the absorption peaks at 699 cm^{-1} and 966 cm^{-1} has a certain relationship with the SBS modifier content. Thus, the characteristic peak area at 1377 cm^{-1} can be used as a reference to respectively establish the linear regression curve of SBS content and the ratio of the absorption peak area at 699 cm^{-1} and 966 cm^{-1} to that at 1377 cm^{-1}. It can well predict the content of the modifier in SBS modified asphalt.

(3) In the orthogonal tests, MANOVA is used to analyze the effect of different additives on the determination results. It is found that at the significance level $\alpha = 0.05$, the additive sulfur and rubber oil have significant effect on the determination of SBS content in modified asphalt by FTIR, and the influence of sulfur content is greater than that of rubber oil. Therefore, the FTIR method for rapid determination of SBS content in modified asphalt needs to be further improved.

Author Contributions: Conceptualization, G.X. and X.C.; methodology, X.G.; validation, X.C., Y.Y.; formal analysis, X.G.; investigation, G.X.; resources, X.C.; data curation, X.G.; writing—original draft preparation, G.X.; writing—review and editing, X.G. and Y.Y.; supervision, X.C.; project administration, X.C.; funding acquisition, X.C. All authors have read and agreed to the published version of the manuscript.

Funding: This research was funded by National Natural Science Foundation of China (No. 51778136), Technology Research and Development Program of China State Railway Group Co., Ltd. (K2020G032). And the APC was funded by Technology Research and Development Program of China State Railway Group Co., Ltd., Beijing, China.

Institutional Review Board Statement: Not applicable.

Informed Consent Statement: Not applicable.

Data Availability Statement: Not applicable.

Conflicts of Interest: The authors declare no conflict of interest.

References

1. Stangl, K.; Jaeger, A.; Lackner, R. The effect of styrene-butadiene-styrene modification on the characteristics and performance of bitumen. *Monatsh. Chem.* **2007**, *138*, 301–307. [CrossRef]
2. Larsen, D.O.; Alessandrini, J.L.; Bosch, A.; Cortizo, M.S. Micro-structural and rheological characteristics of SBS-asphalt blends during their manufacturing. *Constr. Build. Mater.* **2009**, *23*, 2769–2774. [CrossRef]
3. Liang, M.; Liang, P.; Fan, W.; Qian, C.; Xin, X.; Shi, J.; Nan, G. Thermo-rheological behavior and compatibility of modified asphalt with various styrene-butadiene structures in SBS copolymers. *Mater. Design.* **2015**, *88*, 177–185. [CrossRef]
4. Lin, P.; Huang, W.; Li, Y.; Tang, N.; Xiao, F. Investigation of influence factors on low temperature properties of SBS modified asphalt. *Constr. Build. Mater.* **2017**, *154*, 609–622. [CrossRef]
5. Jin, H.L.; Gao, G.T.; Zhang, Y.; Zhang, Y.X.; Sun, K.; Fan, Y.Z. Improved properties of polystyrene-modified asphalt through dynamic vulcanization. *Polym. Test.* **2002**, *21*, 633–640. [CrossRef]

6. Huang, W.; Tang, N. Characterizing SBS modified asphalt with sulfur using multiple stress creep recovery test. *Constr. Build. Mater.* **2015**, *93*, 514–521. [CrossRef]
7. Aguirre de Carcer, I.; Masegosa, R.M.; Teresa Vinas, M.; Sanchez-Cabezudo, M.; Salom, C.; Prolongo, M.G.; Contreras, V.; Barcelo, F.; Paez, A. Storage stability of SBS/sulfur modified bitumens at high temperature: Influence of bitumen composition and structure. *Constr. Build. Mater.* **2014**, *52*, 245–252. [CrossRef]
8. Yang, L. Research progress of SBS modified asphalt stabilizer. *Shanxi Transp. Technol.* **2013**, *4*, 1–2.
9. Zhang, F.; Yu, J.; Wu, S. Effect of ageing on rheological properties of storage-stable SBS/sulfur-modified asphalts. *J. Hazard. Mater.* **2010**, *182*, 507–517. [CrossRef]
10. Wen, G.; Zhang, Y.; Zhang, Y.X.; Sun, K.; Chen, Z.Y. Vulcanization characteristics of asphalt/SBS blends in the presence of sulfur. *J. Appl. Polym. Sci.* **2001**, *82*, 989–996. [CrossRef]
11. Li, W.C.; Fan, L.; Lin, L.P. Effect of SBS content on performance of modified asphalt. *Shandong Transp. Technol.* **2009**, *1*, 35–37.
12. Shi, Y.M.; Yang, Q.L.; Pan, C.L.; Cao, S.J.; Liu, Y.L.; Wang, W.L. Feasibility analysis of SBS content determination in modified asphalt based on infrared spectrum. *Transp. Technol.* **2018**, *1*, 127–130.
13. Xiao, P.; Kang, A.H.; Liu, P.P. Study on the quantitative performance of modified asphalt by fluorescence microscopy. *Pet. Asph.* **2005**, *19*, 45–48.
14. Geng, J.G.; Chang, Q.; Yuan, J.A.; Dai, J.L. Study on crosslinking structure and stability of SBS modified asphalt by GPC. *J. Zhengzhou Univ. Eng.* **2008**, *29*, 14–17.
15. Wang, L.H.; Xia, Y. Study on detection method of SBS content in modified asphalt. *Chem. Eng.* **2012**, *2*, 24–25.
16. Wang, K. Application of FTIR spectroscopy and artificial neural networks for the quantification of SBS content in modified asphalt. *Chem. Resea. Appl.* **2018**, *30*, 1938–1942.
17. Zhong, K.; Cao, D.; Luo, S. Determination the modifier content in SBS modified asphalt based on infrared spectroscopy technique. In Proceedings of the International Conference on Mechanical Engineering and Green Manufacturing (MEGM) 2010, Xiangtan, China, 19–22 November 2010; pp. 1129–1134.
18. Zhang, X.Y. Effect of powder-to-gel ratio on properties of SBS modified asphalt binder. *Shandong Transport. Technol.* **2016**, *2*, 109–112.
19. Fu, Y.K.; Zhang, L.; Tan, Y.Q.; Meng, D.Y. *Low-Temperature Properties Evaluation Index of Rubber Asphalt*; Crc Press-Taylor & Francis Group: Boca Raton, FL, USA, 2016; p. 16.
20. Chen, J.S.; Liao, M.C.; Shiah, M.S. Asphalt modified by styrene-butadiene-styrene triblock copolymer: Morphology and model. *J. Mater. Civ. Eng.* **2002**, *14*, 224–229. [CrossRef]
21. Yan, C.Q.; Huang, W.D.; Xiao, F.P.; Wang, L.F.; Li, Y.W. Proposing a new infrared index quantifying the aging extent of SBS-modified asphalt. *Road Mater. Pavement* **2017**, *1*, 1–16. [CrossRef]
22. Masson, J.F.; Pelletier, L.; Collins, P. Rapid FTIR method for quantification of styrene-butadiene type copolymers in bitumen. *Appl. Polym.* **2001**, *79*, 959–1149. [CrossRef]
23. Deng, Z.W.; Yu, P.; Chen, L. Application of SPSS software in orthogonal experimental design and results analysis. *Comput. Learn.* **2009**, *5*, 15–17.
24. Gong, J.; Shi, P.C.; Li, C.Y. Multivariate analysis of variance using SPSS software. *Agric. Net. Infor.* **2012**, *4*, 31–33.
25. Xiao, P.; Kang, A.H.; Li, X.F. Blending mechanism of SBS modified asphalt based on infrared spectroscopy. *J. Jiangsu Uni. Nat. Sci.* **2005**, *26*, 529–532.

applied sciences

MDPI

Article

Evaluating the Rheological, Chemical and Morphological Properties of SBS Modified Asphalt-Binder under Multiple Aging and Rejuvenation Cycles

Xiaobing Chen [1,*], Yunfeng Ning [1,*], Yongming Gu [2], Ronglong Zhao [1], Jinhu Tong [1], Juntian Wang [3], Xiaorui Zhang [1] and Wei Wen [1]

1 School of Transportation, Southeast University, Nanjing 211189, China; along@seu.edu.cn (R.Z.); 101010035@seu.edu.cn (J.T.); zxr@seu.edu.cn (X.Z.); 220203258@seu.edu.cn (W.W.)
2 Suzhou Communications Investment Group Co., Ltd., Suzhou 215004, China; szgtsn@sina.com
3 Hangzhou Transportation Development Support Center, Hangzhou 310030, China; 220183082@seu.edu.cn
* Correspondence: xbchen@seu.edu.cn (X.C.); 220193090@seu.edu.cn (Y.N.)

Featured Application: This paper presents the trends in the rheological, chemical, and morphological properties of SBS-modified asphalt-binders under multiple cycles of aging and rejuvenation. The results from this study may provide guidance for the future development of rejuvenators and application of repeated penetrating rejuvenation.

Abstract: To investigate the influence of multiple cycles of aging and rejuvenation on the rheological, chemical, and morphological properties of styrene–butadiene–styrene (SBS)-modified asphalt-binders, the asphalt-binders were aged using two laboratory simulation methods, namely a rolling thin film oven (RTFO) test for short-term aging and pressure aging vessel (PAV) for long-term aging. The asphalt-binders were then rejuvenated with three types of rejuvenators (Type I, II, and III) with different dosages (i.e., 6%, 10%, and 14% for the first, second, and third rejuvenation, respectively). A dynamic shear rheometer (DSR) was then used to analyze the effect of rejuvenators on the rheological properties of all the asphalt-binders. The changes in the functional groups and microscopic morphology in the process of multiple aging and rejuvenation cycles were studied using Fourier transform infrared (FTIR) and atomic force microscopy (AFM). The results indicated that the three rejuvenators could soften the stiffness and restore the microstructures of the aged asphalt-binders in the process of repeated aging and rejuvenation from DSR and AFM testing. Considering the rutting and fatigue properties, the Type I rejuvenator exhibited the potential to achieve the desired rejuvenation effects under multiple rejuvenation cycles. During the multiple aging and rejuvenation cycles, the aging resistance of SBSMA decreased gradually from the FTIR results. This inherently limited the number of repeated rejuvenation cycles. This research is conducive to promoting the application of repeated penetrating rejuvenation.

Keywords: multiple aging and rejuvenation cycles; SBS modified asphalt; morphological properties; DSR; FTIR; AFM

Citation: Chen, X.; Ning, Y.; Gu, Y.; Zhao, R.; Tong, J.; Wang, J.; Zhang, X.; Wen, W. Evaluating the Rheological, Chemical and Morphological Properties of SBS Modified Asphalt-Binder under Multiple Aging and Rejuvenation Cycles. *Appl. Sci.* **2021**, *11*, 9242. https://doi.org/10.3390/app11199242

Academic Editors: Amir Tabakovic, Jan Valentin and Liang He

Received: 31 August 2021
Accepted: 1 October 2021
Published: 4 October 2021

1. Introduction

Due to the mature production process, cost-effectiveness, and the outstanding performance in terms of rutting, cracking, and moisture resistance, styrene–butadiene–styrene-modified asphalt-binder (SBSMA) is one of the most commonly used asphalt-binders for pavement construction, particularly as a surfacing layer [1]. The principle idea is to take advantage of the good binding and hardening properties of the SBS polymer to achieve effective modification of the bituminous material [2]. Thus, SBSMA has broad market prospects and accounts for more than 50% of the modified asphalt-binder around the world [3].

Nevertheless, SBSMA is inevitably prone to aging during the service life of the pavement as a result of the negative effects of oxygen, heat, and ultraviolet (UV) light, which ultimately lead to the hardening and embrittlement of the asphalt-binder [4,5]. These effects usually result in the degradation of the SBS polymer and oxidation along with the volatilization of light components and the polycondensation of the asphalt-binder, consequently causing a decay in the physico-rheological properties [6–8]. Many researchers have observed an increase in the complex modulus (G^*), elastic modulus (G'') and creep recovery (R), but a reduction in phase angle (δ) and non-recoverable creep compliance (J_{nr}) with the deepening of the aging by DSR, which means that the viscous behavior of the asphalt-binder has partly moved to elastic behavior [9–11].

SBSMA's performance decay accelerates pavement distresses and shortens its service life under traffic loading and fluctuating environmental conditions. It is essential to make a proper decision to restore the physico-rheological properties of the asphalt-binder at an appropriate time during the life cycle of the pavement [12]. Therefore, rejuvenators are often added to aged asphalt-binders to soften them and make the rejuvenated asphalt-binders more fluid. Specifically, the effects of the rejuvenators on the asphalt-binders include an increase in the penetration grade, ductility, phase angle, m-value, and a corresponding decrease in the softening point, viscosity, shear complex modulus, rutting factor, and stiffness, respectively [13,14].

During the aging and rejuvenation processes, the change in the physico-rheological properties of the SBSMA can be attributed to the chemical composition and microstructures. Fourier transform infrared (FTIR), atomic force microscopy (AFM), environmental scanning electron microscope (ESEM), and other new technologies have been successfully applied to analyze the aging mechanism and rejuvenating process of SBSMA at a micro scale [11,15–25]. FTIR was used to investigate the aging and rejuvenation mechanism of SBSMA as well as the relationship between the chemical and rheological characteristics by identifying the differences between the absorption peaks which represent the functional groups and contents in the asphalt-binder based on Lambert–Beer's Law [15–18]. Yang et al. [20] found that the oxygen content of the asphalt-binder increased greatly after thermal aging, leading to an increase in the polar functional groups such as carbonyl and sulfoxide groups in the asphalt-binder. In terms of rejuvenation, rejuvenators regularly decreased the carbonyl and sulfoxide indices [21]. The AFM can obtain the microscopic morphology of the asphalt-binder surface [10]. Guo et al. [22] observed that the bee structures slightly increased after PAV aging, and rejuvenator made the microstructures larger. Ozdemir et al. [23] studied aging effects of varying processing parameters on the phase structure of SBSMA. Cong et al. [24] observed a decrease in the surface roughness with aging. Aghazadeh Dokandari et al. [25] observed that tiny bee structures appeared with the addition of the rejuvenator.

The rejuvenation of aged SBSMA and reuse of asphalt mixtures reduces the environmental impacts, such as land occupation and consumption of nonrenewable resources [4]. Rapid in-place pavement recycling at an appropriate time is a very promising technology for preventive maintenance, and has an advantage of time- and cost-efficiency [26]. However, one cycle of rejuvenation has become inadequate to meet the current environmental and traffic demands. Considering the availability of bitumen and its price uncertainty, the need for multiple instances of maintenance in road construction has become essential. Three or four instances of maintenance has been proposed to extend the service life by 10–15 years with over 50% cost savings [12].

With the above background, it is therefore imperative to conduct a laboratory study on the multiple aging and rejuvenation cycles of asphalt-binder where rejuvenated asphalt-binder is widely used. Furthermore, multiple-cycle rejuvenation is expected to be a common practice in the future. To explore the mechanisms of multiple cycle aging and rejuvenation, various technological methods including FTIR and AFM were applied in this study for quantitatively evaluating the chemical and morphological properties of SBSMA through repeated aging and rejuvenation laboratory testing. Another aim of this study was to investigate the effects of multiple aging and rejuvenation cycles on the rheological parameters of SMSMA and the potential for repeated rejuvenation of flexible pavements, so laboratory experimentations were accomplished by using DSR for investigating.

2. Materials and Test Methods

SBSMAs were separately blended with three types of rejuvenators at different dosages in multiple aging and rejuvenation cycles. Basic performance tests, namely DSR, FTIR, and AFM, were used to test the original, aged, and rejuvenated asphalt-binder samples. The high-temperature rutting parameters and intermediate-temperature fatigue parameters were obtained using the DSR test device [27]. The functional group changes of the asphalt-binder samples were obtained by analyzing the results from FTIR [20]. The microstructures of asphalt-binder samples were observed and characterized with AFM [10].

2.1. Materials

The main materials used in this study were asphalt-binder and three types of rejuvenators, namely Type I, Type II, and Type III, respectively.

2.1.1. Asphalt-Binder

The physical properties of the SBS-modified asphalt-binder (SBSMA) used in this study were acquired from SHELL Co., Ltd. The physical properties of SBSMA are listed in Table 1.

Table 1. Technical indices of SBS modified asphalt-binder (SBSMA).

Technical Index	Unit	Virgin	RTFO Residue	PAV Residue	Test Method
Penetration (25 °C, 100 g, 5 s)	0.1 mm	50.6	35.0	22.4	ASTM D5
Softening point, $T_{R\&B}$	°C	78.9	81.0	83.2	ASTM D36
Ductility (5 °C, 5 cm/min)	cm	25.2	14.7	2.0	ASTM D113

2.1.2. Rejuvenators

The three different types of rejuvenators were asphaltene-free cationic rejuvenating emulsions fabricated from petroleum-based oil. The principal properties and infrared spectra of the three rejuvenators used are listed in Table 2 and shown in Figure 1, respectively, all indicating nearly similar components and chemical structures.

2.2. Experimental Design Plan

Laboratory short-term aging was carried out using the rolling thin-film oven test (RTFO, 163 °C, 1.0 MPa, 85 min) to simulate aging during storage, mixing, transport, and placing. Thereafter, the pressurized aging vessel test (PAV, 100 °C, 2.1 MPa, 20 h) was performed to roughly simulate 5 years of field aging [28] in accordance with ASTM D2872 [29] and ASTM D6521 [30]. The virgin SBSMA was subjected to RTFO and PAV aging to obtain the first aged SBSMA.

Table 2. Technical indices of the rejuvenators.

Technical Index	Unit	Type I	Type II	Type III	Test Method
Viscosity @25 °C	SFS	25~150	20~60	25~150	ASTM D7496
Residue	%	>64	60~65	>65	ASTM D6934
Particle charge	—	Positive	Positive	Positive	ASTM D7402
Asphaltenes	%	<11.5	2~4	<11	ASTM D2006
Maltene distribution ratio $(PC+A_1)/(S+A_2)$ [1]	—	0.8~1.4	0.5~0.9	0.7~1.1	ASTM D2006
PC/S ratio	—	>1.2	>1.2	>0.5	ASTM D2006

[1] PC = Polar Compound; A1 = First Acidaffins; A2 = the Second Acidaffins; S = Saturated Hydrocarbons.

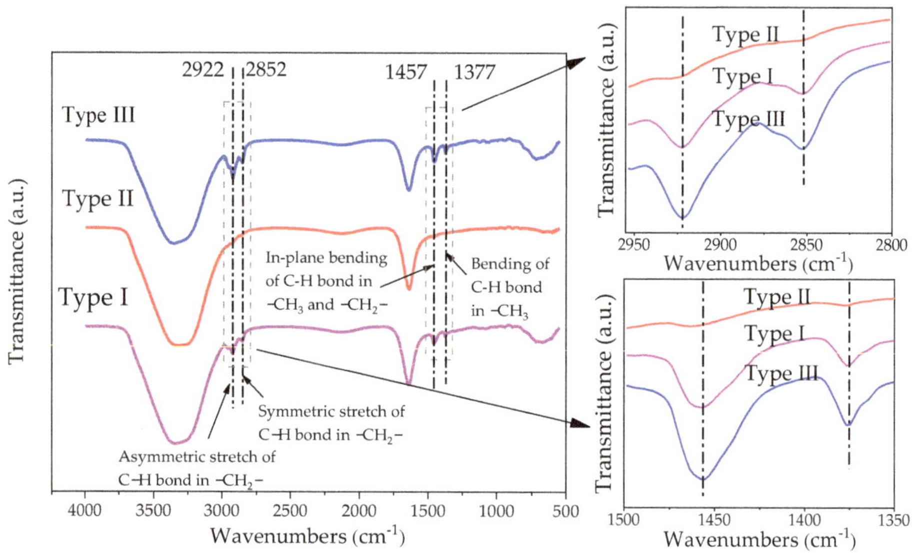

Figure 1. FTIR spectra of the three rejuvenators.

For the purpose of investigating the influence of multiple aging and rejuvenation cycles on the SBSMA in the field, 6%, 10%, and 14% by weight (wt%) of the asphalt-binder were selected as the rationale dosage of the rejuvenator for the first, second, and third rejuvenation, respectively, in accordance with previous research studies [26]. The first aged SBSMA was heated to 140~160 °C for melting and blended with three rejuvenators for 30 min to prepare the first rejuvenated SBSMA. Thereafter, the first rejuvenated SBSMA was directly placed in the PAV (100 °C, 2.1 MPa, 20 h) for long-term aging (rather than RTFO first) to obtain the second aged asphalt-binder on account of the in-place penetrating rejuvenation without the need for heated construction. The second rejuvenated asphalt-binder was prepared by adding 10 wt% of the rejuvenator to the second aged SBSMA. The third aged asphalt-binder and the third rejuvenated asphalt-binder were obtained by the same method. Figure 2 illustrates the experimental design plan for this research study.

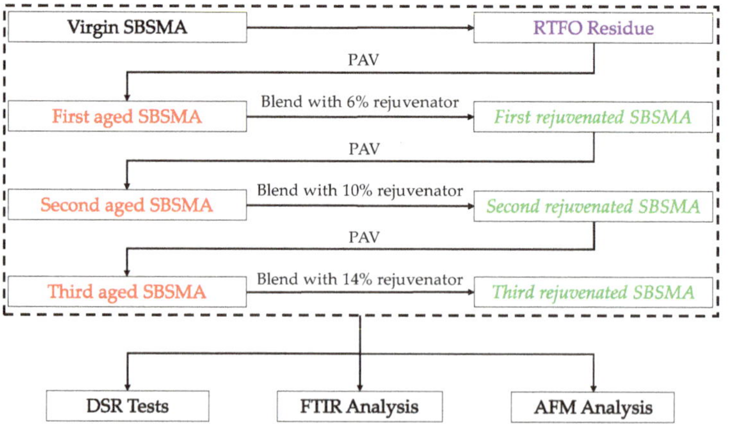

Figure 2. Experimental flow chart.

2.3. Laboratory Testing

2.3.1. Rheological Property Tests

The dynamic rheological properties of the SBSMA at different cycles of aging and rejuvenation were measured using the DSR test device (Kinexus Pro+, Malvern, Wolverhampton, UK).

Temperature sweep tests were carried out under the strain-controlled loading mode at a constant frequency of 10 rad·s^{-1} in accordance with the ASTM D7405 [31]. The temperature range was 58~82 °C for the rejuvenated asphalt-binders, and 22~34 °C for the aged asphalt-binders. The applied strain levels for the temperature sweep tests were controlled in the linear viscoelastic range. The rejuvenated asphalt-binder samples were molded to a diameter of 25 mm by 1 mm thick whilst the aged asphalt-binder samples to a diameter of 8 mm by 2 mm thick.

Multiple Stress Creep Recover (MSCR) tests were conducted at 76 °C in line with the ASTM D7405 [31] test specification at standard stress levels of 0.1 kPa and 3.2 kPa, respectively. After subjecting the asphalt-binder samples to 10 loading cycles, each level with a 1 s loading period followed by a 9 s recovery period, two major rheological parameters, namely the percentage recovery (R) and non-recoverable creep compliance (J_{nr}), were calculated automatically at both stress levels of the DSR test.

Linear Amplitude Sweep (LAS) tests were performed using the DSR test device with 8 mm parallel plates and 2 mm gap to determine the asphalt-binders' resistance to fatigue damage in accordance with the AASHTO T391-20 [32] test specification. The LAS test procedure consisted of two phases. Firstly, the frequency sweep test was performed at 0.1% strain level with a frequency range of 0.1~30 Hz to obtain the undamaged property of the samples (α). Then, in the amplitude sweep, the load was applied from 0 to 30% strain at 10 Hz for 300 s under strain-controlled loading mode. The LAS test data were processed using a Visco-Elastic Continuum Damage (VECD) model approach to determine the fatigue parameters A and B. In this study, all the LAS tests were performed at a room temperature of 25 °C.

2.3.2. FTIR Tests

FTIR tests were performed using a Bruker Alpha FTIR spectrometer (Germany) equipped with a reflection diamond ATR accessory to analyze the functional groups of samples qualitatively or quantitatively with an accumulation of 32 scans in a 400~4000 cm^{-1} range and a resolution of 4 cm^{-1} (Figure 3a). The samples were directly placed on a zinc selenide horizontal ATR crystal, which should be totally cleaned using kerosene and alcohol after each test operation [33]. The raw FTIR data were processed and analyzed using a software program called OPUS.

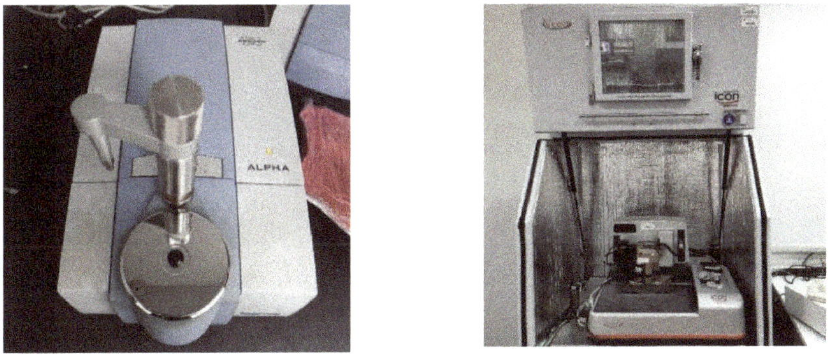

(a) (b)

Figure 3. Microscopic analysis equipment: (a) FTIR-ATR; (b) AFM.

2.3.3. AFM Tests

Atomic force microscopy (AFM, Bruker Dimension ICON, USA) was used to investigate the micro-morphology of the asphalt-binder at different cycles of aging and rejuvenation using the tapping mode (see Figure 3b). Because of the AFM requirement for a smooth surface of the sample molded, all the asphalt-binder samples for the AFM tests were prepared using the thermal method. This was necessary to avoid toxic solvents from dissolving and damaging the microstructures of the asphalt-binder [10].

A hot liquid drop of asphalt-binder (160 °C) was placed on a 10 × 10 × 2 mm glass substrate and put into the oven at a constant temperature of 145 °C for about 5 min and thereafter, cooled to the ambient temperature. The AFM samples with a flat and smooth surface were obtained and then moved to a closed dust-free container for storage before AFM observations.

Height and Z-images were scanned using an etched silicon probe. The probe (NANOSENSORTM, PPP-NCL-20) was used under the tapping mode. All the experiments were conducted at a temperature of 25 °C. The scanning range of 10 × 10 μm was used for the microstructure of the sample surfaces, and the range of 30 × 30 μm for analysis of the morphology was deemed a good statistical representation of the overall asphalt-binder samples.

2.3.4. Sample Replicates

For each rejuvenator type, content, aging condition, and test type, three sample replicates were tested. Overall, a total of 261 asphalt-binder samples (153 samples for the DSR tests, 54 samples for the FTIR tests, and 54 samples for the AFM tests) were prepared and tested.

3. Test Results, Analyses, and Discussion

3.1. Rheological Properties

The rheological properties of the asphalt-binders from the DSR test were used to evaluate the high-temperature rutting resistance and intermediate-temperature fatigue resistance of the modified asphalt-binder in the process of multiple aging and rejuvenation cycles. The specific DSR test results are presented and discussed below.

3.1.1. Temperature Sweep Analysis

The DSR temperature sweep tests were conducted with an isochronal plot to evaluate the temperature sensitivity of the asphalt-binder. The complex shear modulus (G^*) and phase angle (δ) were measured for subsequent computation of the analyzed rutting factor ($G^*/\sin\delta$) and fatigue factor ($G^* \cdot \sin\delta$).

The rutting factor ($G^*/\sin\delta$) reflects the deformation resistance of asphalt-binders at high temperatures [21]. Asphalt-binders with larger rutting factors have better rutting resistance at high temperatures. Figure 4 shows that the $G^*/\sin\delta$ values of the rejuvenated asphalt-binders decreased with an increase in temperature, and that the logarithm of $G^*/\sin\delta$ had a linear correlation with temperature.

After the second rejuvenation cycle, the $G^*/\sin\delta$ values of the asphalt-binders with Type I and Type III rejuvenators were comparatively close to those of the RTFO residue. However, only the third rejuvenated SBSMA with Type I rejuvenator had a $G^*/\sin\delta$ that was approximately equivalent to that of the RTFO residue. The superiority rank order of the rutting factors of the first, second, and third rejuvenated asphalt-binders was Type II > Type I > Type III, respectively. With an increase in the rejuvenation cycles, the rutting factor of the rejuvenated asphalt-binder with the Type II rejuvenator increased steadily, while that with the Type III rejuvenator decreased. Notably, the rutting factor of the aged SBSMA could be rejuvenated close to that of the RTFO residue by adding the Type I rejuvenator.

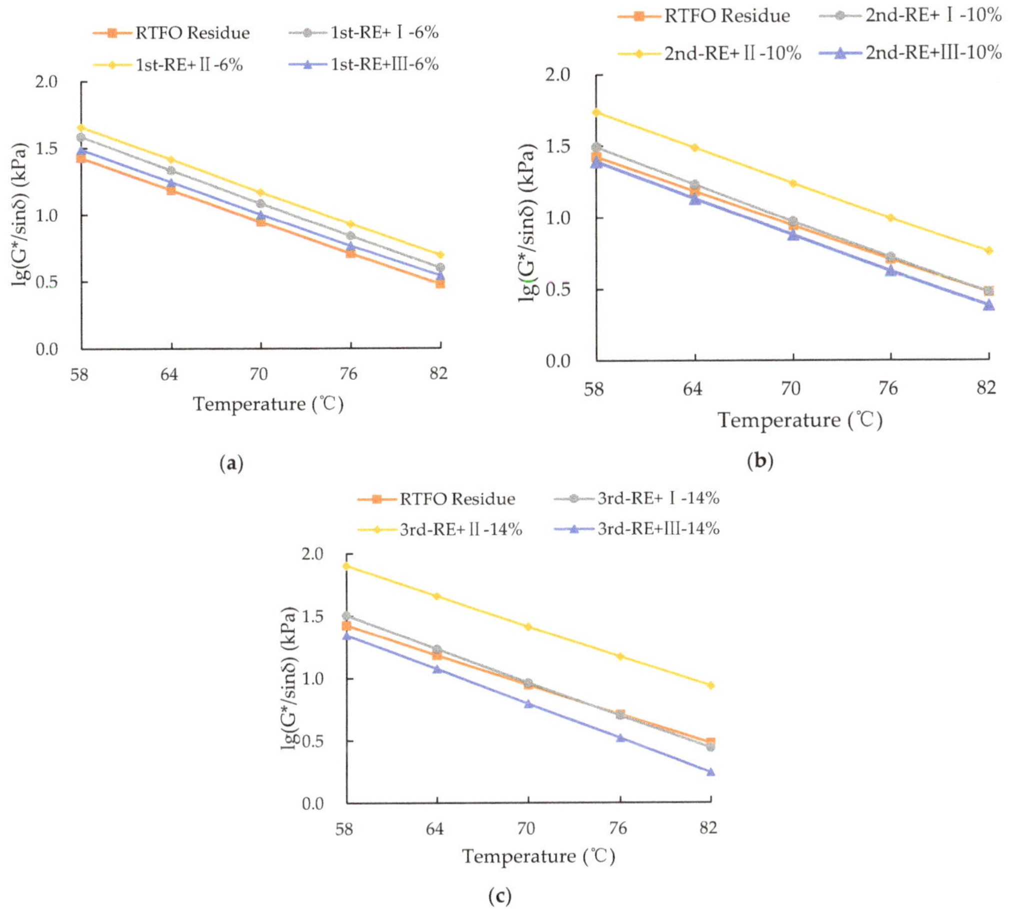

Figure 4. Rutting factor of SBS asphalt-binders at different rejuvenation cycles: (**a**) First cycle; (**b**) Second cycle; (**c**) Third cycle.

As the rejuvenation cycles increased, the dispersion of the rutting factor increased, indicating that repeated rejuvenation had a great influence on asphalt-binder's rutting resistance. In general, the following was observed: (a) Type II rejuvenator enhanced the rutting resistance of the asphalt-binder, (b) Type III rejuvenator exhibited the potential to reduce the rutting resistance, and (c) Type I rejuvenator indicated the potential to help maintain the rutting resistance of the rejuvenated asphalt-binder.

The fatigue factor ($G^{*}\cdot\sin\delta$) reflects the anti-fatigue property of the asphalt-binder at intermediate temperatures [21]. Figure 5a,b show that the fatigue factors of the aged asphalt-binders decreased with an increase in temperature, and that the logarithm of $G^{*}\cdot\sin\delta$ had a linear correlation with temperature. The $G^{*}\cdot\sin\delta$ values of the second aged asphalt-binders with the three rejuvenators were close to that of the first aged asphalt-binder, whereas only the third aged SBSMA with Type I rejuvenator had nearly the same values as that of the PAV residue.

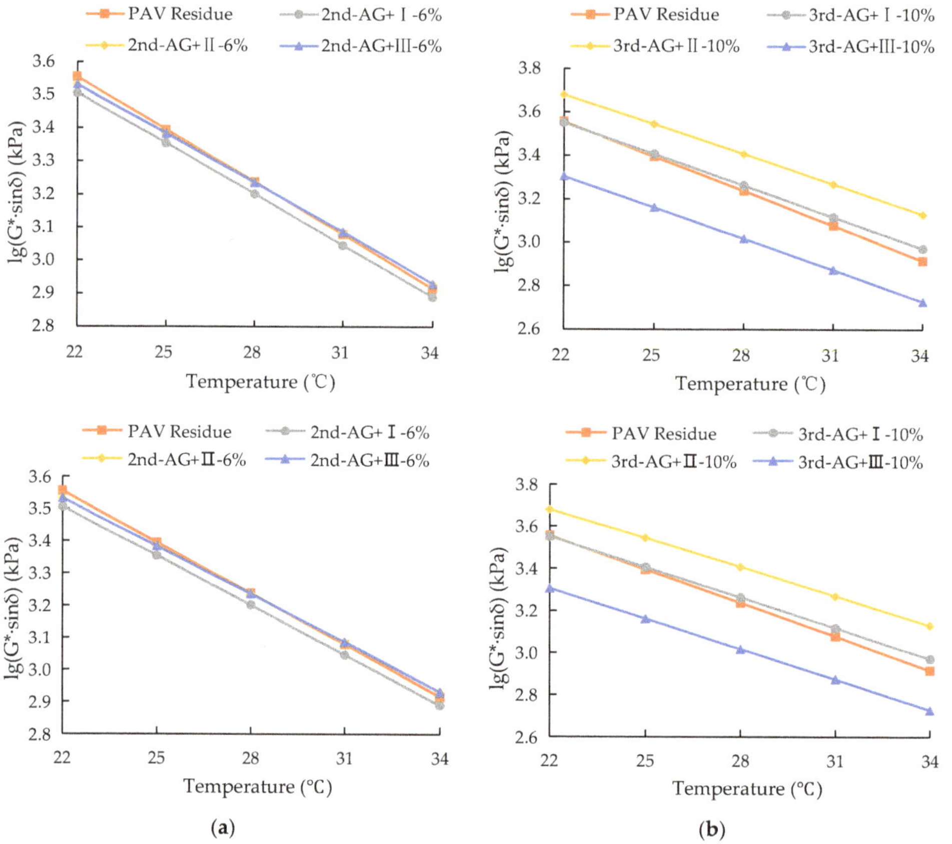

Figure 5. Fatigue factor of SBS asphalt-binders at different aging cycles: (**a**) Second cycle; (**b**) Third cycle.

From Figure 5, the fatigue factors of the third aged asphalt-binders ranked as follows: Type II > Type I > Type III. The fatigue property of the third aged asphalt-binders with the Type I rejuvenator was closest to that of the PAV residue, whilst that with the Type II rejuvenator was deemed as poor. The Type III rejuvenator, on the other hand, was found to be satisfactory and better than the other two types in terms of potential for fatigue resistance.

It was interesting to note that the widest dispersion of the fatigue factor was produced by the third aged asphalt-binders, followed by the second aged asphalt-binders as shown in Figure 5, indicating that the repeated aging process had a great influence on the rheological properties of the SBSMA. By comparison, the fatigue factor of the second and third aged asphalt-binders with the Type I rejuvenator were seen to be close to those of the PAV residue. This observation suggested that the expected rejuvenation performance can be achieved by blending with the Type I rejuvenator.

Theoretically, low continuous grade of the asphalt-binders at high temperature means poor high-temperature rutting resistance. As the number of rejuvenating cycles increased, the high-temperature continuous grade of SBSMA with Type I and III rejuvenators decreased continuously, while the grade of the asphalt-binder with Type II rejuvenator maintained an increasing trend. After the third cycle of rejuvenation, with an increase in high temperature, the grade of the asphalt-binder with Type II demonstrated an insufficient effectiveness of the repeated rejuvenation with Type II. By contrast, Type I and III rejuvenators could be applied to the repeated rejuvenation of SBSMA to reduce its high temperature grade. However, the grade reduction associated with the Type I rejuvenator was relatively low. On the other hand, whilst the third rejuvenated asphalt-binder reached 84.1 °C, the third rejuvenated asphalt-binder with Type III was only 79.8 °C (see Figure 6a).

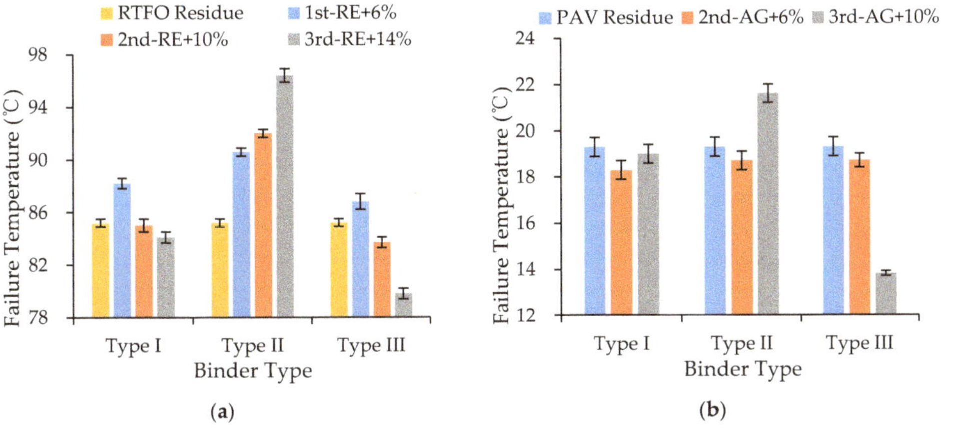

Figure 6. Failure temperature of SBS asphalt-binders at different aging and rejuvenation cycles: (**a**) High-temperature continuous grade; (**b**) intermediate-temperature continuous grade.

For asphalt-binders rejuvenated with Type I and II rejuvenators, the intermediate temperature continuous grade decreased to 18.3 °C and 18.7 °C (Type I and Type II, respectively), and then increased by 19.0 °C and 21.6 °C (Type I and Type II, respectively) with an increase in the number of aging cycles. For the asphalt-binders rejuvenated with Type III, the continuous grade at the intermediate temperature decreased with the aging cycles (see Figure 6b). The continuous grading temperature of the first, second, and third aged asphalt-binders rejuvenated with Type I were extremely close, showing that the fatigue properties of SBSMA rejuvenated with Type I could satisfactorily reach the virgin asphalt-binder level. Based on the results of the rutting and fatigue behavior of the asphalt-binders in different aging and rejuvenation cycles, Type I was considered to achieve the expected rejuvenation effects among the three types of rejuvenators evaluated.

3.1.2. MSCR Analysis

Percentage recovery (*R*), which reflects the recovery deformation for a period of ten creep and recovery cycles at each stress level, represents the elastic recovery performance of an asphalt-binder [34]. The larger percent recovery (*R*) indicates better rutting resistance of asphalt-binders [35]. Figure 7a shows the recovery response of the rejuvenated SBSMA with different rejuvenators at different rejuvenation cycles measured at 0.1 kPa and 3.2 kPa stress levels, respectively.

From Figure 7, it is evident that as the rejuvenation cycles increased, the percentage recovery of the asphalt-binder decreased under the same stress level. This meant a decline in the percentage recovery ability of the asphalt-binders after rejuvenation and the conversion promotion of the rejuvenators to the viscous component of the asphalt-binder. The percentage recovery of the third rejuvenated asphalt-binder with the Type III rejuvenator was 25.62% (3.2 kPa), which compared to the RTFO residue and the third rejuvenated asphalt-binder with Type II, represents a reduction factor of three. At the same stress level, the superiority ranking of the percentage recovery of SBSMA after similar rejuvenation cycles was Type II > Type I > Type III.

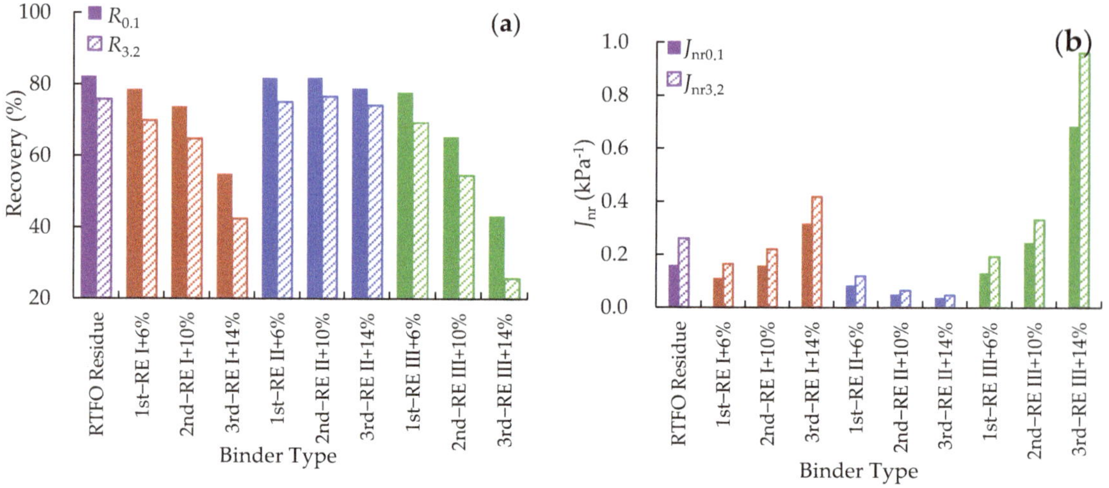

Figure 7. MSCR rutting evaluation of the repetitively rejuvenated asphalt-binders: (**a**) Recovery; (**b**) non-recoverable creep compliance (J_{nr}) (Red bars: the asphalt-binders rejuvenated with Type I; blue bars: the asphalt-binders rejuvenated with Type II; green bars: the asphalt-binders rejuvenated with Type III).

The non-recoverable creep compliance (J_{nr}) reflects the residual strain level of asphalt-binders after the loading-recovery cycle, which represents the rutting resistance of asphalt-binder. With an increased number of rejuvenation cycles, the rejuvenated asphalt-binders with Type I and Type III rejuvenators exhibited an increasing trend in the J_{nr} value and the non-recoverable strain at the same stress level (see Figure 7b). This implies a continuous drop in the rutting resistance. Particularly, the $J_{nr0.1}$ and $J_{nr3.2}$ values of the third rejuvenated asphalt-binder with Type III reached 0.684 kPa^{-1} and 0.959 kPa^{-1}, over 3 times the corresponding values of the RTFO residue. On the contrary, the rejuvenated asphalt-binder with Type II exhibited a continuously decreasing trend, indicating an improvement in the rutting resistance of the asphalt-binder. These findings are consistent with the results from temperature sweep tests and further illustrate that the Type II rejuvenator had a poor repeated rejuvenation effect.

When comparing the differences of the MSCR evaluation indices (R and J_{nr}) under the two stress levels shown in Figure 7, the results showed that the differences in $J_{nr0.1}$ and $J_{nr3.2}$ of the rejuvenated asphalt-binders with the Type I and III rejuvenators increased with an increase in the rejuvenation cycles. The discrepancy between $J_{nr0.1}$ and $J_{nr3.2}$ revealed that the stress sensitivity of the asphalt-binder, namely an increasing sensitivity of the rejuvenated SBSMA to the variation of stress loading as the rejuvenation cycles increased. The trend was mainly attributed to the degradation of the SBS polymer in the SBSMA asphalt-binder during the repeated rejuvenation, which weakened the modification effects of the asphalt-binder.

Comparing Figures 7a and 8a, the percentage recovery of the asphalt-binder slightly increased after aging, pointing to improved resistance to permanent deformation. This increase in percentage recovery indicates a potential increase in elasticity of the asphalt-binders due to aging. Figure 7a also shows that under the same stress level, the creep recovery of the asphalt-binder decreased when the rejuvenation cycles increased. This indicates that the addition of the rejuvenators could supplement the viscous component of the asphalt-binder to effectively compensate and delay the aging of the asphalt-binder [36].

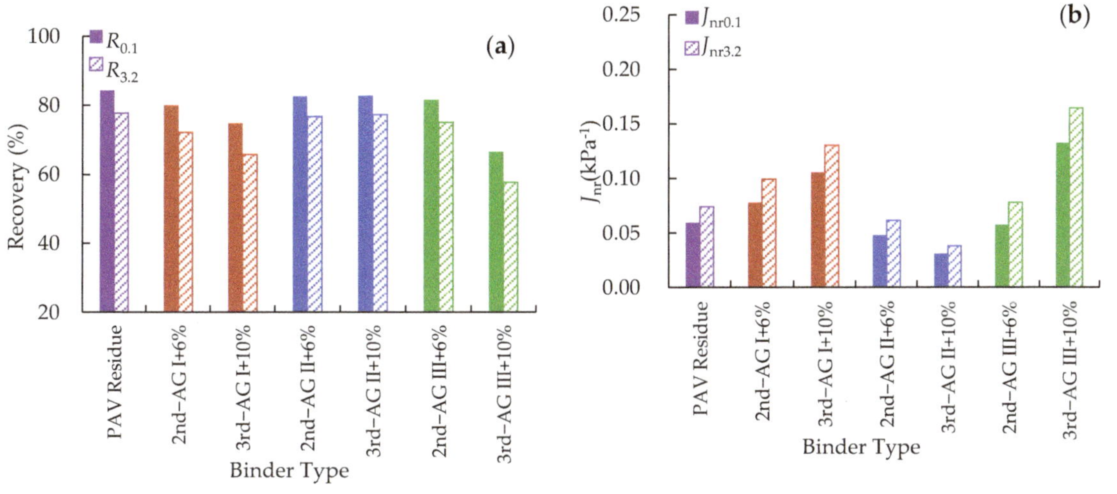

Figure 8. MSCR rutting evaluation of repetitively aged asphalt-binders: (**a**) recovery; (**b**) non-recoverable creep compliance (J_{nr}) (Red bars: the asphalt-binders rejuvenated with Type I; blue bars: the asphalt-binders rejuvenated with Type II; green bars: the asphalt-binders rejuvenated with Type III).

Comparing Figures 7b and 8b, it is observed that the non-recoverable creep compliance of the aged asphalt-binder declined significantly. This indicates that aging partly provided an improvement in elasticity and that the elasticity of the asphalt-binders after rejuvenation was obviously restored. With an increase in the aging cycles, the asphalt-binders with Type I and Type III rejuvenators exhibited an increasing trend in J_{nr}, whilst the aged asphalt-binder with Type II exhibited a decreasing trend at 0.1 kPa and 3.2 kPa, respectively (see Figure 8b). Under the same stress level, the superiority ranking of the non-recoverable creep compliance of SBSMA in the equivalent aging cycle was Type I $\approx$ Type III > Type II, and that of SBSMA in the equivalent rejuvenation cycle was Type III > Type I > Type II, respectively.

3.1.3. LAS Analysis

To evaluate the fatigue life (N_f) of the asphalt-binders with the parameters A and B, strain levels of 2.5%, 5%, and 10% were used in this study. The fatigue behavior of the rejuvenated SBSMA significantly depends on the applied strain loading and the rheological property of asphalt-binder as well as the type of rejuvenators. The results in Figure 9 show a significant decrease in the fatigue life of asphalt-binders at elevated strain levels. That is, high strain loading significantly weakened the fatigue resistance of the SBSMA.

At low strain levels, thermal oxidative aging did not appear to significantly affect the fatigue life of the asphalt-binders. In fact, there was an increase in the fatigue life at lower strain levels. However, this was not the case at the 10% strain level. This may be explained from the perspective that the increase in viscosity and stiffness of the asphalt-binder due to aging promoted the asphalt-binders to convert to more linear-elastic behavior at low strain levels with a logical decline in the fatigue life under the adverse effects of aging at high strain levels [37]. As the aging and rejuvenation cycles increased, both the aged and rejuvenated asphalt-binders roughly showed a decreasing trend in fatigue life at the high strain level.

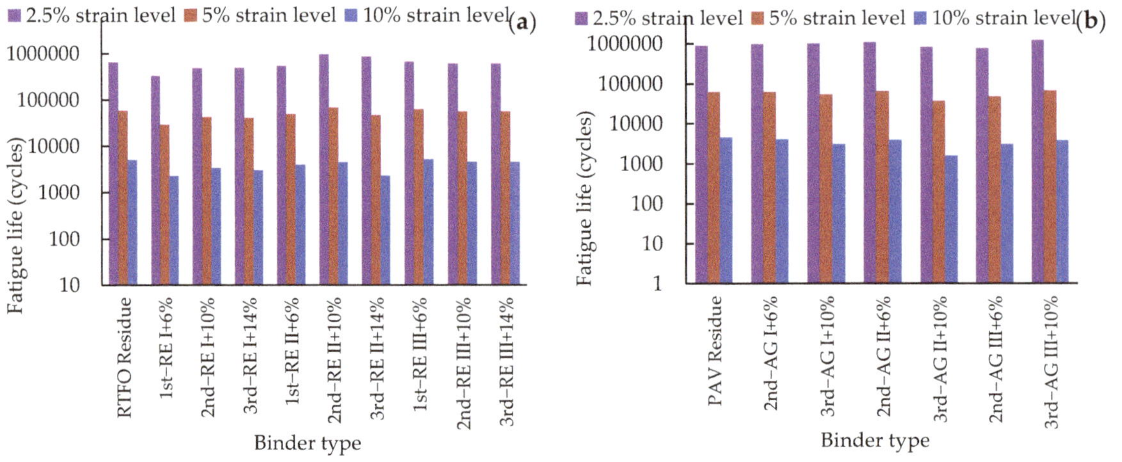

Figure 9. LAS fatigue life results: (**a**) rejuvenated asphalt-binders; (**b**) aged asphalt-binders.

3.2. FTIR Analysis

To quantitatively investigate the influence of multiple aging and rejuvenation cycles on the SBSMA chemical structure and the vibration of the functional groups under repeated aging and rejuvenation, the 600~2000 cm^{-1} functional group peak area was selected as a reference to calculate the butadiene (CH=CH), carbonyl (C=O), and sulfoxide (S=O) relative contents. This helped to eliminate the influence of the thickness and concentration of the asphalt-binders. The chemical aging indices, namely butadiene index (BI), carbonyl index (CI), and sulfoxide index (SI), were determined using Equations (1)–(3) [38]:

$$CI = \frac{\text{Area of the carbonyl centered around } 1700 \text{ cm}^{-1}}{\text{Area of spectral bands between } 2000 \text{ and } 600 \text{ cm}^{-1}} \quad (1)$$

$$SI = \frac{\text{Area of the sulfoxide centered around } 1030 \text{ cm}^{-1}}{\text{Area of spectral bands between } 2000 \text{ and } 600 \text{ cm}^{-1}} \quad (2)$$

$$BI = \frac{\text{Area of the butadiene centered around } 966 \text{ cm}^{-1}}{\text{Area of spectral bands between } 2000 \text{ and } 600 \text{ cm}^{-1}} \quad (3)$$

Figure 10 gives more intuitive information about the changes in the functional groups. As the aging cycles increased, repetitively aged asphalt-binders showed an obviously increasing trend in the CI value. The indistinctive increasing and decreasing trends in the SI and BI values, respectively, are obvious. In the aging process of SBSMA, due to oxidation, dehydrogenation, and cross-linking reactions of the asphalt-binder and SBS polymer, degradation occurred simultaneously, and the spatial network structure was destroyed [39,40]. By comparing the aged and rejuvenated asphalt-binders, it can be seen that the rejuvenating effect was obvious in each aging and rejuvenation cycle, and that the decrease in the BI, CI, and SI values were significant.

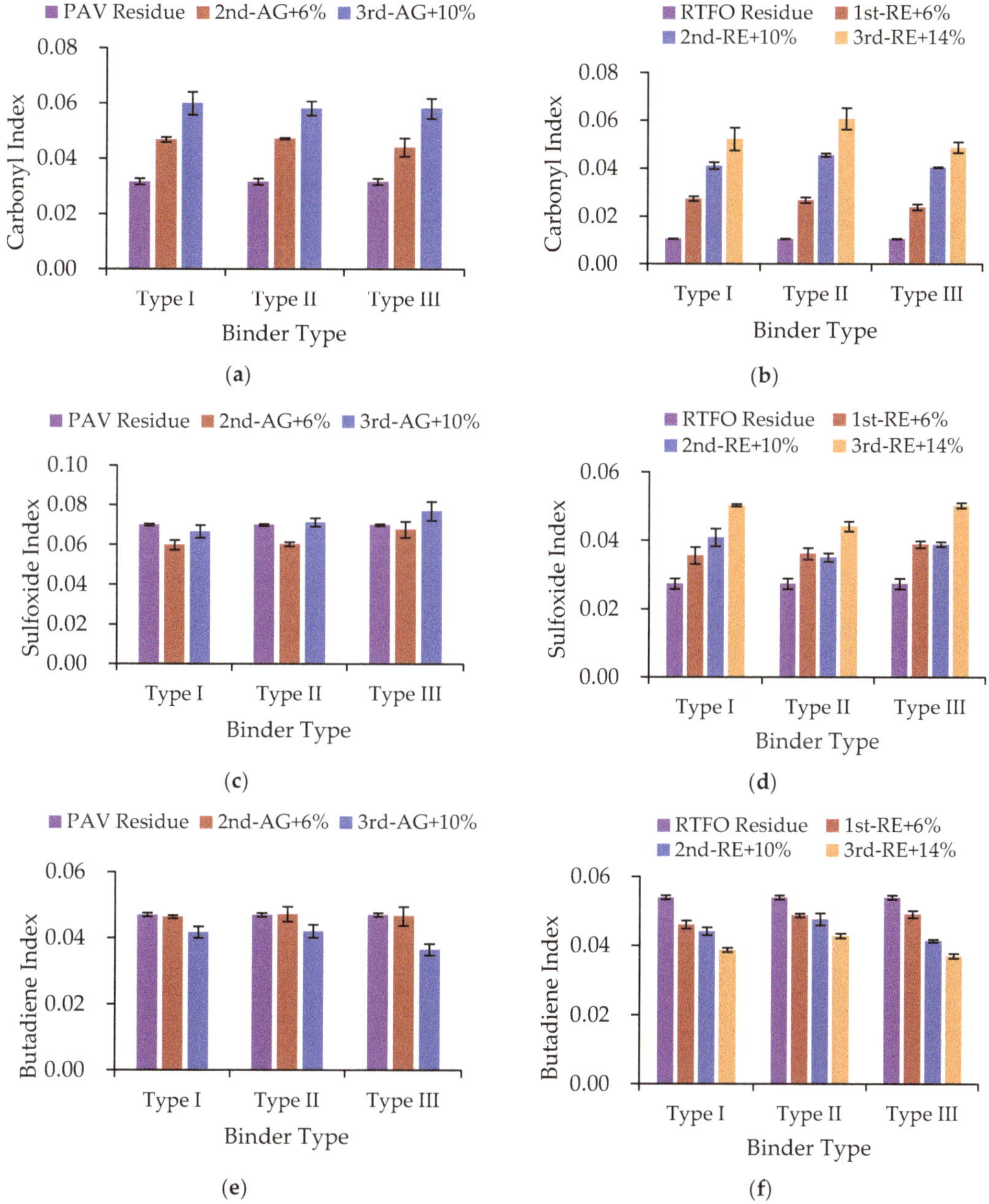

Figure 10. Chemical aging indices of repetitively aged and rejuvenated SBSMA: (**a**) AG−CI; (**b**) RE−CI; (**c**) AG−SI; (**d**) RE−SI; (**e**) AG−BI; (**f**) RE−BI.

Based on the theory of compositional harmonic on rejuvenation mechanisms [41], the rejuvenators rich in aromatics are added to the aged asphalt-binder to dilute the aged asphalt-binders, enhance the proportion of maltenes, and harmonize the different fractions in the asphalt-binder. This was not simply the reversal of the transformation from aromatics into asphaltenes during the aging process, expressing that the aromatics were consumed and asphaltenes increased.

However, as the dosage of rejuvenators increased in each cycle, the CI and SI values of the rejuvenated asphalt-binders also increased, whilst the value of BI decreased during the repeated aging and rejuvenation cycles. In other words, the aging of the rejuvenated asphalt-binder was faster than the asphalt-binder after RTFO aging. This might be explained by the decrease of the aging resistance due to the oxidative degradation of SBS polymer in the multiple aging process [42]. Meanwhile, the oxidation and breaking of polybutadiene (PB) segments of SBS would produce more oxygen-containing groups such as –OH, C=O, –COOH, etc. [43,44]. Hence, the increased aromatic hydrocarbon oil might have also led to the acceleration of the asphalt-binder aging. The addition of the rejuvenator indicated the potential to improve the rheological properties of the aged asphalt-binders but could not restore the chemical aging indices to the same level as the RTFO residue. As a result, the rejuvenating effect was not significant enough to achieve an appreciable repeated rejuvenation.

With an increase in the rejuvenation cycles, the CI value decreased, especially for the Type III rejuvenator. This indicated that all the three types of rejuvenators had certain rejuvenating effects on the aged asphalt-binders, with the Type III rejuvenator being more impactful. On the other hand, the SI value of the third rejuvenated asphalt-binder with Type II was the smallest while the third rejuvenated asphalt-binder with Types I and III had the largest SI value in magnitude. These results indicated that the Type II rejuvenator was relatively more effective at inhibiting aging of the asphalt-binder. By contrast, the BI value of the rejuvenated asphalt-binder with Type II had the lowest rate of decay, whilst that of the asphalt-binder with Type III declined rapidly. A possible explanation for this could be that the Type II rejuvenator had some chemical components to minimize the SBS degradation, whilst Type III had more maltenes to dilute the contents of SBS [45].

3.3. AFM Analysis

3.3.1. Microstructure of the Asphalt-Binder Surface

By means of the NanoScope Analysis software [46], the AFM height images of various SBSMA asphalt-binders after being subjected to multiple aging and rejuvenation cycles are shown in Figure 11. The roughness indices of the asphalt-binders determined from these images are shown in Figure 12.

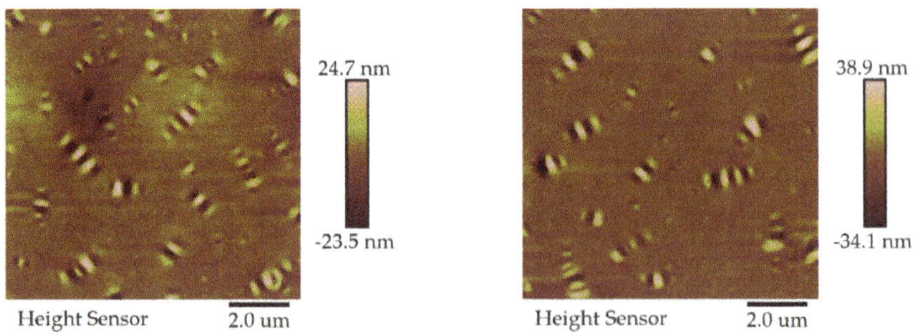

(a) (b)

Figure 11. *Cont.*

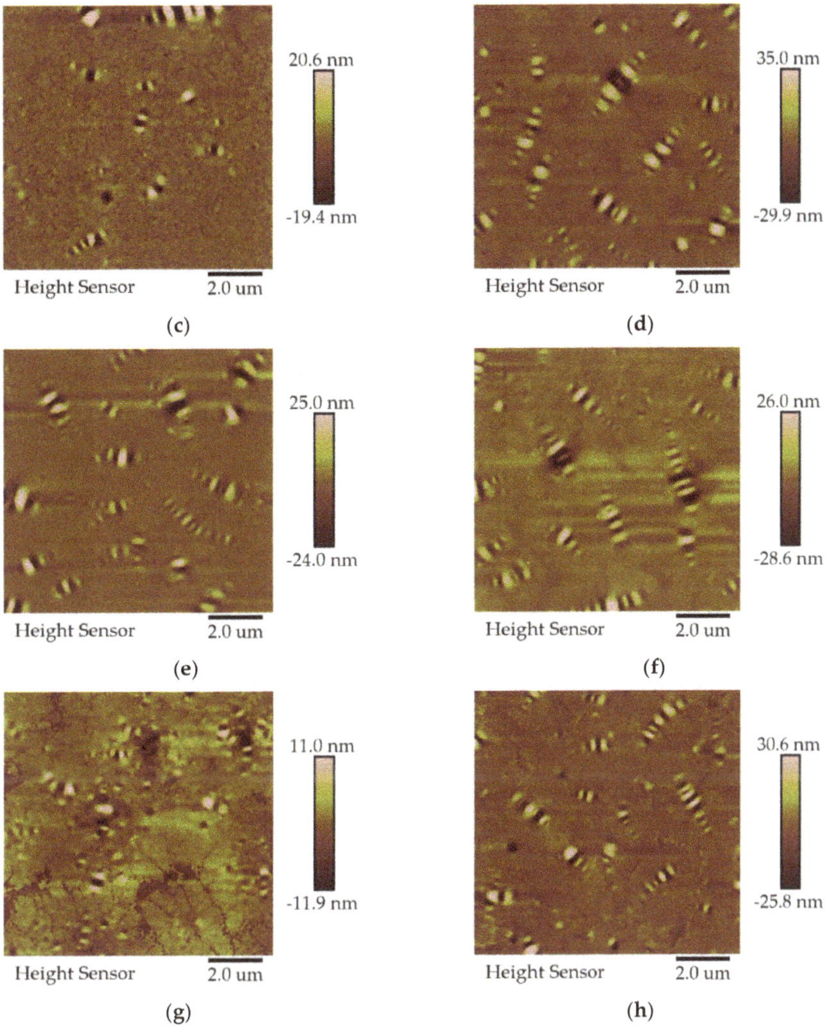

Figure 11. Microstructures obtained from AFM images of the asphalt-binders: (**a**) Virgin; (**b**) RTFO Residue; (**c**) PAV Residue; (**d**) 1st−RE I + 6%; (**e**) 2nd−AG I + 6%; (**f**) 2nd−RE I + 10%; (**g**) 3rd−AG I+10%; (**h**) 3rd−RE I + 14%.

Figure 11 shows the two-dimensional (2-D) AFM images of different asphalt-binders subjected to multiple aging and rejuvenation cycles to illustrate the influence of multiple aging and rejuvenation cycles on the microstructure evolution of the asphalt-binder's surface. The appearance of many beelike structures with different sizes could be clearly seen on the surface of the virgin SBSMA (see Figure 11a). After aging, the number of large-sized beelike structures in the PAV residue declined noticeably.

With the addition of the rejuvenator, tiny beelike structures appeared on the rejuvenated asphalt-binder samples with increased morphological undulations. Compared to the asphalt-binder before aging, the size of the beelike structures in the second and third aged asphalt-binders decreased considerably, with the height of the beelike structures increasing as well. By contrast, the quantity, size, and height of the beelike structures of the second and third rejuvenated asphalt-binder increased substantially. However, the quantity declined in contrast to the asphalt-binder before rejuvenation (see Figure 11d–h).

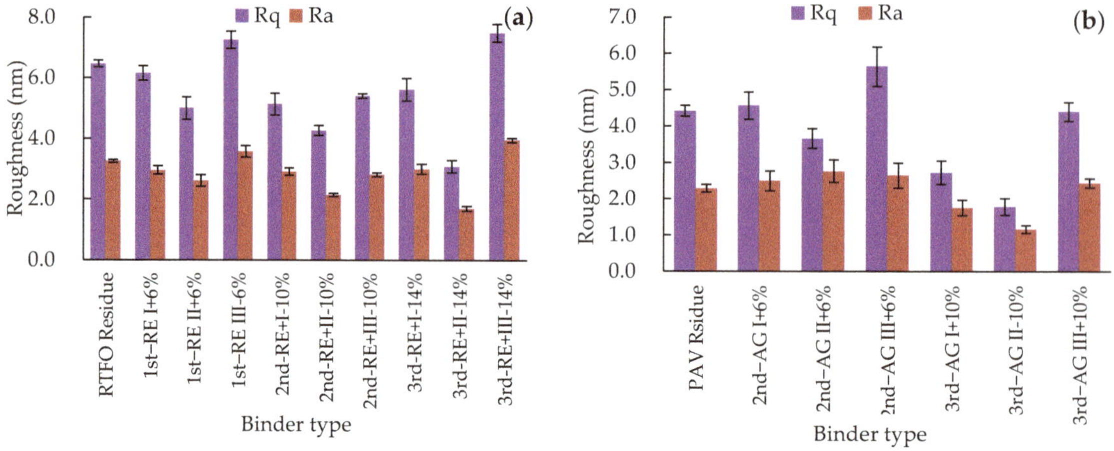

Figure 12. AFM roughness analysis: (**a**) rejuvenated asphalt-binders; (**b**) aged asphalt-binders.

3.3.2. Morphological Features of the Asphalt-Binder Surface

The surface roughness indices reflected the degree of development of the beelike structure. Figure 12 shows the roughness indices that were calculated from the height of 30 × 30 μm AFM images. The roughness indices selected for analysis were the average roughness (*Ra*) and root-mean-square roughness (*Rq*).

After multiple aging and rejuvenation cycles, the *Ra* and *Rq* values of the asphalt-binders with different rejuvenators had a similar response trend. After aging, the *Ra* and *Rq* values of the asphalt-binder surface decreased. By contrast, the roughness indices appeared to have been restored after rejuvenation, which is similar to the findings reported by Yu et al. [47]. This could be because the molecular movement of the asphalt-binders decreased after aging, with the accumulation and nucleation of the polar components being hindered. This led to the reduction in the size and height of the beelike structures. The addition of the rejuvenator to supplement maltenes could promote the mobility of the asphalt-binder molecules. Therefore, the size and height of the beelike structures of rejuvenated asphalt-binder were greater than those of the aged asphalt-binder. This helps to explain the mechanism responsible for the formation of the beelike structures during the rejuvenation process.

However, the three rejuvenators showed different rejuvenation effects. The surface of the asphalt-binders with the Type III rejuvenator were the roughest and had the greatest beelike structures in height. The asphalt-binders with the Type II rejuvenator had the least roughness, with the lowest height in the beelike structures. After three rejuvenation cycles, the roughness of the rejuvenated asphalt-binder with Type III was more than twice that of the asphalt-binder with Type II. This might be because the Type III rejuvenator replenished most of the light components and greatly improved the mobility of the maltene molecules. By contrast, Type II only slightly promoted the accumulation of the polar components.

4. Conclusions and Recommendations

This study investigated the effects of the three aging and rejuvenation cycles with three types of rejuvenators designated as Type I, II, and III with different dosages (6%, 10%, and 14%, respectively). The rheological (thermal, rutting, and fatigue resistance), chemical, and morphological characteristics of SBSMA were evaluated using the DSR, LAS, FTIR, and AFM tests. From this study, the following conclusions were drawn:

- The superiority ranking for the rutting factor of the repetitively aged asphalt-binders and fatigue factor of the repetitively rejuvenated asphalt-binders was Type II > Type I > Type III, respectively. Multiple rejuvenation cycles reduced the percentage recovery of the SBSMA with varying rejuvenators and made the asphalt-binder depend more sensitively on the strain levels. The fatigue life of the rejuvenated SBSMA with the three rejuvenators increased at low stress level and decreased at high strain level as the rejuvenation cycles were increased. Considering the high- and intermediate-temperature performance, the Type I rejuvenator exhibited a satisfactory rejuvenation effect.
- From FTIR analysis, the carbonyl and sulfoxide groups of SBSMA increased whilst the chain segments of butadiene decreased after repeated aging and rejuvenation cycles. The addition of the rejuvenator could not restore the chemical aging indices to the same level as the RTFO residue. During the multiple aging and rejuvenation cycles, the aging resistance of SBSMA gradually decreased. This inherently limited the expected rejuvenation effects under the repeated rejuvenation cycles.
- From AFM morphological analysis, PAV aging decreased the quantity of the beelike structures and the surface roughness of the asphalt-binders. The rejuvenators significantly changed the microscopic morphology and phase dispersion of the aged asphalt-binders. The three rejuvenators showed different rejuvenation effects, but generally increased the morphological undulations and surface roughness.

Overall, plausible results were obtained in this laboratory study with promising potential for using rejuvenators to improve the rheological properties of asphalt-binders. The study has also shown that rejuvenators have the potential of achieving satisfactory repeated aging and rejuvenation in terms of performance characteristics (rutting, thermal, and fatigue cracking resistances) and morphological features of SBSMA. Nonetheless, the conclusions drawn are not exhaustive nor exclusive as the test results presented herein pertain to the specific materials evaluated in this study. Future research should include a wide array of different materials including high-activity rejuvenators to further supplement this study's findings including field application trials and validation.

Author Contributions: Conceptualization, X.C. and Y.N.; methodology, Y.N. and J.W.; validation, X.Z. and Y.G.; formal analysis, Y.N. and W.W.; investigation, X.Z.; resources, X.C. and R.Z.; writing—original draft preparation, Y.N. and J.W.; writing—review and editing, X.C. and X.Z.; visualization, Y.N.; supervision, X.C. and J.T.; funding acquisition, Y.G. All authors have read and agreed to the published version of the manuscript.

Funding: This research was funded by the National Science Foundation of China (Grant No. 51778140 and No. 52078130), Suzhou Transport Investment, Planning Design and Management Co., Ltd. (Grant No. 8521008637), and Kunshan Transportation Development Holding Group Co., Ltd. (Grant No. 8521002599 and No. 8521008420).

Institutional Review Board Statement: Not applicable.

Informed Consent Statement: Not applicable.

Data Availability Statement: No new data were created or analyzed in this study. Data sharing is not applicable to this article.

Acknowledgments: The authors would like to acknowledge sincerely the financial support provided by the National Science Foundation of China (Grant No. 51778140 and No. 52078130), Suzhou Transport Investment, Planning Design and Management Co., Ltd. (Grant No. 8521008637), and Kunshan Transportation Development Holding Group Co., Ltd. (Grant No. 8521002599 and No. 8521008420) for this study. Special thanks and due gratitude go to all those who offered the administrative and technical support during the course of this study.

Conflicts of Interest: The authors declare no conflict of interest.

References

1. Lesueur, D.; Teixeira, A.; Lázaro, M.M.; Andaluz, D.; Ruiz, A. A simple test method in order to assess the effect of mineral fillers on bitumen ageing. *Constr. Build. Mater.* **2016**, *117*, 182–189. [CrossRef]
2. Ren, S.; Liu, X.; Xu, J.; Lin, P. Investigating the role of swelling-degradation degree of crumb rubber on CR/SBS modified porous asphalt binder and mixture. *Constr. Build. Mater.* **2021**, *300*, 124048. [CrossRef]
3. Zhu, C.; Zhang, H.; Zhang, D.; Chen, Z. Influence of Base Asphalt and SBS Modifier on the Weathering Aging Behaviors of SBS Modified Asphalt. *J. Mater. Civ. Eng.* **2018**, *30*, 04017306. [CrossRef]
4. Cong, P.; Guo, X.; Mei, L. Investigation on rejuvenation methods of aged SBS modified asphalt binder. *Fuel* **2020**, *279*, 118556. [CrossRef]
5. Zeng, W.; Wu, S.; Wen, J.; Chen, Z. The temperature effects in aging index of asphalt during UV aging process. *Constr. Build. Mater.* **2015**, *93*, 1125–1131. [CrossRef]
6. Ma, T.; Huang, X.; Zhao, Y.; Yuan, H. Aging Behaviour and Mechanism of SBS-Modified Asphalt. *J. Test. Eval.* **2012**, *40*, 20120150. [CrossRef]
7. Ouyang, C.; Wang, S.; Zhang, Y.; Zhang, Y. Improving the aging resistance of styrene–butadiene–styrene tri-block copolymer modified asphalt by addition of antioxidants. *Polym. Degrad. Stab.* **2006**, *91*, 795–804. [CrossRef]
8. Tauste, R.; Moreno-Navarro, F.; Sol-Sánchez, M.; Rubio-Gámez, M.C. Understanding the bitumen ageing phenomenon: A review. *Constr. Build. Mater.* **2018**, *192*, 593–609. [CrossRef]
9. Wang, Z.; Ye, F. Experimental investigation on aging characteristics of asphalt based on rheological properties. *Constr. Build. Mater.* **2020**, *231*, 117158. [CrossRef]
10. Xu, J.; Sun, L.; Pei, J.; Xue, B.; Liu, T.; Li, R. Microstructural, chemical and rheological evaluation on oxidative aging effect of SBS polymer modified asphalt. *Constr. Build. Mater.* **2021**, *267*, 121028. [CrossRef]
11. Cuciniello, G.; Leandri, P.; Filippi, S.; Lo Presti, D.; Polacco, G.; Losa, M.; Airey, G. Microstructure and rheological response of laboratory-aged SBS-modified bitumens. *Road Mater. Pavement Des.* **2021**, *22*, 372–396. [CrossRef]
12. Huang, M.; Dong, Q.; Ni, F.; Wang, L. LCA and LCCA based multi-objective optimization of pavement maintenance. *J. Clean. Prod.* **2021**, *283*, 124583. [CrossRef]
13. Jiang, H.; Zhang, J.; Sun, C.; Liu, S.; Liang, M.; Yao, Z. Experimental assessment on engineering properties of aged bitumen incorporating a developed rejuvenator. *Constr. Build. Mater.* **2018**, *179*, 1–10. [CrossRef]
14. Behnood, A. Application of rejuvenators to improve the rheological and mechanical properties of asphalt binders and mixtures: A review. *J. Clean. Prod.* **2019**, *231*, 171–182. [CrossRef]
15. Yao, H.; Dai, Q.; You, Z. Fourier Transform Infrared Spectroscopy characterization of aging-related properties of original and nano-modified asphalt binders. *Constr. Build. Mater.* **2015**, *101*, 1078–1087. [CrossRef]
16. Yan, C.; Huang, W.; Ma, J.; Xu, J.; Lv, Q.; Lin, P. Characterizing the SBS polymer degradation within high content polymer modified asphalt using ATR-FTIR. *Constr. Build. Mater.* **2020**, *233*, 117708. [CrossRef]
17. Zhao, X.; Wang, S.; Wang, Q.; Yao, H. Rheological and structural evolution of SBS modified asphalts under natural weathering. *Fuel* **2016**, *184*, 242–247. [CrossRef]
18. Feng, Z.; Cai, F.; Yao, D.; Li, X. Aging properties of ultraviolet absorber/SBS modified bitumen based on FTIR analysis. *Constr. Build Mater.* **2021**, *273*, 121713. [CrossRef]
19. Lin, P.; Liu, X.; Apostolidis, P.; Erkens, S.; Zhang, Y.; Ren, S. ESEM observation and rheological analysis of rejuvenated SBS modified bitumen. *Mater. Des.* **2021**, *204*, 109639. [CrossRef]
20. Yang, Z.; Zhang, X.; Zhang, Z.; Zou, B.; Zhu, Z.; Lu, G.; Xu, W.; Yu, J.; Yu, H. Effect of Aging on Chemical and Rheological Properties of Bitumen. *Polymers* **2018**, *10*, 1345. [CrossRef]
21. Zhu, H.; Xu, G.; Gong, M.; Yang, J. Recycling long-term-aged asphalts using bio-binder/plasticizer-based rejuvenator. *Constr. Build. Mater.* **2017**, *147*, 117–129. [CrossRef]
22. Guo, M.; Liang, M.; Jiao, Y.; Tan, Y.; Yu, J.; Luo, D. Effect of aging and rejuvenation on adhesion properties of modified asphalt binder based on AFM. *J. Microsc.* **2021**. [CrossRef] [PubMed]
23. Kaya Ozdemir, D.; Topal, A.; McNally, T. Relationship between microstructure and phase morphology of SBS modified bitumen with processing parameters studied using atomic force microscopy. *Constr. Build. Mater.* **2021**, *268*, 121061. [CrossRef]
24. Cong, P.; Liu, N.; Tian, Y.; Zhang, Y. Effects of long-term aging on the properties of asphalt binder containing diatoms. *Constr. Build. Mater.* **2016**, *123*, 534–540. [CrossRef]
25. Aghazadeh Dokandari, P.; Topal, A.; Kaya Ozdemir, D. Rheological and Microstructural Investigation of the Effects of Rejuvenators on Reclaimed Asphalt Pavement Bitumen by DSR and AFM. *Int. J. Civ. Eng.* **2021**, *19*, 749–758. [CrossRef]
26. Chen, X.; Wang, J.; Zhang, X.; Liu, H.; Tong, J.; Zhao, R. Evaluating the Physical and Rheological Properties of Rejuvenated Styrene-Butadiene-Styrene-Modified Asphalt Binders. *Adv. Mater. Sci. Eng.* **2020**, *2020*, 1–14. [CrossRef]
27. Grilli, A.; Gnisci, M.I.; Bocci, M. Effect of ageing process on bitumen and rejuvenated bitumen. *Constr. Build. Mater.* **2017**, *136*, 474–481. [CrossRef]

28. Qian, Y.; Guo, F.; Leng, Z.; Zhang, Y.; Yu, H. Simulation of the field aging of asphalt binders in different reclaimed asphalt (RAP) materials in Hong Kong through laboratory tests. *Constr. Build. Mater.* **2020**, *265*, 120651. [CrossRef]
29. *ASTM D2872: Standard Test Method for Effect of Heat and Air on a Moving Film of Asphalt (Rolling Thin-Film Oven Test)*; ASTM International: West Conshohocken, PA, USA, 2019.
30. *ASTM D6521: Standard Practice for Accelerated Aging of Asphalt Binder Using a Pressurized Aging Vessel*; ASTM International: West Conshohocken, PA, USA, 2019.
31. *ASTM D7405: Standard Test Method for Multiple Stress Creep and Recovery (MSCR) of Asphalt Binder Using a Dynamic Shear Rheometer*; ASTM International: West Conshohocken, PA, USA, 2015.
32. *AASHTO T391-20: Standard Method of Test for Estimating Fatigue Resistance of Asphalt Binders Using the Linear Amplitude Sweep*; American Association of State Highway and Transportation Office: Washington, DC, USA, 2020.
33. Ren, R.; Han, K.; Zhao, P.; Shi, J.; Zhao, L.; Gao, D.; Zhang, Z.; Yang, Z. Identification of asphalt fingerprints based on ATR-FTIR spectroscopy and principal component-linear discriminant analysis. *Constr. Build. Mater.* **2019**, *198*, 662–668. [CrossRef]
34. Zhang, J.; Simate, G.S.; Hu, X.; Souliman, M.; Walubita, L.F. Impact of recycled asphalt materials on asphalt binder properties and rutting and cracking performance of plant-produced mixtures. *Constr. Build. Mater.* **2017**, *155*, 654–663. [CrossRef]
35. Kataware, A.V.; Singh, D. A study on rutting susceptibility of asphalt binders at high stresses using MSCR test. *Innov. Infrastruct. Solut.* **2017**, *2*, 4. [CrossRef]
36. Zhong, H.; Huang, W.; Yan, C.; Zhang, Y.; Lv, Q.; Sun, L.; Liu, L. Investigating binder aging during hot in-place recycling (HIR) of asphalt pavement. *Constr. Build. Mater.* **2021**, *276*, 122188. [CrossRef]
37. Ziari, H.; Amini, A.; Goli, A. The effect of different aging conditions and strain levels on relationship between fatigue life of asphalt binders and mixtures. *Constr. Build. Mater.* **2020**, *244*, 118345. [CrossRef]
38. Zhao, Y.; Gu, F.; Xu, J.; Jin, J. Analysis of Aging Mechanism of SBS Polymer Modified Asphalt based on Fourier Transform Infrared Spectrum. *J. Wuhan Univ. Technol. Mater. Sci. Ed.* **2010**, *25*, 1047–1052. [CrossRef]
39. Cortizo, M.S.; Larsen, D.O.; Bianchetto, H.; Alessandrini, J.L. Effect of the thermal degradation of SBS copolymers during the ageing of modified asphalts. *Polym. Degrad. Stab.* **2004**, *86*, 275–282. [CrossRef]
40. Li, Y.; Li, L.; Zhang, Y.; Zhao, S.; Xie, L.; Yao, S. Improving the aging resistance of styrene-butadiene-styrene tri-block copolymer and application in polymer-modified asphalt. *J. Appl. Polym. Sci.* **2010**, *116*, 754–761. [CrossRef]
41. Kuang, D.; Liu, W.; Xiao, Y.; Wan, M.; Yang, L.; Chen, H. Study on the rejuvenating mechanism in aged asphalt binder with mono-component modified rejuvenators. *Constr. Build. Mater.* **2019**, *223*, 986–993. [CrossRef]
42. Hao, G.; Huang, W.; Yuan, J.; Tang, N.; Xiao, F. Effect of aging on chemical and rheological properties of SBS modified asphalt with different compositions. *Constr. Build. Mater.* **2017**, *156*, 902–910. [CrossRef]
43. Xu, X.; Yu, J.; Xue, L.; Zhang, C.; Zha, Y.; Gu, Y. Investigation of Molecular Structure and Thermal Properties of Thermo-Oxidative Aged SBS in Blends and Their Relations. *Materials* **2017**, *10*, 768. [CrossRef] [PubMed]
44. Cao, Z.; Chen, M.; Yu, J.; Han, X. Preparation and characterization of active rejuvenated SBS modified bitumen for the sustainable development of high-grade asphalt pavement. *J. Clean. Prod.* **2020**, *273*, 123012. [CrossRef]
45. Cao, X.; Wang, H.; Cao, X.; Sun, W.; Zhu, H.; Tang, B. Investigation of Rheological and Chemical Properties Asphalt Binder Rejuvenated with Waste Vegetable Oil. *Constr. Build. Mater.* **2018**, *180*, 455–463. [CrossRef]
46. Ding, L.; Wang, X.; Zhang, M.; Chen, Z.; Meng, J.; Shao, X. Morphology and properties changes of virgin and aged asphalt after fusion. *Constr. Build. Mater.* **2021**, *291*, 123284. [CrossRef]
47. Yu, X.; Zaumanis, M.; dos Santos, S.; Poulikakos, L.D. Rheological, microscopic, and chemical characterization of the rejuvenating effect on asphalt binders. *Fuel* **2014**, *135*, 162–171. [CrossRef]

applied sciences

MDPI

Article

A Multiscale Study of Moisture Influence on the Crumb Rubber Asphalt Mixture Interface

Lan Wang [1,2,*], Yang Liu [3] and Le Zhang [3]

1 School of Civil Engineering, Inner Mongolia University of Technology, Hohhot 010051, China
2 Key Laboratory of Civil Engineering Structure and Mechanics, Inner Mongolia University of Technology, Hohhot 010051, China
3 College of Science, Inner Mongolia University of Technology, Hohhot 010051, China; liuyang_imut@foxmail.com (Y.L.); zhangle218@foxmail.com (L.Z.)
* Correspondence: wanglan661018@163.com

Abstract: In order to study the influence of moisture on the interface of crumb rubber–asphalt (CR) mixture, the interface bonding performance and crack resistance of a crumb rubber–asphalt mixture under dry and wet conditions were studied at three scales. At the macroscale, the characteristics of medium temperature fatigue cracking and low temperature fracture were studied by semi-circular bending tensile test (SCB) on the example of digital image correlation (DIC) technique. At the microscale, the surface energy of CR with basalt and limestone was measured using the contact angle measurement test, and then the adhesion work was calculated and analyzed. At the molecular scale, the model of CR, the model of basalt representative mineral (augite) and limestone representative mineral (calcite) were studied by molecular dynamics simulation. The relationship between these three scales was further explored to reveal the mechanism of the damage of moisture on the interface deterioration of the CR mixture. The results show that moisture has a certain effect on the interface of the CR mixture, which is characterized by macroscopically reducing the crack resistance of the asphalt mixture, microscopically reducing the adhesion ability between the asphalt and the aggregate and weakening the interaction between the asphalt and aggregate molecules at the molecular scale. Molecular dynamics can accurately simulate the deterioration of micro asphalt-aggregate adhesion under the damage of moisture. The decrease in microadhesion leads to the decrease in the crack resistance of the macro-CR mixture.

Keywords: moisture; interface cracking; digital image correlation (DIC); molecular dynamics; crumb rubber asphalt (CR)

Citation: Wang, L.; Liu, Y.; Zhang, L. A Multiscale Study of Moisture Influence on the Crumb Rubber Asphalt Mixture Interface. *Appl. Sci.* **2022**, *12*, 6940. https://doi.org/10.3390/app12146940

Academic Editor: Leonid Burakovsky

Received: 14 June 2022
Accepted: 6 July 2022
Published: 8 July 2022

1. Introduction

The study of a crumb rubber–asphalt (CR) mixture proved that it had good road performance and diverse functions [1–6]. However, in the service process of crumb rubber–asphalt pavement, the looseness, spalling, cracks and other diseases have seriously affected the service performance and service life of crumb rubber–asphalt pavement [7–9]. The reason is that many issues are related to moisture. The performance of the interface between CR and aggregate decreases under the condition of moisture, resulting in moisture damage to the CR mixture, thus affecting the service life of asphalt pavement [10,11]. Therefore, clarifying the interaction mechanism between the CR and the aggregate in the presence of moisture is the core issue to ensure the service durability of a CR mixture and provides a reference for the application and promotion of a crumb asphalt mixture.

The interface behavior of asphalt mixtures under the action of moisture is usually evaluated by macro-indicators such as the crack strength and spalling rate [12–15]. In addition, some scholars have used microcharacterization technology to study the asphalt mixture, including the surface morphology, structure, mechanical properties and other microproperties of the material [16–20]. Molecular simulation [21,22] is a computer numerical simulation test, which is not limited by the detection means, specimen preparation or other conditions. It effectively complements the shortcomings of macro tests and can reveal the molecular-scale mechanism of many macro tests. In recent years, it has also been more widely applied to the research of asphalt mixtures. From the current research, the research on the interface effect of asphalt mixtures mainly focuses on single-scale research and analyses of macroperformance and materials' microcharacteristics, but the multiscale and cross-scale research idea can better solve the mechanism problem of asphalt mixture performance. Many scholars have also carried out multiscale studies on different properties of asphalt and asphalt mixtures, including asphalt aging performance [23,24], the combination of old and new materials in recycled asphalt mixtures [25–27], the diffusion of modifiers [28,29], and the interaction between asphalt and aggregates [30,31]. A large number of studies have confirmed that the multiscale idea is feasible in the research of asphalt and asphalt mixtures.

Therefore, in order to study the interface adhesion and failure mechanism of CR mixtures under the action of moisture using the multiscale method, this study carried out a cracking resistance test of a CR mixture at the macroscale and compared the effects of different cracking forms on the CR mixture. At the microscale, the surface free energies of asphalt and aggregate were calculated and analyzed by the surface energy theory, and the damage from moisture on the surface adhesion of asphalt and aggregate was studied. At the molecular scale, the interaction mechanism between CR and aggregate under the action of moisture was studied by constructing a 12-component CR model, a representative molecular aggregate model, and an interface model of CR and aggregate. Finally, the interface adhesion mechanism and failure behavior of the CR mixture under the action of moisture was revealed by correlating the interface properties under different scales.

2. Materials and Models

2.1. Preparation of Test Materials and SCB Specimens

2.1.1. Asphalt

The penetration degree of matrix asphalt used in this study was 90, and the modifier was 0.25 mm rubber powder particles. The rubber powder particle content was 20% by weight of asphalt. The quantitative matrix asphalt was weighed, the quality accuracy was controlled at 0.01 g, the temperature of the matrix asphalt was controlled at 140 °C, and the rubber powder particles preheated and dried in advance were slowly added to the matrix asphalt many times, and gradually heated up and stirred. Finally, the temperature of asphalt and rubber powder mixture was controlled at 180 °C, and the mixture was stirred for 30 min. The evenly mixed mixture of asphalt and rubber powder was put into the high-speed shear emulsifier to maintain the shear rate of 6000~7000 r/min and the temperature of 160 °C for 30 min, and then it was kept in the oven at 120 °C for 2 h to prepare the CR. The technical matrix asphalt and CR indices indices are shown in Table 1.

Table 1. Technical indices of CR and matrix asphalt.

Technical Index	CR	Matrix Asphalt
Penetration (25,100 g, 5 s) (0.1 mm)	68.2	91.5
Softening Point/°C (Ring and ball test)	57.3	51.5
Ductility (5 cm/min,10 °C)/cm	17.2	9.4

Asphalt is a complex mixture of macromolecules. In the study of the chemical composition of asphalt, the asphalt was divided into several parts according to different solubility and physicochemical properties by using the different solubilities of various organic compounds in asphalt. According to the four-component analysis method stipulated in *Standard Test Methods of Bitumen and Bituminous Mixtures for Highway Engineering* (JTG E20-2011) [32], the four-component analysis test of matrix asphalt was carried out. The mass percentages of the four components of matrix asphalt are shown in Table 2.

Table 2. Mass percentages of four components of matrix asphalt.

Asphalt Components	Asphaltene	Saturate	Aromatic	Resin
Mass percentage (w/%)	9.4	25.2	28.8	36.6

2.1.2. Aggregate and Gradation of CR Mixture

The aggregate used in this study was basalt machine-made crushed aggregate, and relevant tests were carried out according to *Test Methods of Aggregate for Highway Engineering* (JTG E42-2005) [33]. The technical indices of the coarse aggregate and fine aggregate are shown in Tables 3 and 4, respectively.

Table 3. Technical indices of coarse aggregate.

Item	Unit	Specification	Test Result			Method
			10~20 mm	5~10 mm	3~5 mm	
Crushing value of stone	%	≤26	13.0		11.0	T0316
LA abrasion value	%	≤28	9.5		10.5	T0317
Soundness	%	≤12	1.5		1.6	T0314
Soft stone content	%	≤2	1.1	1.0	2.0	T0320
Water absorption	%	≤2	0.8	0.9	0.9	T0304
Water washing method <0.075 content	%	≤1	0.1	0.3	0.4	T0310

Table 4. Technical indices of fine aggregate.

Item	Unit	Specification	Test Result	Method
Apparent relative density	g/cm^3	≥2.5	2.734	T0328
Soundness (>0.3)	%	≥12	25	T0340
Mud content (<0.075 mm)	%	≤3	0.4	T0333
Sand equivalent	%	≥60	88	T0334
Methylene blue value	g/kg	≤25	1.8	T0346

The filler was lime powder ground from limestone. Relevant tests were carried out according to *Test Methods of Aggregate for Highway Engineering* (JTG E42-2005). The technical indices are shown in Table 5.

Table 5. Technical indices of filler.

Item		Unit	Specification	Test Result	Method
Appearance		—	No agglomerate	No agglomerate	—
Water content		%	≤1	0.5	T0103
Apparent density		t/m^3	≥2.5	2.686	T0352
Hydrophilic coefficient		—	<1	0.5	T0353
Plasticity index		%	<4	2.1	T0354
Heating stability		—	Actual observation record	No deterioration	T0355
Particle size range	<0.6 mm	%	100	100	T0351
	<0.15 mm	%	90~100	100	
	<0.075 mm	%	75~100	99.6	

The gradation of the CR mixture was AC-16, and the gradation was designed in accordance with the requirements of *Technical Specifications for Construction of Highway Asphalt Pavement* (JTG F40-2004) [34]. The gradation composition is shown in Table 6.

Table 6. Gradation composition of CR mixture.

Particle size (mm)	19	16	13.2	9.5	4.75	2.36	1.18	0.6	0.3	0.15	0.075
Synthetic gradation	100	94.2	83.6	71.8	48.5	33.6	25.4	18.6	12.9	9.6	6.6

2.1.3. Preparation of SCB Specimen

The SCB specimen was simple in molding and had strong repeatability. In addition, the stress state of the specimen was more in line with the actual use of road surface and better simulates the wheel load [35].

The CR mixture was formed by a rotary compaction instrument. The diameter of the specimen was 150 mm, the compressive stress was 600 kPa, the deviation angle was 1.16°, the speed was 30 r/min, and the compaction control method was 170 mm height control. The target void ratio was 5.0%. A 100 mm diameter core machine was used to core the cylindrical CR mixture. The core sample was cut into a semicircle specimen with a thickness of 40 mm ± 0.5 mm. Then, a pre-cut with a depth of 1 cm and width of 4 mm was made in the middle of the semicircle span. Preparation process of SCB specimen is shown in Figure 1.

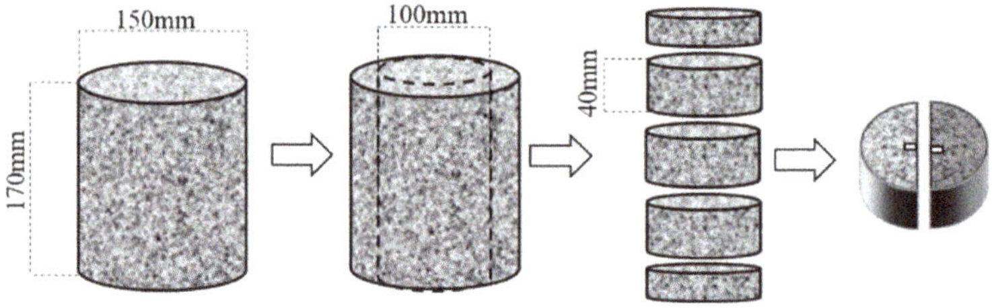

Figure 1. SCB specimen preparation.

Half of the SCB specimens were saturated with water for 15 min under the vacuum degree of 97.3 kpa and then soaked in water for 48 h to achieve the purpose of soaking. The water temperature was 25 °C. The dried specimen was used as a control for reserve.

2.2. Preparation of Contact Angle Measurement Sample

The CR was heated to a molten state and placed on the glass slide. The glass slide and asphalt were put into the oven and heated at 120 °C for 30 min to make the asphalt on the surface of the glass slide fully leveled. The leveled asphalt and the glass slide were removed and put into the drying chamber for cooling. The basalt and limestone aggregates were cut into cubes of 60 mm × 60 mm × 60 mm and then polished with lithographic polishing machine to remove surface particles to avoid their impact on experimental results. The polished test blocks were put into the drying chamber for standby. Contact angle measurement samples are shown in Figure 2.

(a) (b)

Figure 2. Contact angle measurement sample: (**a**) Crumb rubber sample, (**b**) Basalt and limestone blocks.

2.3. Construction of Models

2.3.1. Model of CR

In order to better study the physical, mechanical and rheological properties of asphalt, Li et al. [36] proposed a 12-component model of asphalt, including 3 asphaltenes, 5 resins, 2 saturates and 2 aromatics. The matrix asphalt molecular model was constructed according to the four-component ratio of asphalt obtained in the experiment (as shown in Table 2) and the 12-component asphalt model. The molecular ratio of each component was listed in Table 7.

Table 7. Ratio of matrix asphalt component model.

Components		Number of Molecules	Mass Ratio	Simulated Mass Ratio	Experimental Mass Ratio
Asphaltene	AS-1	1	3.9%	9.5%	9.4%
	AS-2	1	2.5%		
	AS-3	1	3.1%		
Resin	R-1	4	9.7%	36.7%	36.6%
	R-2	3	7.3%		
	R-3	4	5.1%		
	R-4	3	7.3%		
	R-5	4	7.3%		
Aromatic	Ar-1	8	15.8%	28.3%	28.8%
	Ar-2	7	12.5%		
Saturate	S-1	6	12.7%	25.7%	25.2%
	S-2	7	13%		

Representative molecular models of asphaltenes, aromatics, saturates and resins were added according to the ratio calculated in Table 7 to construct the matrix asphalt molecular group model, as shown in Figure 3.

The rubber components of waste tire rubber powder were mainly natural rubber, butadiene rubber and styrene butadiene rubber. The main rubber components of radial truck tires were natural rubber and styrene–butadiene rubber, and the mass ratio of natural rubber to styrene–butadiene rubber was 7:3. Natural rubber was the homopolymer of the natural rubber repeat unit, while styrene–butadiene was formed by the emulsion polymerization of styrene and butadiene at low temperature. The content of each structural unit is shown in Table 8 below.

The natural rubber was polymerized into a single natural rubber chain according to 15 repeating units, the styrene–butadiene rubber was polymerized into a single styrene–butadiene rubber chain according to the content in Table 8, and then the structure and energy of the single natural rubber chain and the single styrene–butadiene rubber chain were optimized according to the ratio of 7:3. After optimization, the Amorphous Cell module was used to construct the rubber powder molecular group model. The construction process of the rubber powder molecular group model is shown in Figure 4.

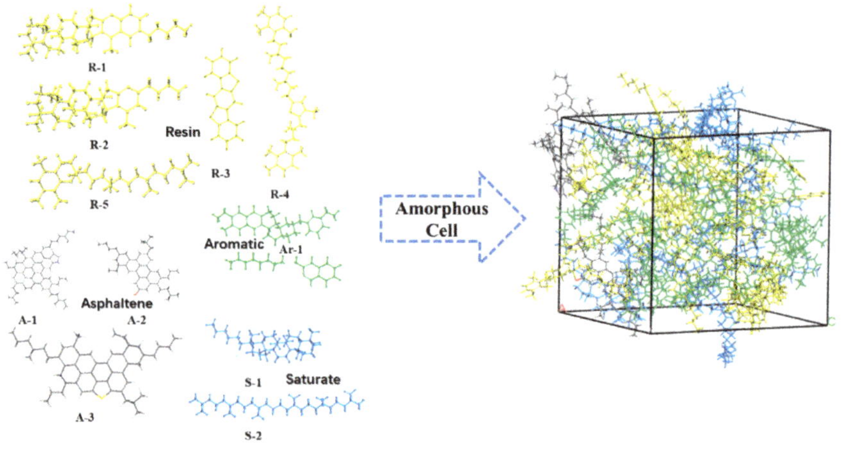

Figure 3. Matrix asphalt model composition.

Table 8. Content of styrene butadiene rubber polymer structure repeat unit (w%).

Styrene	Trans 1-4 Butadiene	Cis 1-4 Butadiene	1,2-Butadiene	Others
23.5	58	5.5	12	1

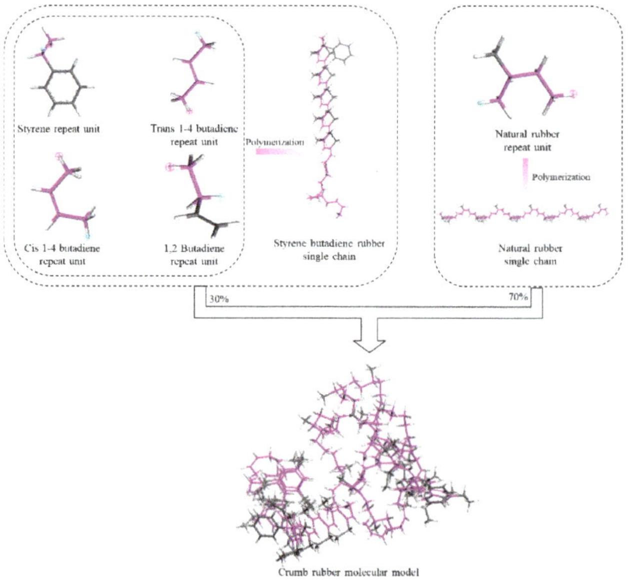

Figure 4. Construction process of the crumb rubber molecular group model.

After the molecular model of asphalt and rubber powder was established, the molecular model of CR was constructed by combining the two molecular group models according to a 20%mass ratio of rubber powder to asphalt, as shown in Figure 5.

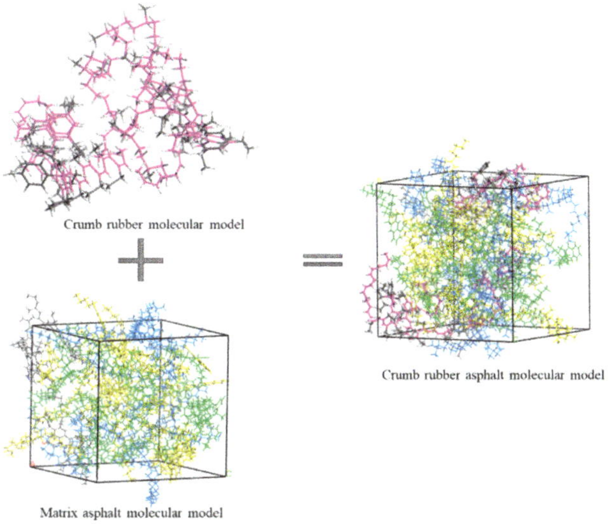

Figure 5. Composition of molecular CR model.

2.3.2. Model of Aggregate Mineral

Basalt was the most commonly used aggregate in engineering applications in Inner Mongolia. Therefore, basalt was selected as the representative aggregate in this study, and limestone was selected as the control group. The main mineral composition of basalt was augite, and the main mineral composition of limestone was calcite. The crystal cells were treated with supercell, and the crystal surface was cut along the direction of (0 0 1) to form a two-dimensional supercell system. A 10 Å vacuum layer was established on the surface of the supercell along the Z direction, which had a three-dimensional periodic boundary condition. It was used as the representative molecular model of the aggregate. The molecular model and information are listed in Table 9.

Table 9. The representative aggregate molecular models and information.

Mineral Type	Cell Model	Lattice Parameters	Molecular Model
Augite	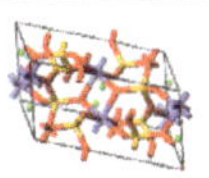	a = 9.7 Å b = 8.9 Å c = 5.3 Å α = 90° β = 106.8° γ = 90°	
Calcite	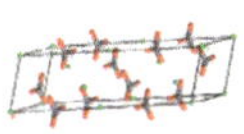	a = 5.0 Å b = 5.0 Å c = 17.1 Å α = 90° β = 90° γ = 120°	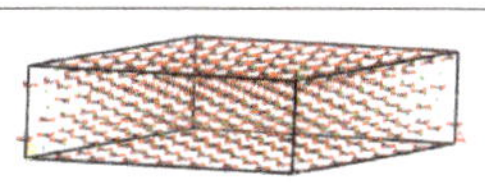

Note: a, b and c are lattice lengths, α, β, γ are the lattice angles.

2.3.3. Interface Model of CR-Aggregate

The CR-aggregate interface model was constructed by attaching the asphalt layer to the aggregate model, a 50 Å vacuum layer was established along the Z direction on the asphalt layer and the aggregate model was fixed to eliminate the basement effect of three-dimensional periodic boundary conditions. Water molecules were added between the asphalt and aggregate interface to form the asphalt–aggregate interface model under water immersion conditions. The asphalt–aggregate interface model is shown in Figure 6.

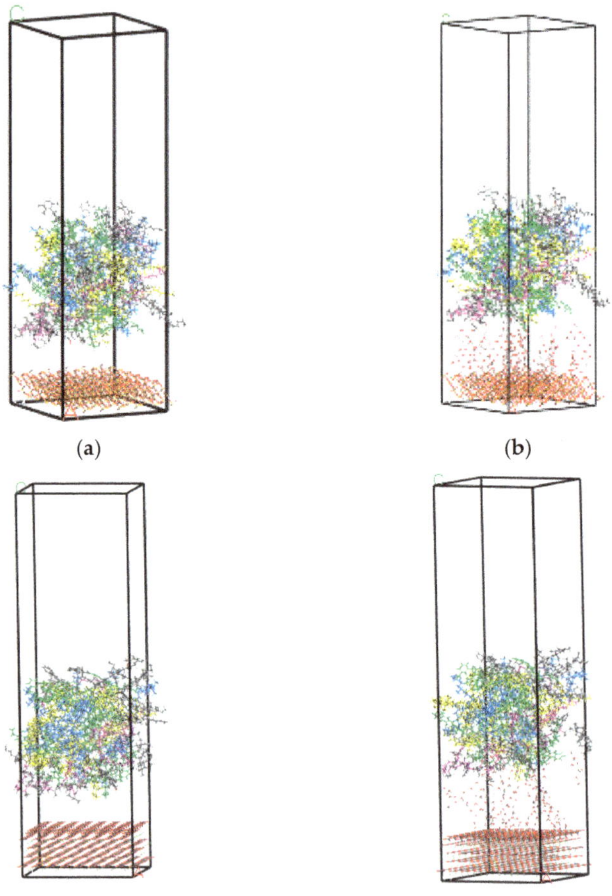

(**a**) (**b**) (**c**) (**d**)

Figure 6. (**a**) Interface model of CR–augite. (**b**) Interface model of CR–water–augite. (**c**) Interface model of CR-calcite. (**d**) Interface model of CR–water–calcite.

3. Results and Analysis

3.1. SCB Test and Analysis

3.1.1. SCB Test

The loading equipment was a universal material testing (UTM) machine, and the distance between the two fulcrum points at the bottom of the semicircular specimen was 80 mm. The two failure modes adopted different loading methods. Sinusoidal wave was selected for the loading waveform of the medium-temperature fatigue failure test, and the loading frequency was determined by simulating the actual road speed. Considering the highway driving speed (60–120 km/h), when the driving speed was 60 km/h, the road surface was unfavorable, and the corresponding frequency was 10 Hz. According to the preliminary semi-circular bending (SCB) test, the failure load of the specimen with a pre-cut slot was 2500 N and the selected stress ratio was 0.2, so the loading range was 50 to 500 N. The test termination condition was that the specimen was completely fractured. In the low-temperature fracture test, the specimens were subjected to vertical uniform load at a loading rate of 1 mm/min. The test termination condition was that the specimen was completely fractured. The loading conditions are shown in Table 10.

Table 10. Loading conditions of SCB.

	Temperature/°C	Control Mode	Loading Mode
Medium-temperature fatigue failure test	20	Stress control	Repeated load: 10 Hz sine wave, loading range 50–500 N
Low-temperature fracture test	−10	Displacement control	Uniform load: 1 mm/min

3.1.2. Digital Image Measurement System

The digital image measurement system consisted of two fixed light sources, image acquisition equipment (two CCD cameras), calibration equipment, image processing equipment, a loading system, etc. The basic principle of digital image correlation (DIC) technology is shown in Figure 7. Speckle images before and after the surface deformation of the measured object were recorded by image acquisition equipment. The reference subregion with displacement point (x, y) as the center and side length of (2 M + 1) × (2 M + 1) was selected in the speckle images before deformation. By searching the speckle image after deformation, the target subregion with the greatest correlation with the reference subregion was found. The center point of the target subregion was defined as (x′, y′). The horizontal and vertical displacement components u and v of the subregion before and after deformation were obtained by the changes of the center points of the reference subregion and the target subregion after deformation.

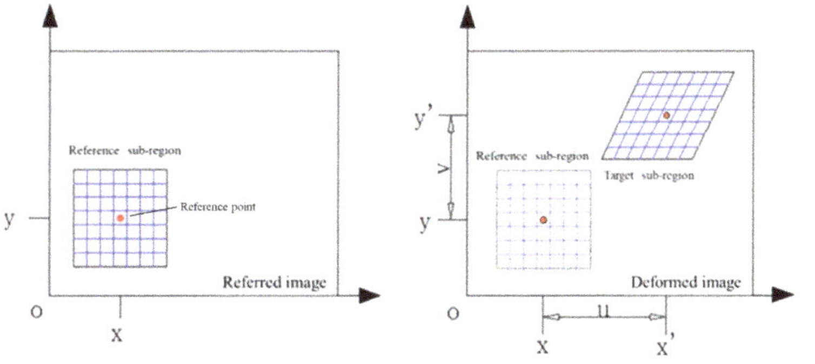

Figure 7. Displacement diagram before and after deformation.

3.1.3. Strain Energy and Interfacial Damage Factor

The horizontal strain density D_E was the area enclosed under the *Exx-t* curve before the horizontal strain reached the maximum value (macrofracture), namely, the cumulative value of horizontal strain of asphalt mixture from the beginning of loading to the macrofracture. The higher the horizontal strain density D_E value was, the better the cracking resistance of the asphalt mixture was, and its function expression [37] was as follows:

$$D_E = \int_0^{t_0} Exx(t)dt \tag{1}$$

where *Exx (t)* is a horizontal strain–time curve function; t_0 is the time corresponding to the maximum horizontal strain; and *dt* is the derivative of time.

The interfacial damage factor can represent the damage degree of the interface, thus reflecting the aggregate–asphalt interface bonding strength. The greater the interface damage factor D_I was, the greater the damage degree of the aggregate–asphalt interface was and the weaker the interface bonding strength was. The interfacial damage factor was defined as D_I [38]:

$$D_I = Exx/Exx_{\max} \tag{2}$$

where $Exx_{\max}$ is the maximum value of Exx, that is, Exx at the time of specimen failure.

The medium-temperature fatigue fracture strain density D_E was the loading time obtained according to the loading times of repeated load, and the horizontal strain before the cumulative horizontal strain reached the maximum value (macrofracture) was integrated with the loading time to obtain the area surrounded under the *Exx-t* curve, that is, the cumulative value of horizontal strain of asphalt mixture from the beginning of loading to the macrofracture. In the interfacial damage factor D_I, $Exx_{\max}$ selected the maximum horizontal strain before crack penetration. Taking Figure 8 as an example, the variation curve of the horizontal strain of the medium-temperature fatigue fracture with the loading time/times of repeated load under dry condition, the dark area in the figure was the integral area of Exx to the loading time, that is, the horizontal strain energy density D_E. The peak value of the Exx curve in the figure was the maximum horizontal strain $Exx_{\max}$.

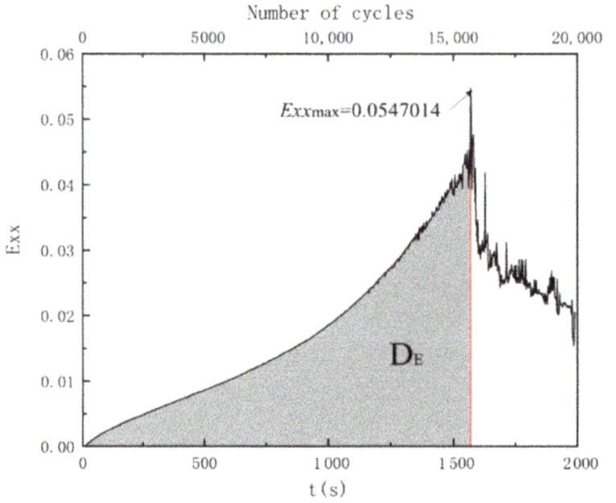

Figure 8. Variation curve of horizontal strain with loading time/times of repeated load.

3.1.4. Analysis of Influence of Moisture on Cracking Resistance of the CR Mixture

The cracking of the asphalt mixture was mainly characterized by low-temperature fracture and medium-temperature fatigue cracking. Based on DIC technology, the strain energy density (D_E) and interfacial damage factor (D_I) of the CR mixture before and after immersion with different failure forms were analyzed, as shown in Figures 9–12.

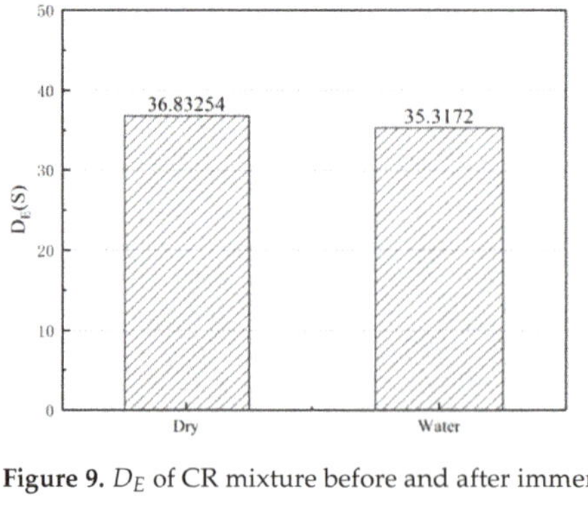

Figure 9. D_E of CR mixture before and after immersion under medium temperature fatigue failure.

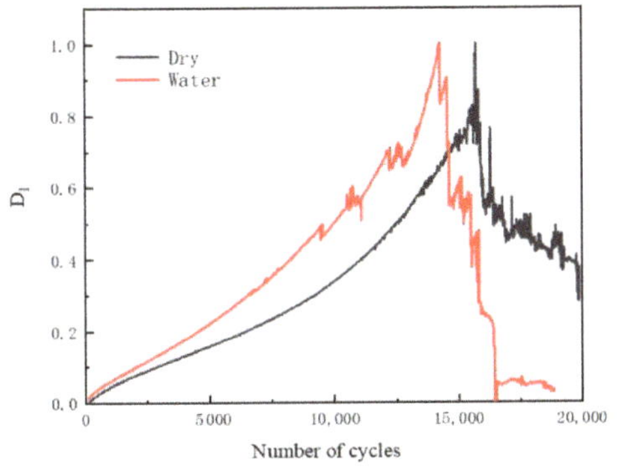

Figure 10. D_I of CR mixture before and after immersion under medium temperature fatigue failure.

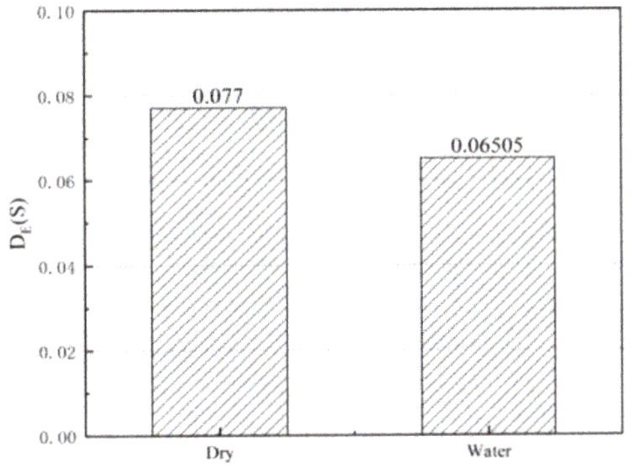

Figure 11. D_E of CR mixture before and after immersion under low temperature fracture.

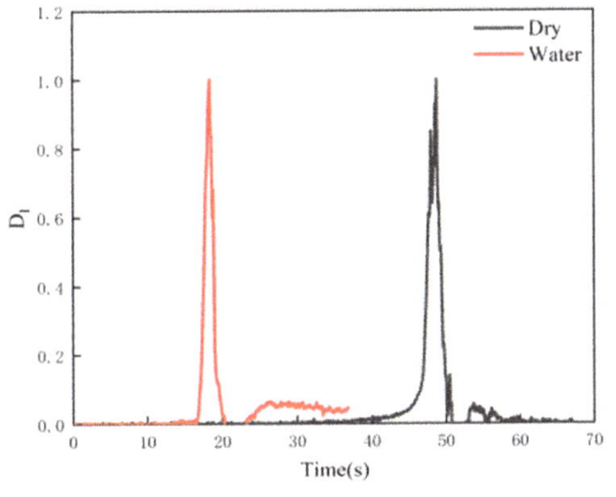

Figure 12. D_I of CR mixture before and after immersion under low-temperature fracture.

It can be seen from Figures 9–12 that the D_E of CR decreases after immersion, and the D_I occurs earlier than that of the dry environment. In addition, it can be seen that the D_I increases faster in the environment with water, indicating that the interface is easier to be destroyed in the environment with water.

As a kind of viscoelastic–plastic composite material, the temperature and loading mode have great significance to the performance of the asphalt mixture. By comparing Figures 9 and 11, the strain energy density of the asphalt mixture at low temperature is significantly reduced by three orders of magnitude compared to that under repeated medium-temperature load. This indicates that when the temperature is lower, the loading time is shorter, the force is larger and the accumulated microstrain of the asphalt mixture before fracture is smaller; that is, the specimen fracture occurs when there is no obvious deformation. Although the strain energy density of the specimen decreases due to both failure modes, it can be found that the strain energy density of the specimen after immersion is 15.5% lower than that of the dry specimen under low-temperature fracture failure, and the strain energy density of the specimen after immersion is 4.1% lower than that of the dry specimen under medium temperature repeated load. Obviously, the damage from water is more serious under the low-temperature fracture. This is due to the decrease in asphalt adhesion caused by water immersion in the asphalt mixture, which makes the asphalt peel off from the aggregate surface. Under the low-temperature condition, the temperature stress generated in the asphalt layer exceeds the tensile strength of the asphalt mixture, and cracks occur at the adhesion interface between the asphalt and the aggregate. Therefore, the low-temperature condition intensifies the damage effect from water on the interface between the asphalt and the aggregate.

Comparing Figures 10 and 12, it can be seen that the curve in Figure 10 shows that the damage factor of repeated medium-temperature load increases gradually with time, which can be divided into three stages. The first stage is the rapid growth stage, in which *Exx* increases rapidly, but the duration is short. The specimen as a whole resists the applied repeated load, but there is no cracking, and the strain increases rapidly, indicating that the crumb rubber–asphalt mixture has a certain deformation resistance. The second stage is the accelerated growth stage, in which *Exx* and the growth rate gradually increase until reaching the peak value, which is caused by the gradual generation and full development of cracks until they are penetrated. The third stage is the abrupt fluctuation decline stage. In this stage, *Exx* drops suddenly after the peak value. Then, the fluctuation drops to the lowest value because the specimen is completely cracked, and the horizontal strain gradually decreases with the release of horizontal stress. However, in Figure 12, the peak in the damage factor value of low-temperature fracture failure occurs suddenly. The results show that the cracks of specimens under medium temperature repeated load are a cumulative process from damage to micro-cracks and then to macro-cracks, while the cracks in specimens under low-temperature fracture failure appear suddenly, and the damage process is very short. The interfacial damage factors of the two failure modes reach the peak value in advance after immersion, while the dry one lags behind. However, the change in the interfacial damage factors of the two failure modes with time is also significantly different. It is not difficult to find that the distance between the two peaks before and after immersion under low-temperature fracture failure is greater than that before and after immersion under repeated medium-temperature load, which is consistent with the conclusion of the above-mentioned strain energy density. That is to say, the damage from water is more serious under low-temperature fracture.

3.2. Contact Angle Measurement Test and Analysis

3.2.1. Contact Angle Measurement Test

According to material surface theory, surface free energy in a vacuum was the energy that separated a solid or liquid to create a new interface. Gibbs free energy refers to the difference between the initial and final free energy of a certain system, which was represented by ΔG. According to Gibbs free energy, the asphalt–aggregate model in the dry condition and the asphalt–water–aggregate adhesion-cracking model in the wet condition were established, as shown in Figure 13.

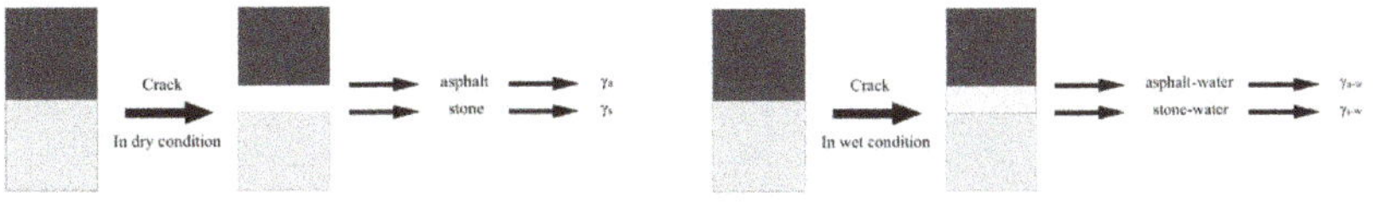

(a) (b)

Figure 13. Schematic diagram of asphalt–aggregate adhesion-cracking model: (**a**) Dry condition; (**b**) Wet Condition.

Fowkes et al. [39] assumed that the dispersion force between the liquid and solid interface can be expressed by the geometric mean of the dispersion components of the free energy on liquid and solid surfaces. On this basis, Owens et al. [40] further developed this method to make it applicable to polar components. Adhesion work between asphalt and aggregate is calculated as follows:

$$W_{as} = 2\sqrt{\gamma_a^d \gamma_s^d} + 2\sqrt{\gamma_a^+ \gamma_s^-} + 2\sqrt{\gamma_a^- \gamma_s^+}$$

$$W_{asw} = 2\begin{pmatrix} \sqrt{\gamma_a^d \gamma_s^d} - \sqrt{\gamma_a^d \gamma_w^d} - \sqrt{\gamma_s^d \gamma_w^d} + \gamma_w^d + \sqrt{\gamma_w^+}\left(\sqrt{\gamma_w^-} - \sqrt{\gamma_a^-} - \sqrt{\gamma_s^-}\right) \\ +\sqrt{\gamma_w^-}\left(\sqrt{\gamma_w^+} - \sqrt{\gamma_a^+} - \sqrt{\gamma_s^+}\right) + \sqrt{\gamma_a^+ \gamma_s^-} + \sqrt{\gamma_a^- \gamma_s^+} \end{pmatrix} \quad (3)$$

where W_{as} is the adhesion work between asphalt and aggregate under dry conditions (mJ/m^2); W_{asw} is the adhesion work between asphalt and aggregate under wet conditions (mJ/m^2); γ_a^W is the surface free energy between the asphalt and water interface (mJ/m^2); γ_s^w is the surface free energy between the aggregate and water interface (mJ/m^2); γ_a^s is the surface free energy between the asphalt–aggregate interface (mJ/m^2); γ_w^d is the dispersion component of the surface free energy of water (mJ/m^2); γ_S^d is the dispersion component of the surface free energy of aggregate (mJ/m^2); γ_a^d is the dispersion component of the surface free energy of asphalt (mJ/m^2); γ_S^+ is the Lewis acid component of the surface free energy of aggregate (mJ/m^2); γ_S^- is the Lewis basic component of the surface free energy of aggregate (mJ/m^2); γ_a^+ is the Lewis acid component of the surface free energy of asphalt (mJ/m^2); γ_a^- is the Lewis basic component of the surface free energy of asphalt (mJ/m^2); γ_w^+ is the Lewis acid component of the surface free energy of water (mJ/m^2); and γ_w^- is the Lewis basic component of the surface free energy of water (mJ/m^2).

The water stability index W' [41] was defined, and its calculation is shown in Equation (4). The larger W' was, the less the adhesion work of the material was affected by water, and the better its surface adhesion ability was:

$$W' = \left|\frac{W_{asw}}{W_{as}}\right| \times 100\% \quad (4)$$

where W' is the water stability index; W_{as} is the adhesion work between the asphalt and aggregate under dry conditions (mJ/m^2); and W_{asw} is the adhesion work between the asphalt and aggregate under wet conditions (mJ/m^2).

3.2.2. Analysis of the Contact Angle Measurement Results

The surface free energy parameters obtained through the contact angle measurement test were listed in Table 11.

Table 11. Surface free energy parameters of different materials (mJ/m^2).

	γ	γ^d	γ^p	γ^+	γ^-
CR	22.80	21.94	0.86	0.17	1.10
Basalt	32.39	24.64	6.75	1.08	10.56
Limestone	37.74	31.77	5.97	0.75	11.89
Distilled water	72.8	21.8	51.0	25.5	25.5

The surface free energy parameters listed in Table 11 were substituted into Equation (3) to obtain the adhesion work W_{as} and W_{asw} between CR and basalt and limestone in dry and wet conditions, respectively. The water stability indices of the adhesion work of different aggregates and CR were further calculated, and the results are plotted in Figures 14 and 15.

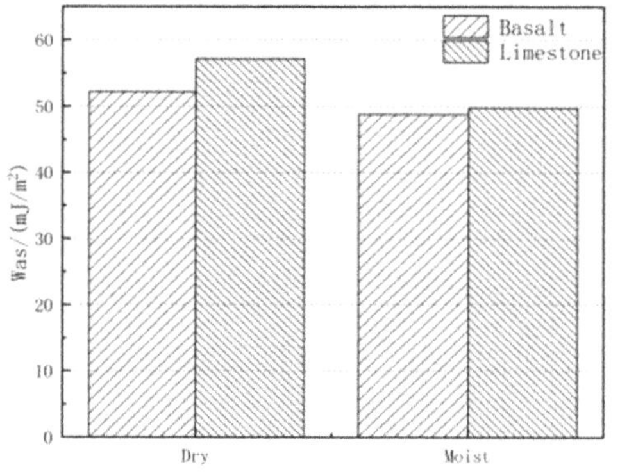

Figure 14. Influence of water on adhesion work of CR.

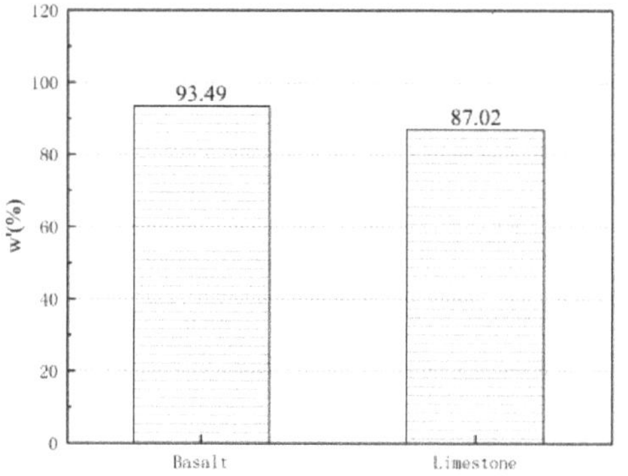

Figure 15. Influence of different aggregates and CR on W'.

It can be seen from Figure 14 that the wet surface reduces the adhesion work between the aggregate and the CR. Although the adhesion work between the CR and the basalt and limestone aggregates decreases, it is obvious that water has a greater impact on the limestone. Under the wet condition, the adhesion work between the limestone and CR decreases more than that between the basalt and CR on the wet surface. It can be seen from Figure 15 that the W' between CR and basalt is greater than that between CR and limestone, indicating that the adhesion between limestone and CR decays faster under wet conditions. Although limestone has better adhesion with CR under dry conditions, limestone does not have good water stability.

3.3. *Molecular Dynamics Simulation and Analysis*

3.3.1. Optimization and Calculation of Molecular Dynamics Model

After the construction of the molecular dynamics model, it is necessary to optimize the model to ensure its accuracy, rationality and stability. (a) In order to eliminate the unreasonable configurations generated in the process of molecular model construction, the molecular model needed to be geometrically optimized first. The molecular model was iterated 100,000 times under the Compass II force field. At this time, the molecular group energy decreased and tended to be stable with the increase in the number of iterations. (b) In order to eliminate the unreasonable energy in the process of molecular model construction, an annealing treatment was carried out on the geometrically optimized molecular model. The NVE ensemble was selected at 300~500 K, and the number of steps was set to 50,000 for annealing. (c) In order to make the molecular model achieve energy and volume stability, the molecular model after annealing was optimized dynamically. The NVT ensemble was set at 1500 K, one configuration was output every 1000 steps and the dynamics calculation time was 100 ps. (d) At this time, the molecular model reached the stability of configuration, energy and volume, and the dynamics calculation can be carried out. The NVT ensemble was selected to set the temperature at 298 K, and one configuration was output every 1000 steps. The dynamics calculation time was 200 ps, and the dynamics calculation results can be obtained.

3.3.2. Intermolecular Interaction Energy

In order to quantify the adhesion between the aggregate and the asphalt binder, the interaction energy of the asphalt aggregate system was calculated. The interface interaction energy represented the energy absorbed or released by the formation of a new interface when the interface was separated. The interaction energy between terms A and B in system $(A-B)$ was calculated by Formula (5) [42]:

$$E_{interaction} = E_{(A-B)} - (E_A + E_B) \quad (5)$$

where $E_{interaction}$ is the interaction energy between system A and system B; E_A is the potential energy of term A; E_B is the potential energy of term B; and $E_{(A-B)}$ is the total potential energy of system $(A-B)$.

Considering the influence of the interface's molecular weight on the adhesion energy of the crystal model, the influence of the molecular weight difference of the crystal interface model can be eliminated by converting the interface interaction energy into interface interaction energy density. The interface interaction energy density is the energy required or released per unit area during interface peeling. Because of the linear correlation between the interaction energy density and the interaction energy, the interaction energy density can also be used to evaluate the bonding degree of the interface. The interaction energy density was calculated by Equation (6):

$$E_\rho = E_{interaction} / A \quad (6)$$

where E_ρ is the interaction energy density; $E_{interaction}$ is the interaction energy between system A and system B; and A is the area of the model interface.

3.3.3. Analysis of Molecular Dynamics Results

It can be seen from Figure 16 that there are great differences in the interaction energy between different aggregates and CR. It is obvious that limestone has a better interaction relationship with CR, while basalt is worse. At the same time, it can be seen that the addition of water reduces the interaction energy between aggregates and CR. In order to eliminate the influence of contact area on the interaction energy between the aggregate and the asphalt in the molecular model, the interaction energy per unit area between the aggregate and CR in dry and wet conditions was further compared, as shown in Figure 17. It can be seen from Figure 17 that the addition of water has a significant impact on the interaction between aggregate and CR. By comparing the two aggregates, it can be found that the interaction energy density between basalt and CR decreases by only 38.2% after the addition of water, while the interaction energy density between limestone and CR decreases by 64.5% after the addition of water. It can be seen that the interaction relationship of basalt, as an aggregate mineral, decays more slowly after it is combined with CR. From the perspective of interaction energy, the interaction performance between basalt and CR is not as good as that of limestone, but the interaction energy density decreases less after immersion, indicating that basalt has better adhesion to asphalt under the action of water, and it is an aggregate mineral with better resistance to water damage of asphalt mixture. The reason why the performance of limestone is inferior to that of basalt in the presence of water is that the main component of limestone is calcium carbonate, and a small part of it will be hydrolyzed and ionized after encountering water, which destroys its original crystal structure and destroys the interaction relationship between limestone and CR, so it shows instability after encountering water. In addition, the combination of asphalt and aggregate is mainly in the form of hydrogen bond. When water molecules appear at the interface between asphalt and aggregate, the aggregate will be bonded with some water molecules in the form of hydrogen bonds, thus reducing the binding ability between the asphalt and the aggregate. The interaction between the asphalt and the aggregate is weakened. Limestone, as a more polar molecule than basalt, is more likely to bond with water molecules, so its interaction with asphalt decreases more dramatically in the presence of water.

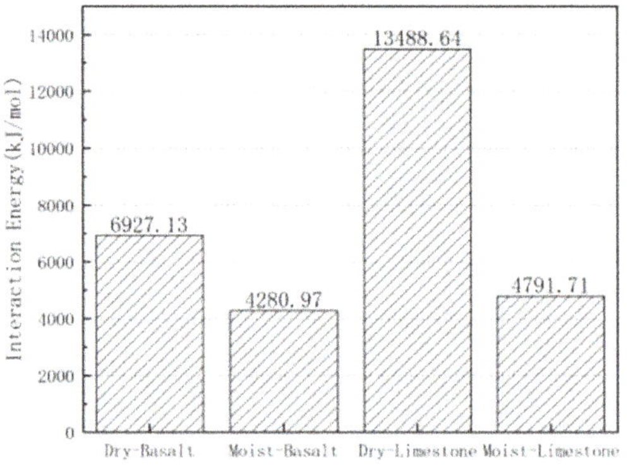

Figure 16. Influence of water on the asphalt–aggregate interaction energy.

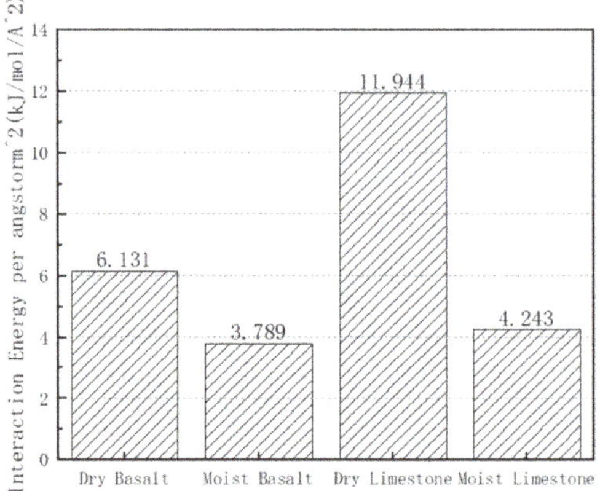

Figure 17. Influence of water on interaction energy density.

3.4. Multiscale Association Analysis

For the unified research object, observation and analysis at multiple scales can provide basis and method at micro and molecular scales to solve the macroscopic performance problem. The relationship between multiple scales is shown in Figure 18. The influence indices of water at multiple scales on the interface of crumb rubber–asphalt mixture are shown in Table 12.

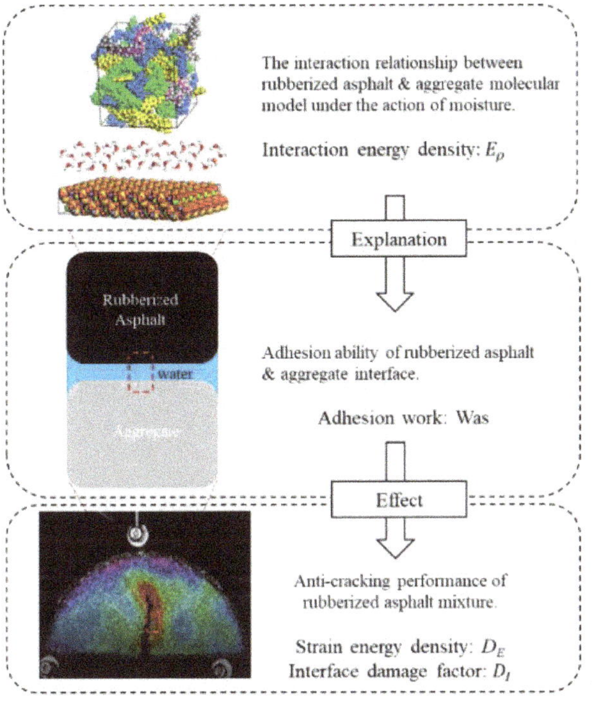

Figure 18. Multiscale association analysis.

Table 12. Multiscale indices of the influence of water.

Scales	Indexes	Unit	Dry	Wet
Molecular scale	Basalt-Asphalt E_ρ	(kJ/mol/A^2)	6.131	3.789
	Limestone-Asphalt E_ρ	(kJ/mol/A^2)	11.944	4.243
Micro scale	Basalt-Asphalt W'	(%)	100	93.49
	Limestone-Asphalt W'	(%)	100	87.02
Macro scale	Medium temperature fatigue failure D_E	-	36.83254	35.3172
	Low temperature fracture D_E	-	0.077	0.06505

Through molecular dynamics simulation calculation, it can be known that the interaction energy and interaction energy density of the asphalt and aggregate at the molecular scale decrease with water participation; the interaction energy density of basalt and CR decreases by 38.2%, and the interaction energy density of limestone and CR decreases by 64.5%. Due to the decrease in the interaction energy between CR and the aggregate, the adhesion ability between the asphalt and the aggregate at the microscale decreases obviously under the wet condition, the W' of basalt and CR decreases by 6.51%, and the W' of limestone and CR decreases by 12.98%. Since the adhesion ability between the crumb rubber asphalt and aggregate is an influencing factor on the crack resistance of the crumb rubber–asphalt mixture, the decrease in adhesion ability between the asphalt and aggregate at the microscale leads to the decrease in the crack resistance of the macro crumb rubber–asphalt mixture. The D_E of medium-temperature fatigue failure decreases by 4.11%, and the D_E of low-temperature fracture failure decreases by 15.52%. It can be seen from the above that the influence of water on the interface between the crumb rubber–asphalt and the aggregate is basically the same at the molecular scale, microscale and macroscale. The macrocracking of the asphalt mixture can further be analyzed for its internal mechanism and changes caused at the micro and molecular scale, so as to provide a certain theoretical basis for further improving the performance of asphalt mixtures. However, due to the influence of the size effect, the correlations of the three scales need to be further studied.

4. Conclusions

The influence of water on the interface of the crumb rubber–asphalt mixture was studied using the multiscale method. At the macroscale, the performance of the crumb rubber–asphalt mixture before and after water immersion under the two failure modes of medium-temperature fatigue failure and low-temperature fracture failure were analyzed. At the microscale, the adhesion work and water stability index between crumb rubber asphalt and basalt and limestone under dry and wet conditions were analyzed. A molecular dynamics simulation was carried out on the molecular model of crumb rubber asphalt and a representative molecular model of basalt and limestone at the molecular scale. The interaction energy and interaction energy density between the asphalt model and the mineral model were calculated and analyzed. The research conclusions are as follows:

(1) At macroscale, water damage is more serious under low-temperature fracture damage than that of medium-temperature fatigue failure;
(2) At the microscale, the adhesion energy between the crumb rubber asphalt and aggregate decreases under wet conditions, indicating that the participation of water has a certain damage to the adhesion ability between the crumb rubber asphalt and the aggregate;
(3) The adhesion ability between limestone and crumb rubber asphalt is more affected by water than by basalt;
(4) The results of the molecular-scale simulation are consistent with those of the micro-adhesion test and the macrofracture test, indicating that molecular simulation has certain significance in simulating the adhesion ability of water to the interface.

Author Contributions: L.W.: resources, writing—review and editing, supervision. Y.L.: writing—original draft, conceptualization, methodology, software, validation. L.Z.: formal analysis, writing—review and editing, visualization. All authors have read and agreed to the published version of the manuscript.

Funding: This research was supported by a grant (No. 2019GG031) from the Key Science-Technology Project of Inner Mongolia, China.

Data Availability Statement: The data presented in this research are available on request from the corresponding author. The data are not publicly available due to privacy.

Conflicts of Interest: The authors declare that they have no known competing financial interests or personal relationships that could have appeared to influence the work reported in this paper.

References

1. Wang, L.; Shan, M.; Chang, C.; Zhou, X. The macro-and meso-cracking characteristics of warm mix crumb rubber asphalt mastics before and after aging. *Constr. Build. Mater.* **2020**, *262*, 120724. [CrossRef]
2. Guo, Z.; Wang, L.; Feng, L.; Guo, Y. Research on fatigue performance of composite crumb rubber modified asphalt mixture under freeze thaw cycles. *Constr. Build. Mater.* **2022**, *323*, 126603. [CrossRef]
3. Ding, X.; Chen, L.; Ma, T.; Ma, H.; Gu, L.; Chen, T.; Ma, Y. Laboratory investigation of the recycled asphalt concrete with stable crumb rubber asphalt binder. *Constr. Build. Mater.* **2019**, *203*, 552–557. [CrossRef]
4. Ma, T.; Wang, H.; He, L.; Zhao, Y.; Huang, X.; Chen, J. Property characterization of asphalt binders and mixtures modified by different crumb rubbers. *J. Mater. Civ. Eng.* **2017**, *29*, 04017036. [CrossRef]
5. Liu, Q.; Liu, J.; Yu, B.; Zhang, J.; Pei, J. Evaluation and optimization of asphalt binder and mixture modified with high activated crumb rubber content. *Constr. Build. Materials.* **2022**, *314*, 125676. [CrossRef]
6. Asdollah-Tabar, M.; Heidari-Rarani, M.; Aliha, M.R.M. The effect of recycled PET bottles on the fracture toughness of polymer concrete. *Compos. Commun.* **2021**, *25*, 100684. [CrossRef]
7. Liu, Z.; Li, S.; Wang, Y. Characteristics of asphalt modified by WEO/PPA: Conventional, high-temperature rheological, and mechanism properties. *J. Clean. Prod.* **2022**, *330*, 129844. [CrossRef]

8. Ziari, H.; Divandari, H.; Akbar, S.M.S.A.; Hosseinian, S.M. Investigation of the effect of crumb rubber powder and warm additives on moisture resistance of SMA mixtures. *Adv. Civ. Eng.* **2021**, *4*, 6653594. [CrossRef]
9. Zhang, H.; Gong, M.; Gao, D.; Yang, T.; Huang, Y. Comparative analysis of mechanical behavior of composite modified asphalt mixture based on PG technology. *Constr. Build. Mater.* **2020**, *259*, 119771. [CrossRef]
10. Aigner, E.; Lackner, R.; Pichler, C. Multiscale prediction of viscoelastic properties of asphalt concrete. *J. Mater. Civ. Eng.* **2009**, *21*, 771–780. [CrossRef]
11. Yao, H.; Dai, Q.; You, Z. Chemo-physical analysis and molecular dynamics (MD) simulation of moisture susceptibility of nano hydrated lime modified asphalt mixtures. *Constr. Build. Mater.* **2015**, *101*, 536–547. [CrossRef]
12. Chen, Z.; Xie, J.; Xiao, Y.; Chen, J.; Wu, S. Characteristics of bonding behavior between basic oxygen furnace slag and asphalt binder. *Constr. Build. Mater.* **2014**, *64*, 60–66. [CrossRef]
13. Wang, L.; Shen, A.; Yao, J. Effect of different coarse aggregate surface morphologies on cement emulsified asphalt adhesion. *Constr. Build. Mater.* **2020**, *262*, 120030. [CrossRef]
14. Wei, J.; Dong, F.; Li, Y.; Zhang, Y. Relationship analysis between surface free energy and chemical composition of asphalt binder. *Constr. Build. Mater.* **2014**, *71*, 116–123. [CrossRef]
15. Azarhoosh, A.R.; Nejad, F.M.; Khodaii, A. Using the surface free energy method to evaluate the effects of nanomaterial on the fatigue life of hot mix asphalt. *J. Mater. Civ. Eng.* **2016**, *28*, 04016098. [CrossRef]
16. Guo, M.; Liang, M.; Jiao, Y.; Tan, Y.; Yu, J.; Luo, D. Effect of aging and rejuvenation on adhesion properties of modified asphalt binder based on AFM. *J. Microsc.* **2021**, *284*, 244–255. [CrossRef]
17. Zhang, W.G.; Luan, Y.C.; Ma, T.; Wang, S. Multilevel analysis of the aging mechanisms and performance evolution of rubber-modified asphalt. *J. Mater. Civ. Eng.* **2021**, *33*, 04021365. [CrossRef]
18. Cong, P.H.; van Thom, D.; Duc, D.H. Phase field model for fracture based on modified couple stress. *Eng. Fract. Mech.* **2022**, *269*, 108534. [CrossRef]
19. Heidari-Rarani, M.; Bashandeh-Khodaei-Naeini, K. Micromechanics based damage model for predicting compression behavior of polymer concretes. *Mech. Mater.* **2018**, *117*, 126–136. [CrossRef]
20. Doan, D.H.; Van Do, T.; Nguyen, N.X.; Van Vinh, P.; Trung, N.T. Multi-phase-field modelling of the elastic and buckling behaviour of laminates with ply cracks. *Appl. Math. Model.* **2021**, *94*, 68–86. [CrossRef]
21. Wang, H.; Lin, E.; Xu, G. Molecular dynamics simulation of asphalt-aggregate interface adhesion strength with moisture effect. *Int. J. Pavement Eng.* **2015**, *18*, 414–423. [CrossRef]
22. Wang, L.; Zhang, L.; Liu, Y. Study on compatibility of rubber powder and asphalt in rubber powder modified asphalt by molecular dynamics. *J. Build. Mater.* **2018**, *21*, 689–694. (In Chinese)
23. Wang, L.; Zhang, L.; Liu, Y. Molecular dynamics study on compatibility of asphalt and rubber powders before and after aging. *J. Build. Mater.* **2019**, *22*, 474–479. (In Chinese)
24. Wang, L.; Liu, Y.; Zhang, L. Micro/nanoscale study on the effect of aging on the performance of crumb rubber modified asphalt. *Math. Probl. Eng.* **2020**, *2020*, 1924349. [CrossRef]
25. Rinaldini, E.; Schuetz, P.; Partl, M.N.; Tebaldi, G.; Poulikakos, L.D. Investigating the blending of reclaimed asphalt with virgin materials using rheology, electron microscopy and computer tomography. *Compos. Part B* **2014**, *67*, 579–587. [CrossRef]
26. Nazzal, M.D.; Mogawer, W.; Austerman, A.; Abu Qtaish, L.; Kaya, S. Multi-scale evaluation of the effect of rejuvenators on the performance of high RAP content mixtures. *Constr. Build. Mater.* **2015**, *101*, 50–56. [CrossRef]
27. Nazzal, M.D.; Holcombe, E.; Kim, S.S.; Abbas, A.; Kaya, S. Multi-scale evaluation of the effect of ras on the fracture properties of asphalt mixtures. *Constr. Build. Mater.* **2018**, *175*, 126–133. [CrossRef]
28. Samieadel, A.; Oldham, D.; Fini, E.H. Multi-scale characterization of the effect of wax on intermolecular interactions in asphalt binder. *Constr. Build. Mater.* **2017**, *157*, 1163–1172. [CrossRef]
29. Yang, S.Q.; Sun, S.; Qin, L.S.; Li, Q. Microstructure and meso-mechanical properties of asphalt mixture modified by rubber powder under a multi-scale effect. *Coatings* **2021**, *11*, 1321. [CrossRef]
30. Zhang, L.; Long, N.; Liu, Y.; Wang, L. Cross-scale study on the influence of moisture-temperature coupling conditions on adhesive properties of rubberized asphalt and steel slag. *Constr. Build. Mater.* **2022**, *332*, 127401. [CrossRef]
31. Wang, L.; Zhang, L.; Liu, Y. Molecular dynamics study on the effect of mineral composition on the interface interaction between rubberized asphalt and aggregate. *J. Mater. Civ. Eng.* **2022**, *34*, 04022032. [CrossRef]
32. *JTG E20-2011*; Standard Test Methods of Bitumen and Bituminous Mixture for Highway Engineering. People Transportation Press: Beijing, China, 2011.
33. *JTG E42-2005*; Highway Engineering Aggregate Test Regulations. People Transportation Press: Beijing, China, 2005.
34. *JTG F40-2004*; Technical Specification for Construction of Asphalt Pavement in Highway Engineering. People Transportation Press: Beijing, China, 2004.
35. Aliha, M.R.M.; Heidari-Rarani, M.; Shokrieh, M.M.; Ayatollahi, M.R. Experimental determination of tensile strength and KIc of polymer concretes using semi-circular bend (SCB) specimens. *Struct. Eng. Mech.* **2012**, *43*, 823–833. [CrossRef]
36. Li, D.D.; Greenfield, M.L. Chemical compositions of improved model asphalt systems for molecular simulations. *Fuel* **2014**, *115*, 347–356. [CrossRef]
37. Wang, L.; Shan, M.; Li, C. The cracking characteristics of the polymer-modified asphalt mixture before and after aging based on the digital image correlation technology. *Constr. Build. Mater.* **2020**, *260*, 119802. [CrossRef]

38. Wang, L.; Cui, S.; Feng, L. Research on the influence of ultraviolet aging on the interfacial cracking characteristics of warm mix crumb rubber modified asphalt mortar. *Constr. Build. Mater.* **2021**, *281*, 122556. [CrossRef]
39. Fowkes, F.M. Dispersion force contributions to surface and interfacial tensions, contact angles, and heats of immersion. *Adv. Chem. Ser.* **1964**, *43*, 99–111.
40. Owens, D.K.; Wendt, R.C. Estimation of the surface free energy of polymers. *Apply Polym. Sci.* **1969**, *13*, 1741–1747. [CrossRef]
41. Bhasin, A.; Masad, E.; Little, D.; Lytton, R. Limits on adhesive bond energy for improved resistance of hot-mix asphalt to moisture damage. *Transp. Res. Rec.* **2006**, *1970*, 2–13. [CrossRef]
42. Golzar, K. Molecular simulation study of penetrant gas trans-port properties into the pure and nanosized silica particles filled poly-sulfone membranes. *J. Membr. Sci.* **2014**, *451*, 117–134. [CrossRef]

applied sciences

MDPI

Article

Dynamic Characteristic Master Curve and Parameters of Different Asphalt Mixtures

Shijie Ma [1,2], Liang Fan [1,*], Tao Ma [3], Zhao Dong [1], Yuzhen Zhang [4] and Xiaomeng Zhang [1,*]

1 Shandong Transportation Institute, Jinan 250012, China; mashijie@sdjtky.cn (S.M.); dongzhao@sdjtky.cn (Z.D.)
2 School of Transportation, Shandong Jianzhu University, Jinan 250101, China
3 School of Transportation, Southeast University, Nanjing 211189, China; matao@seu.edu.cn
4 College of Chemistry and Chemical Engineering, China University of Petroleum, Qingdao 266580, China; zhangyuzhen@sdjtky.cn
* Correspondence: fanliang@sdjtky.cn (L.F.); zhangxiaomeng@sdjtky.cn (X.Z.)

Abstract: Using an AMPT tester and based on laboratory tests, this paper performed a comparative study on the dynamic characteristics of different asphalt mixtures, analyzed the influence of different asphalt binders on the characteristic parameters of the dynamic modulus master curve and the phase angle master curve of asphalt mixture, and expounds the evaluation function of the phase angle master curve for mixture relaxation characteristics. The results show that the modulus master curve parameters of the asphalt mixture are closely related to voids in the mineral aggregate, mixture density, and asphalt content of the asphalt mixture. For the same kind of asphalt mixture, because the gradation of mineral aggregate is fixed and the volume parameters are almost the same, the ultimate modulus of the mixture at different temperatures is unique; when the temperature changes or the asphalt changes, the shape parameter β of the modulus master curve changes regularly, which brings different dynamic responses, and the lower β will show the characteristics of a higher modulus. Asphalt is the source of the viscoelasticity of the asphalt mixture. Although the influence of particle gradation of the mixture will bring about the change of modulus, the phase angle of the mixture depends on the viscoelastic properties of asphalt, and the initial phase angle in the main curve is positively correlated with asphalt penetration and negatively correlated with the softening point and viscosity, while the peak phase angle A is negatively correlated with penetration, and the softening point viscosity is positively correlated. The viscoelastic interval, represented by ω, is negatively correlated with penetration but positively correlated with the softening point and viscosity. The peak position, parameter ωc, of the phase angle master curve can evaluate the relaxation characteristics of the mixture, and the crack resistance of different mixtures can be compared without complex model calculation. In the comparison of the relaxation time of asphalt mixture, the relaxation time of foam cold-recycled mixture is the largest, which is significantly higher than that of other forms of cement mixture; the emulsified asphalt cold-recycled mixture is equivalent to AC20 and LSPM30 mixtures; the SBS-modified asphalt mixture has the best relaxation characteristics.

Keywords: asphalt mixture; master curve; characteristic parameter; relaxation characteristics; cold recycling; foamed asphalt; emulsified asphalt

Citation: Ma, S.; Fan, L.; Ma, T.; Dong, Z.; Zhang, Y.; Zhang, X. Dynamic Characteristic Master Curve and Parameters of Different Asphalt Mixtures. *Appl. Sci.* **2022**, *12*, 3619. https://doi.org/10.3390/app12073619

Academic Editor: Luís Picado Santos

Received: 22 March 2022
Accepted: 1 April 2022
Published: 2 April 2022

1. Introduction

In the 1970s, the dynamic performance master curve was first used to study polymer damping materials and was used as the main content of the product specification of damping materials [1]. The master curve contains the basic dynamic parameters of viscoelastic materials, including dynamic modulus and phase angle, both of which express the material properties as a whole, and the phase angle can better express the relaxation characteristics of the material. Materials with the same dynamic modulus and different phase angles are different materials [2,3]. However, in the research and application of master curves, it is not appropriate to pay more attention to dynamic modulus and ignore the study of phase angle master curves.

At present, the dynamic performance of the asphalt mixture is also characterized by the method of the master curve [4–6]. The actual state of the asphalt pavement under dynamic load is complex, which can be regarded as the superposition of multiple sine waves. The response of asphalt pavement materials to a single sine wave is the basic dynamic property. Because asphalt pavement material is a viscoelastic system, its dynamic modulus and phase angle parameters constitute the basis for understanding pavement load response [7]. Generally speaking, the mechanical response of asphalt pavement is closely related to driving speed; that is, the stiffness modulus is dependent on the loading time, and the shorter the stress pulse time is, the greater the modulus of the asphalt mixture is, which has the characteristics of high frequency and high elasticity [8–18]. However, the current research pays more attention to the dynamic modulus of asphalt pavement and does not perform an in-depth study on the phase angle, which is an important parameter reflecting the viscoelastic structure. Based on laboratory tests, this paper studied the dynamic response characteristics of different asphalt mixtures and discussed the influence of the asphalt binder on the characteristic parameters of the dynamic modulus master curve and phase angle master curve of the asphalt mixture. The evaluation functions of phase angle master curve and mixture relaxation characteristics are expounded.

2. Acquisition of Dynamic Principal Curve

The asphalt mixture was tested by an AMPT performance tester, and the modulus and phase angle at 20, 30, 40, and 50 °C under nine frequency conditions in the range of 0.1 to 25 Hz were obtained, and the modulus and phase angle at different temperatures were transformed into master curves at the same reference temperature by using the principle of time–temperature equivalence. Based on the existing research, the Sigmoid model was used to fit the dynamic modulus master curve, and the improved Gauss model was used to fit the phase angle master curve [9,19].

(1) Dynamic modulus master curve

The Sigmoid model of the modulus master curve is shown in Equation (1).

$$lg(E^*) = E_0 + \frac{\alpha}{1 + e^{\beta + \gamma(lg\omega_r)}} \tag{1}$$

Among them, E^* is the dynamic modulus, ω_r is the load frequency at the reference temperature, E_0 represents the minimum value of dynamic modulus, $E_0 + \alpha$ represents the maximum value of dynamic modulus, and β and γ are the shape parameters of the model curve.

(2) Phase angle master curve

The phase angle master curve was fitted according to the improved Gauss model with the same displacement factor as the modulus principal curve. The Gauss model is shown in Equation (2).

$$\delta = \delta_0 + Ae^{\frac{(lg\omega_r - \omega_c)^2}{2\omega^2}} \tag{2}$$

In the formula, δ is the phase angle, ω_r is the frequency at the reference temperature, ω_c is the frequency corresponding to the peak of the phase angle, and δ_0, A is the eigenvalues of the master curves.

Figure 1 shows the master curve of the AC-20 type mixture. Among them, the modulus master curve in Figure 1A shows a smooth S-shaped curve, which reflects the mechanical properties of the full frequency range. The asphalt mixture has the characteristics of high frequency and high elasticity, and the dynamic modulus increases and tends to have a stable value at high frequency; under the condition of low frequency, the dynamic modulus decreases and tends to a minimum.

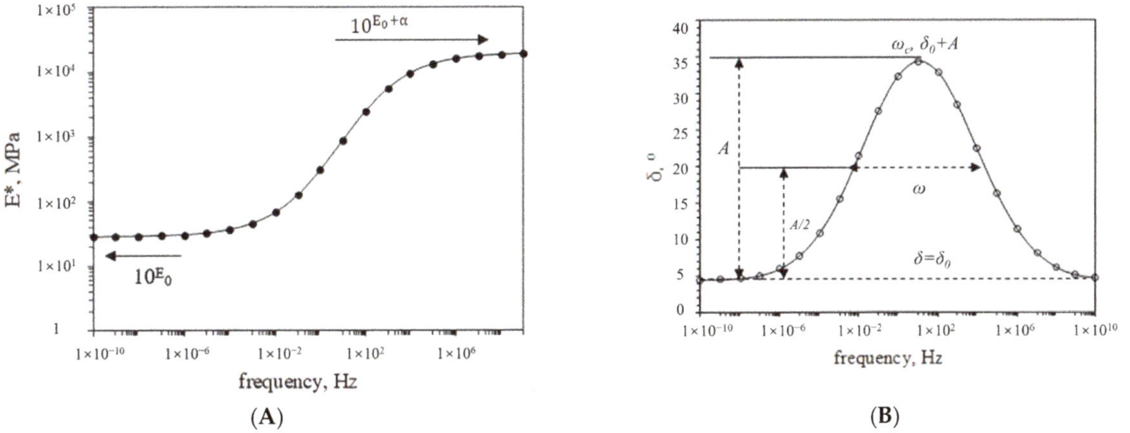

Figure 1. Example of master curves of the asphalt mixture. (**A**) Dynamic modulus master curve, (**B**) Phase angle master curve.

Figure 1B is the phase angle master curve. It can be seen that parameter δ_0 represents the initial phase angle, parameter A represents the variation amplitude of the phase angle, parameter ω represents the full width at half maxima (FWHM) of the phase angle curve, and ω_c represents the position of the phase peak. According to the theory of polymer physics, the FWHM parameter of the phase angle master curve covers more of the viscoelastic state of the material, the left range tends to the rubber state, and the right range tends to the glass state [10,20,21]. In this paper, the division between the amplitude of phase angle A and FWHM (A/ω) is used to represent the significant degree of viscoelasticity of the material; the higher the A/ω value, the greater the sensitivity of the material to frequency, and vice versa.

3. Characteristic Parameters of Master Curves

Taking the same hot-mix asphalt mixture AC-20 as the research object, three kinds of asphalt mixtures were prepared according to different petroleum asphalts. The volume index of the mixture is shown in Table 1. The dynamic modulus and phase angle of the mixture were obtained on the AMPT tester, and the main curve was drawn by model fitting according to Equations (1) and (2). The characteristic parameters of the main curve are shown in Table 2, and the master curves are shown in Figure 2.

Table 1. Asphalt and asphalt mixture indexes.

Mixture Type	ZH50-AC20	ZH70-AC20	ZH90-AC20
Penetration grade	50	70	90
Asphalt content (Pb) %	4.3	4.4	4.3
Mixture density (γ_f) g/cm^3	2.42436	2.42498	2.4296
Mixture void percent (Vv) %	4.59	4.42	4.4
Aggregate void ratio (VMA) %	14.15	14.1	14.4
Asphalt saturation (VFA) %	67.32	68.54	68.47

Table 2. Master curve parameters for three asphalt mixtures.

Type	Temperature, °C	Fitting Parameters for E* Master Curve					Fitting Parameters for δ Master Curve				
		E_0	α	β	γ	R^2	δ_0	A	ω_c	ω	R^2
ZH50-AC20	20	1.4522	2.844	−1.537	−0.664	0.997	4.436	29.961	−2.084	2.893	0.989
	30			−0.743					−0.889		
	40			−0.046					0.161		
	50			0.561					1.074		
ZH70-AC20	20	1.4926	2.7866	−1.339	−0.694	0.995	15.259	19.253	−1.607	2.063	0.990
	30			−0.607					−0.551		
	40			0.149					0.536		
	50			0.924					1.654		
ZH90-AC20	20	1.3762	2.9551	−0.998	−0.627	0.996	18.223	16.309	−1.216	2.025	0.987
	30			−0.461					−0.364		
	40			0.156					0.624		
	50			0.846					1.725		

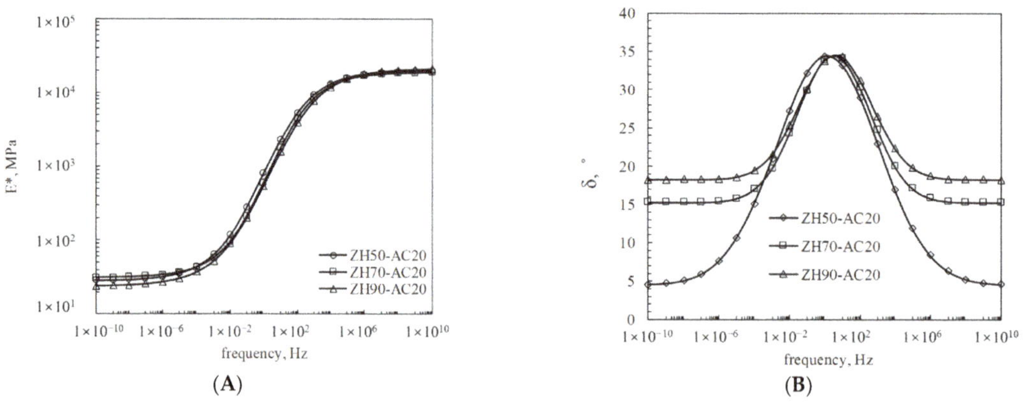

Figure 2. Master curves of three kinds of asphalt mixtures. (**A**) Dynamic modulus master curve, (**B**) Phase angle master curve.

3.1. Modulus Master Curve Parameters

Figure 2A shows that the modulus master curve of the asphalt mixture tends to a stable modulus value—the minimum modulus value (10^{E_0}) and the maximum modulus value ($10^{E_0+\alpha}$) under the condition of extremely low frequency and extremely high frequency. This shows that the dynamic modulus of the asphalt mixture at very low frequency (or high temperature) and very high frequency (low temperature) is not affected by load frequency and only shows obvious frequency sensitivity in the general range of 1×10^{-5} to 1×10^{5} Hz. According to the related research, the ultimate modulus of the asphalt mixture is only related to the mixture gradation, which depends on the mixture volume parameters and presents uniqueness [9,11]. Due to the use of the same gradation, the volume indexes of the three kinds of asphalt mixtures are almost the same. The ultimate modulus is also close to each other, and the difference is small. The parameters α and γ of the principal curve of modulus are also stable.

The linear correlation analysis was carried out by using the master curve parameters of three kinds of mixture modulus (E_0, α, γ) and the volume index of the mixture. Figure 3 shows that the correlation coefficients of the aggregate void ratio (VMA), mixture density (γ_f), and asphalt content (Pb%) are the highest, while the effects of the asphalt saturation (VFA) and void percentage (Vv%) are not significant.

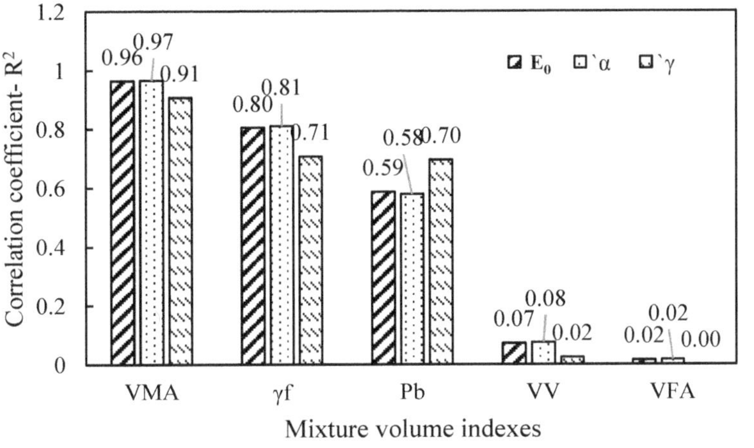

Figure 3. Correlation between the modulus master curve parameters and mixture volume indexes.

However, the parameter β of the master curve increases monotonously with the increase of the reference temperature and presents the linear relationship shown in Figure 4. This phenomenon is related to the shift factor of the principal curve and is essentially related to the temperature sensitivity of asphalt. A good power-law relationship can be established by using the principal curve parameter β at 20, 30, 40, and 50 °C and the complex modulus (G*) of asphalt at the same temperature (test conditions: stress 100 Pa, frequency 5 rad/s). Figure 5 shows that the shape parameter β of the main curve of mixture modulus decreases with the increase of complex modulus G* of asphalt (when the temperature decreases or the asphalt hardens). Under the same gradation condition, a low β value is more likely to lead to a higher modulus. In the range of 10^{-5} to 10^{5} Hz, the modulus of mixture made by penetration grade 50 asphalt is slightly higher than that of penetration grade 70 and grade 90 asphalt. Relative to the influence of modulus brought by gradation, the influence of asphalt is weak.

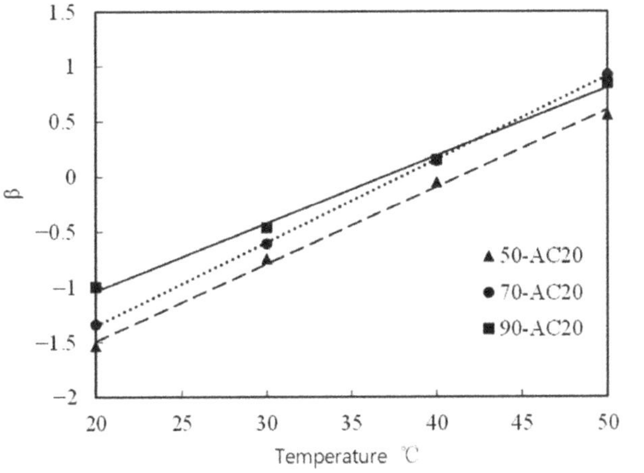

Figure 4. Relationship between shape parameter β and temperature.

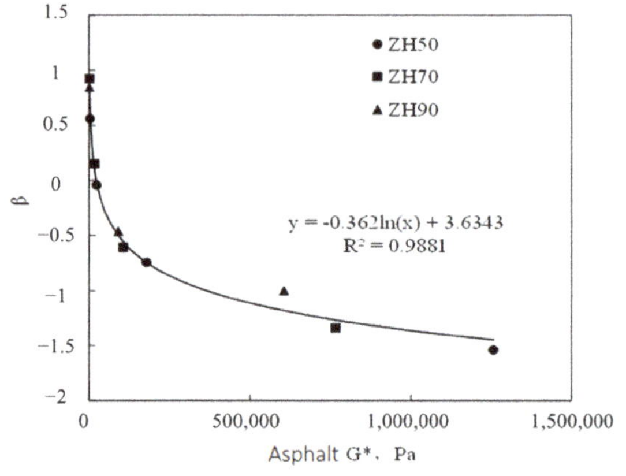

Figure 5. Relationship between asphalt modulus and shape parameter β.

It can be said that for the same kind of asphalt mixture, due to the fixed gradation of mineral aggregate and little difference in volume parameters, the ultimate modulus of the mixture at different temperatures is unique; when the temperature changes or the asphalt changes, the shape parameter β of the modulus master curve changes regularly, which brings different dynamic responses, and the lower β will show the characteristics of higher modulus.

3.2. Phase Angle Master Curve Parameters

3.2.1. δ_0, A, and ω

In the same mixture, different asphalt will create a difference in the viscoelasticity of the mixture, which is mainly reflected in the change of the master curve parameters of the phase angle. Table 1 and Figure 2B show that the phase angle master curve parameter δ_0 increases significantly with the increase of asphalt penetration grade, and the height parameter A of the master curve decreases with the increase of asphalt penetration grade. The half-width parameter ω decreases with the increase of asphalt penetration grade. This shows that different asphalt brings different viscoelastic characteristics to the same mixture; the viscous component of high-grade asphalt is large, so the viscous component of asphalt mixture is also high, resulting in a higher initial phase angle (δ_0), which brings a narrower phase angle change range (A value) and a narrow viscoelastic response frequency range, which is reflected in the reduction of FWHM value (ω).

Figure 6 shows the correlation analysis between the phase angle curve parameters and the asphalt indexes. The master curve of the phase angle of the mixture is closely related to the asphalt properties. The initial phase angle (δ_0) is positively correlated with penetration and negatively correlated with softening point and viscosity; phase angle height A is negatively correlated with penetration; softening point viscosity is positively correlated; and the viscoelastic interval, represented by FWHM (ω), is negatively correlated with penetration, but positively correlated with softening point and viscosity. It can be considered that asphalt is the source of the viscoelasticity of the asphalt mixture. Although the influence of particle gradation of the mixture will bring about the change of the modulus, the phase angle of mixture depends on the viscoelastic properties of asphalt, which is consistent with the previous studies; that is, there is a good sinusoidal relationship between the phase angle of asphalt and the asphalt mixture [11,22–24]. Moreover, the phase angles of different asphalt mixtures from previous studies, stone mastic asphalt (SMA-13), large-stone porous asphalt mixture (LSPM-25), F fine asphalt mixture AC-13F, and asphalt mixture AC-25, are shown in Figure 7.

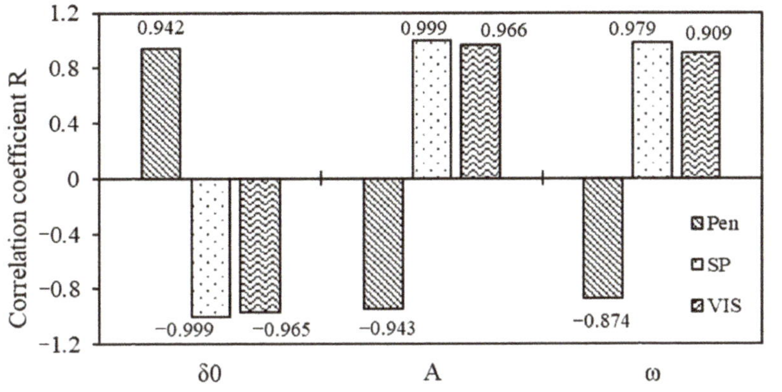

Figure 6. Correlation between master curve parameters of phase angle and asphalt indexes.

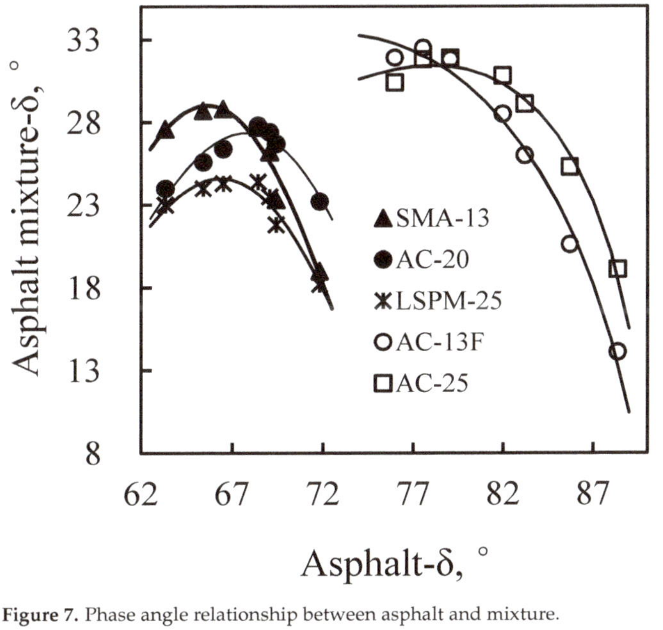

Figure 7. Phase angle relationship between asphalt and mixture.

3.2.2. Phase Angle Peak Position Parameter-ω_c

ω_c is a parameter that varies with temperature. Figure 8 shows that the ω_c value of the phase angle master curve moves to the high-frequency position with the increase of temperature, which shows that the viscoelasticity of the asphalt mixture changes with temperature, which has nothing to do with aggregate but is only caused by the temperature sensitivity of asphalt. Figure 9 shows the relationship between ω_c and the phase angle of the asphalt binder at the same temperature, with R^2 greater than 95%, which is sufficient to illustrate this temperature sensitivity issue.

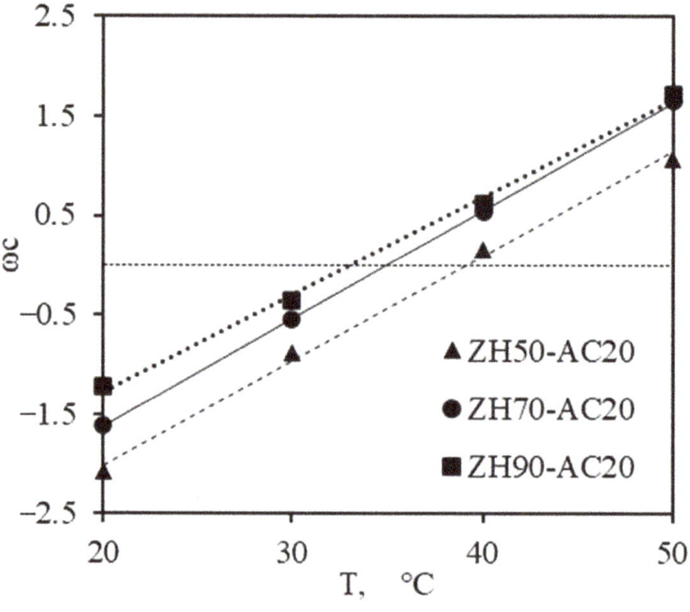

Figure 8. Relationship between ω_c and temperature.

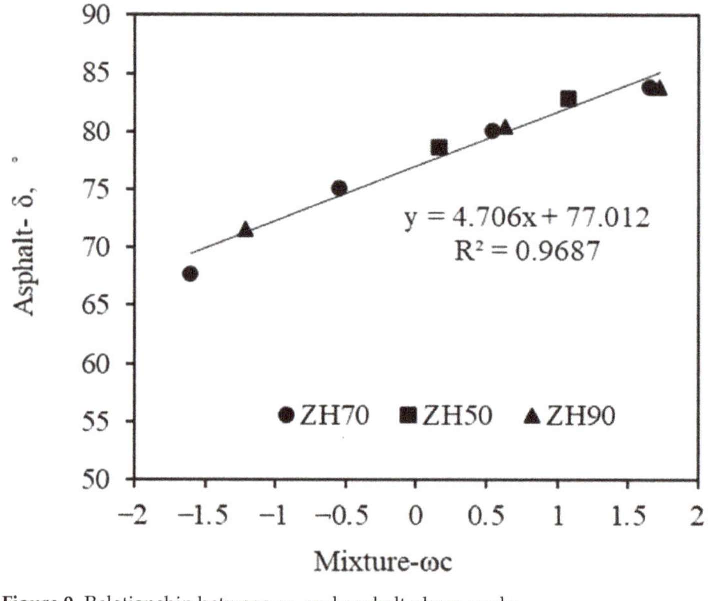

Figure 9. Relationship between ω_c and asphalt phase angle.

In the theory of polymer physics, the ω_c position is related to the relaxation characteristics of polymer materials, which is defined as the relaxation peak [10]. Its significance is to characterize the response ability of the viscoelasticity of materials under external action. Only when the load time (t) is close to the material relaxation time (τ), that is, the action frequency (ω_r) is the same as the reciprocal of the relaxation time (τ^{-1}) ($\omega_r = \tau^{-1}$), is the energy consumption of the material the highest, which reflects the peak value of the phase angle or loss angle tangent; when the frequency is high, the action time of external force is short, and the viscous pot effect of the material is too late to respond, reflecting the elastic characteristics and glass state. When the frequency is low, the action time of the external force is long, the spring effect of the material is completely restored, which shows the full sticky pot effect, and the material shows more viscous characteristics and a rubber state.

Because the asphalt mixture will not appear with absolute flow behavior, there will not be multiple relaxation characteristic peaks like polymer materials, and there is only one phase angle peak or loss angle tangent peak. Therefore, at a certain temperature, the peak position of the phase angle of the asphalt mixture can be used as a reflection of its relaxation ability, which is also important information given by the phase angle master curves, which is often ignored by researchers. According to the peak position of the phase angle master curve, the relaxation time and ability of the mixture at a certain temperature can be calculated and investigated according to the relationship $\omega = \tau^{-1}$. The relaxation times of three kinds of asphalt mixtures at different temperatures are shown in Table 3.

Table 3. Peak position of phase angle and relaxation time.

Mixture Type	Temperature, °C	ω_c	10^{ω_c}, Hz	τ, s
ZH50-AC20	20	−2.084	0.008	121.34
	30	−0.889	0.129	7.74
	40	0.161	1.449	0.69
	50	1.074	11.858	0.08
ZH70-AC20	20	−1.607	0.025	40.46
	30	−0.551	0.281	3.56
	40	0.536	3.436	0.29
	50	1.654	45.082	0.02
ZH90-AC20	20	−1.216	0.061	16.44
	30	−0.364	0.433	2.31
	40	0.624	4.207	0.24
	50	1.725	53.088	0.02

Figure 10 shows the relationship between relaxation time and temperature. It shows that in the range of 20 to 50 °C, the higher the temperature is, the smaller the relaxation time of the asphalt mixture is, and it shows an exponential attenuation relationship. For penetration grading 50 asphalt, the relaxation time of AC20 asphalt mixture is 121.3 s at 20 °C, but it can be reduced to 0.08 s at 50 °C. This explains the reason for the cracking disease caused by the stress concentration of the asphalt mixture under the condition of low temperatures. In addition, for the same kind of asphalt mixture, the high-grade asphalt has less relaxation time; for example, at 20 °C, the relaxation times of grades 50, 70, and 90 asphalt mixtures are 121.3, 40.5, and 16.4 s, respectively. This shows that high-grade asphalt is helpful for improving the cracking resistance of the mixture. At the same time, the relaxation time is of great significance to the characterization of crack resistance of the asphalt mixture. Under the condition of low temperatures, cooling will bring greater temperature stress; if the relaxation time of the asphalt mixture is short, it can relax most of the temperature stress and greatly reduce the possibility of pavement cracking. This is consistent with the conclusion of the existing study using the relaxation time spectrum [12,13].

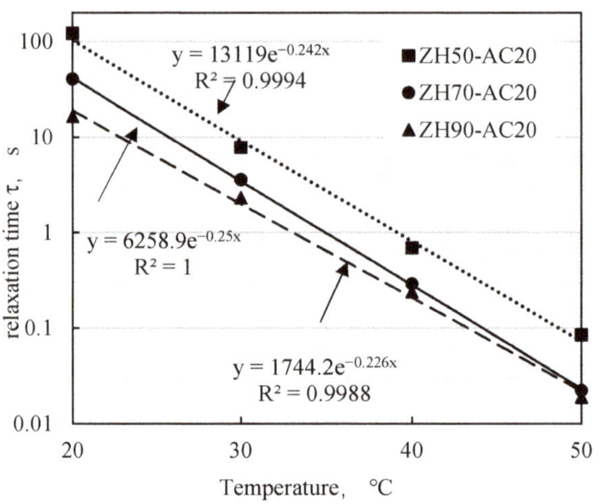

Figure 10. Relationship between relaxation time (τ) and temperature.

4. Master Curves of Different Asphalt Mixture

Table 4 shows the material types and volume indexes of several kinds of asphalt mixtures. The data of the dynamic modulus and phase angle were obtained on an AMPT tester. The master curve parameters were analyzed by Equations (1) and (2), and the main curve diagram of the mixture at 40 °C was drawn.

Table 4. Material types and volume indices of several asphalt mixtures.

Code	Asphalt Type	Mixture Type	Pb, %	Vv, %	VMA, %	VFA, %
ZH50-AC20	Pen-grade 50	HMA, AC20	4.3	4.6	14.2	67.32
ZH70-AC20	Pen-grade 70	HMA, AC20	4.4	4.4	14.1	68.54
ZH90-AC20	Pen-grade 90	HMA, AC20	4.3	4.4	14.4	68.47
SBS-SMA13	SBS modified asphalt	HMA, SMA13	6	4.2	17.1	75.6
MAC-AC20	MAC modified asphalt	HMA, AC20	4.4	4.3	13.4	68.3
ZH70-AC-25	Pen-grade 70	HMA, AC25	4.1	4.1	12.6	68
MAC-LSPM30	MAC modified asphalt	HMA, LSPM30	3.2	15.2	23.5	/
SBS-AC13	SBS modified asphalt	HMA, AC13	6.2	2.4	15.2	82.9
E-CRM	Emulsified asphalt	Cold recycling	3.5 (plus 1.5%cement)	10.2	/	/
F-CRM1	Formed asphalt	Cold recycling	3 (plus 1.5%cement)	11.5	/	/
F-CRM2	Formed asphalt	Cold recycling	3 (plus 1.5%cement)	12.0	/	/

(1) Figure 11A shows the modulus master curve of the asphalt mixtures. Due to the difference in aggregate gradation, asphalt content, and volume index of asphalt mixtures, the difference in the dynamic modulus is obvious; due to the different form and properties of the asphalt binder, the shape of the modulus master curve is obviously different, and the shape parameters (β) are also different. Figure 12 shows that grade 50, grade 70, and grade 90 petroleum asphalt is obviously different from SBS and MAC-modified asphalt, and the modulus is more sensitive to frequency. At the same time, it is also different from the binder form of emulsified asphalt (containing cement) and foamed asphalt (containing cement). The existence of emulsified asphalt and foamed asphalt combined with cement can improve the low-frequency ultimate modulus of the mixture. The dynamic characteristics of different asphalt mixtures are determined by the factors of mixture and binder.

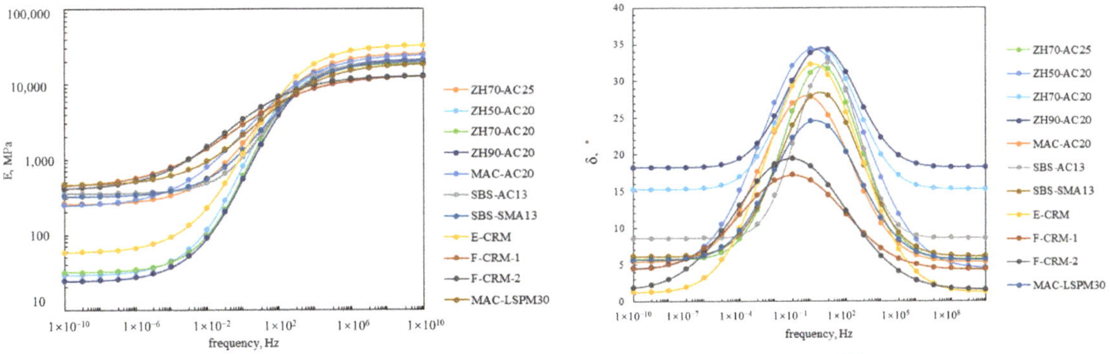

(**A**) (**B**)

Figure 11. Comparison of master curves of different asphalt mixtures. (**A**) Dynamic modulus master curve, (**B**) Phase angle master curve.

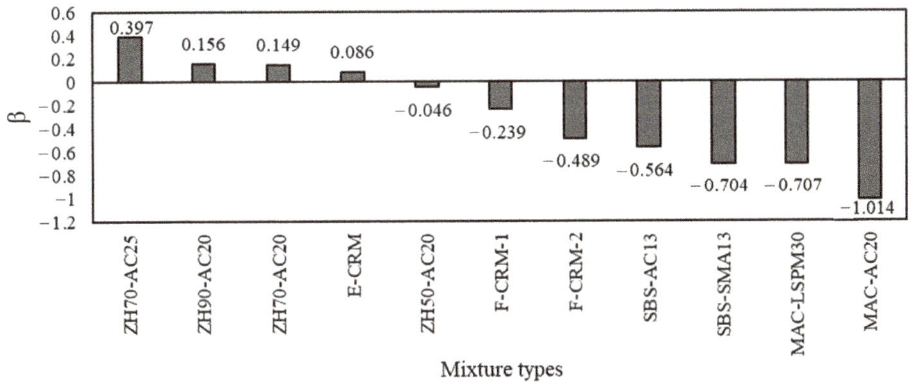

Figure 12. Comparison of shape parameters (β) of modulus master curves.

(2) Figure 11B shows the phase angle master curve of the mixture. Table 5 shows the phase angle master curve parameters. Due to the different forms and properties of the binder, the phase angle master curve of the mixture is obviously different. Among them, the change range of phase angle (A) of foam cold-recycled asphalt mixture is much smaller than that of other asphalt mixtures and emulsified asphalt mixtures. Figure 13 shows that its A/ω value is the smallest, reflecting the weakest viscoelastic characteristics, and tends to be rigid, which is related to the cement cementitious state and the non-continuous distribution of foamed asphalt on the aggregate surface. Although the cold-recycled mixture of emulsified asphalt has a lower initial phase angle (δ_0), it shows more significant viscoelasticity and the variation range of FWHM (ω) and phase angle height (A) is larger than that of foamed asphalt. This is related to the continuous distribution of emulsified asphalt among aggregate particles, shown in Figure 14. For the binder of the hot-mix asphalt mixture, SBS-modified asphalt mixture SBS-AC13 has the best viscoelastic performance, and its A/ω ratio is the highest.

Table 5. Fitting parameters of phase angle master curves.

Mixture Type	δ_0, °	A, °	ω_c, Hz	ω, Hz	A/ω
F-CRM-2	1.53	17.94	0.08	1233.81	0.01
F-CRM-1	4.29	12.92	0.10	800.49	0.02
MAC-AC20	5.35	22.63	0.57	303.40	0.07
ZH50-AC20	4.44	29.96	1.45	781.38	0.04
E-CRM	1.20	31.10	1.53	432.83	0.07
MAC-LSPM30	5.64	19.05	1.86	259.68	0.07
SBS-SMA13	6.16	22.34	3.37	187.31	0.12
ZH70-AC20	15.26	19.25	3.44	115.51	0.17
ZH70-AC25	5.70	26.40	3.92	148.42	0.18
ZH90-AC20	18.22	16.31	4.21	105.94	0.15
SBS-AC13	8.62	23.98	8.96	56.52	0.42

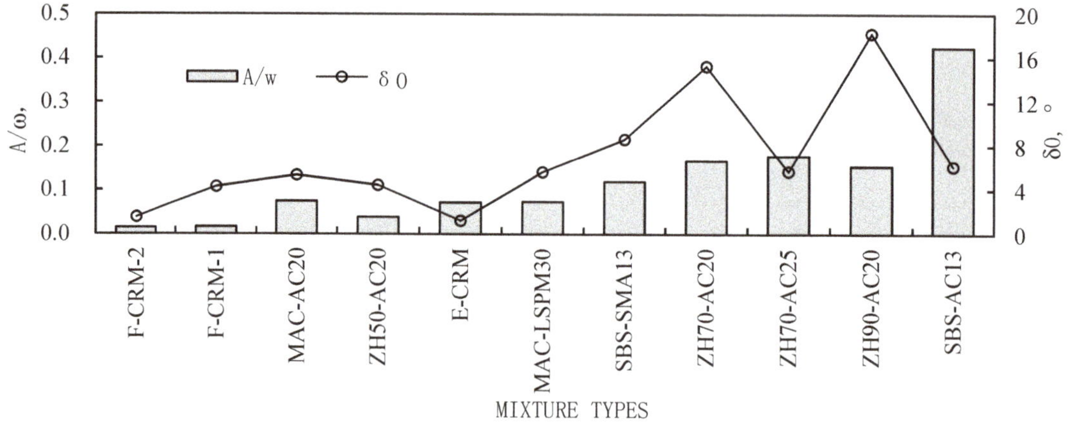

Figure 13. Comparison of master curve parameters (A/ω, δ_0) of phase angle.

Figure 14. Cold recycling asphalt mixture (**A**) Emulsified asphalt cold recycling, (**B**) Foam cold recycling.

(3) Due to the different phase compositions and viscoelasticities of different asphalt mixtures, they reflect different stress relaxation behavior and relaxation times under specific temperatures and alternating stress environments. According to the phase angle master curve parameter ω_c, the relaxation time of different mixtures can be calculated according

to the same calculation method in Table 2. Figure 15 shows the comparison of relaxation time of different asphalt mixtures. Among them, the relaxation time of foam cold-recycled mixture is the longest, which is significantly higher than that of other forms of asphalt mixture; the relaxation time of E-CRM cold-recycled mixture is similar to that of AC20 and LSPM30 mixtures; SBS-modified asphalt mixture SBS-AC13 has the best relaxation characteristics. In the current cold-recycling technology for a flexible base, it can be predicted that foam cold-recycling cannot achieve the effect of the LSPM30 mixture, but emulsified asphalt cold-recycling can.

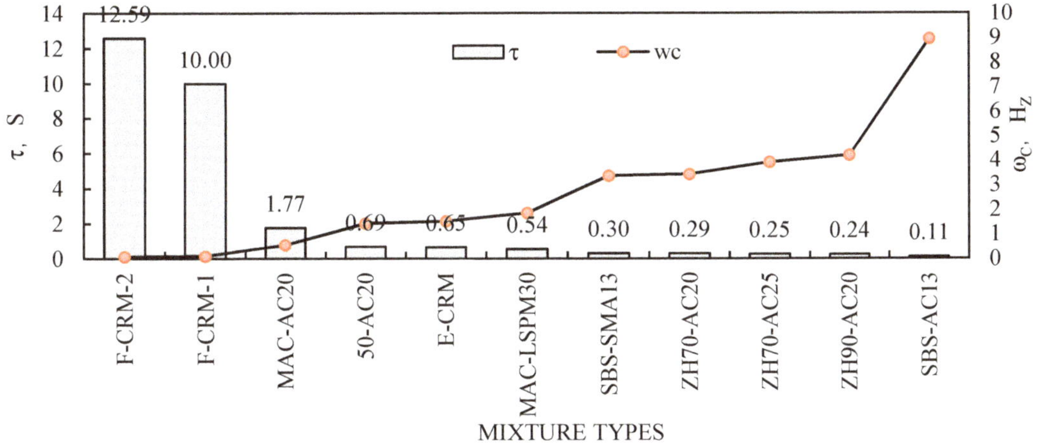

Figure 15. Comparison of relaxation time of different asphalt mixtures.

In this sense, the peak position parameter ω_c of the phase angle master curve can evaluate the relaxation characteristics of the mixture.

5. Conclusions

(1) The modulus master curve parameters (E_0, α, γ) of asphalt mixtures are closely related to the mineral voidage (VMA), mixture density (γ_f), and asphalt content (Pb%) of the asphalt mixture. For the same kind of asphalt mixture, because the gradation of mineral aggregate is fixed, the volume parameters are almost the same, and the ultimate modulus of the mixture at different temperatures is unique. When the temperature changes or the asphalt changes, the shape parameter β of the modulus master curve changes regularly, which brings different dynamic responses, and the lower β will show higher modulus characteristics.

(2) Asphalt is the source of the viscoelasticity of the asphalt mixture. Although the influence of particle gradation of the mixture will bring about the change of modulus, the phase angle of the mixture depends on the viscoelastic properties of the asphalt. The initial phase angle (δ_0) in the main curve is positively correlated with asphalt penetration and negatively correlated with softening point and viscosity, the peak height A is negatively correlated with penetration and softening point viscosity, and the viscoelastic interval represented by FWHM (ω) is negatively correlated with penetration, but positively correlated with the softening point and viscosity.

(3) The peak position parameter ω_c of the phase angle master curve can evaluate the relaxation characteristics of the mixture and avoid the complex process of model calculation. In the comparison of the relaxation time of the asphalt mixture, the relaxation time of the foam cold-recycled mixture is the largest, which is significantly higher than that of other forms of asphalt mixture; emulsified asphalt cold-recycled mixture is equivalent to AC20 and LSPM30 mixtures; SBS-modified asphalt mixture

has the best relaxation characteristics; different relaxation times indicate different crack resistance, the longer the relaxation time, the worse the crack resistance. However, less relaxation time (less viscosity) leads to a large viscous flow in a very short time, creating a large rutting, which is one of the most serious issues in asphalt pavement on heavy-duty roadways.

Author Contributions: Investigation, S.M. and T.M.; Methodology, L.F. and Y.Z.; Experiment, Z.D.; Writing—original draft, S.M. and L.F.; Writing—review and editing, X.Z. All authors have read and agreed to the published version of the manuscript.

Funding: This research was funded by the Applied Basic Research Project of the Ministry of Transport, grant number 2013319781140, and project ZR2020QE271 supported by Shandong Provincial Natural Science.

Institutional Review Board Statement: Not applicable.

Informed Consent Statement: Not applicable.

Data Availability Statement: Not applicable.

Conflicts of Interest: The authors declare no conflict of interest.

References

1. Song, R.Y.; Xie, X.Z. Dynamic property master curves of visco-elastical material. *Noise Vib. Control* **1985**, *S1* (Suppl. S1), 25–32.
2. Liu, F.Q.; Tang, X.Y. *Polymer Physics*; Higher Education Press: Bejing, China, 2004.
3. Menglan, Z.; Xinzhe, H.; Jie, L.; Qun, L. Relationship between dynamic properties of asphalt binders and mixtures. *Cent. South Highw. Eng.* **2003**, *4*, 11–15.
4. Jin, L. Experimental Study on Master Curve of Failure Strength of Asphalt Mixture. Master's Thesis, Dalian University of Technology, Dalian, China, 2019.
5. Wang, H.P.; Yang, J.; Shi, X.Q.; Zhou, W.Z. Research of dynamic modulus and master curve of high modulus asphalt mixture. *J. Highw. Trans. Res. Dev.* **2015**, *32*, 12–17.
6. Hu, L.Y.; Zhen, Q.H.; Yu, F.; Chen, J.L.; Luo, R. Master Curve Drawing of Axial Transverse Dynamic Modulus Based on the Same Time-temperature Equivalent Factor. *J. Wuhan Univ. Technol. Transp. Sci. Eng.* **2019**, *43*, 146–152.
7. Song, X.J.; Zeng, M.L.; Fan, L. Relationship between back-calculated modulus of asphalt pavement and dynamic modulus of asphalt mixture. *China Civ. Eng. J.* **2019**, *52*, 121–128.
8. Song, X.J.; Fan, L. Durations of Stress Response in Asphalt Pavement under Moving Load. *J. Hunan Univ. Nat. Sci.* **2018**, *45*, 71–76.
9. Guo, L.; Xu, Q.; Zeng, G.; Wu, W.; Zhou, M.; Yan, X.; Zhang, X.; Wei, J. Comparative Study on Complex Modulus and Dynamic Modulus of High-Modulus Asphalt Mixture. *Coatings* **2021**, *11*, 1502. [CrossRef]
10. He, M.J.; Zhang, H.D.; Chen, W.X.; Dong, X.X. *Polymer Physics*; Fudan University Press: Shanghai, China, 2007.
11. Song, X.; Zeng, M.; Fan, L. Relationship among dynamic properties of pavement asphalt materials. *J. Build. Mater.* **2018**, *21*, 920–925.
12. Xiao, C.F.N.; Gui, Z.L. Continuous relaxation time spectrum of asphalt mixture based on master curve of dynamic modulus. *J. S. China Univ. Technol. Nat. Sci. Ed.* **2008**, *10*, 76–80.
13. Wang, Z.C.; Guo, N.S.; Zhao, Y.H.; Chen, Z.D. Determination and conversion of relaxation and retardation time spectrum of asphalt mixtures. *J. Beijing Univ. Technol.* **2019**, *45*, 168–176.
14. Zhu, J.; Ahmed, A.; Said, S.; Dinegdae, Y.; Lu, X. Experimental analysis and predictive modelling of linear viscoelastic response of asphalt mixture under dynamic shear loading. *Constr. Build. Mater.* **2022**, *328*, 12095.
15. Han, D.; Yuan, M.; Hu, H. Establishing prediction master curve of dynamic modulus of asphalt mixture considering randomness of aggregate morphology. *Constr. Build. Mater.* **2021**, *294*, 123575.
16. Shi, Y.; Liu, H.; Wang, G. Modeling of asphalt mixture-screed interaction: A nonlinear dynamic vibration model for improving paving density. *Constr. Build. Mater.* **2021**, *311*, 125296.
17. Pablo, M.V.; Silvio, L.S.; Pedro, O.B.A.; Cléber, F.; Luciano, P.S.; Deividi, S.P. Dynamic modulus master curve construction of asphalt mixtures: Error analysis in different models and field scenarios. *Constr. Build. Mater.* **2021**, *301*, 124343.
18. Zhang, J.; Zhang, X.; Liang, M.; Jiang, H.; Wei, J.; Yao, Z. Influence of different rejuvenating agents on rheological behavior and dynamic response of recycled asphalt mixtures incorporating 60% RAP dosage. *Constr. Build. Mater.* **2020**, *238*, 117778.
19. Bueno, L.D.; Schuster, S.L.; Specht, L.P.; da Pereira, D.S.; do Nascimento, L.A.H.; Kim, Y.R.; Brenner, M.G.B. Asphalt pavement design optimisation: A case study using viscoelastic continuum damage theory. *Int. J. Pavement Eng.* **2020**, *23*, 1070–1082. [CrossRef]
20. Ling, M.; Luo, X.; Gu, F.; Lytton, R.L. Time-temperature-aging-depth shift functions for dynamic modulus master curves of asphalt mixtures. *Constr. Build. Mater.* **2017**, *157*, 943–951. [CrossRef]

21. Specht, L.P.; de Babadopulos, L.F.A.L.; di Benedetto, H.; Cedric Sauzeat, S.J.B. Application of the theory of viscoelasticity to evaluate the resilient modulus test in asphalt mixes. *Constr. Build. Mater.* **2017**, *149*, 648–658. [CrossRef]
22. Su, N.; Xiao, F.; Wang, J.; Amirkhanian, S. Precision analysis of sigmoidal master curve model for dynamic modulus of asphalt mixtures. *J. Mater. Civ. Eng.* **2018**, *30*, 04018290. [CrossRef]
23. Eric, L.T.; Vaillancourt, M.; Perraton, D.; di Benedetto, H. Linear viscoelastic (LVE) properties of asphalt mixtures with different glass aggregates and hydrated lime content. *Int. J. Pavement Eng.* **2020**, *21*, 1170–1179.
24. Mangiafico, S.; Sauzeat, C.; Di Benedetto, H. 2S2P1D model calibration error from user panel for one bitumen and one bituminous mixture. *Adv. Mater. Sci. Eng.* **2019**, *2019*, 6547025. [CrossRef]

applied sciences

MDPI

Article

Experimental Study of Low Temperature Performance of Porous Asphalt Mixture

Jian Wang [1,*], Pui-Lam Ng [2,3,*], Yuhua Gong [4], Han Su [5] and Jinsheng Du [4]

1 School of Civil Engineering, Sun Yat-sen University, Guangzhou 510275, China
2 Faculty of Civil Engineering, Vilnius Gediminas Technical University, LT-10223 Vilnius, Lithuania
3 Department of Civil and Environmental Engineering, The Hong Kong University of Science and Technology, Hong Kong 999077, China
4 School of Civil Engineering, Beijing Jiaotong University, Beijing 100044, China; 15121017@bjtu.edu.cn (Y.G.); jshdu@bjtu.edu.cn (J.D.)
5 Key Laboratory of Civil Engineering Safety and Durability of China Education Ministry, Tsinghua University, Beijing 100084, China; suhan@mail.tsinghua.edu.cn
* Correspondence: wangj833@mail.sysu.edu.cn (J.W.); irdngpl@gmail.com (P.-L.N.)

Abstract: Porous asphalt mixture can be used as a road surface paving material with the remarkable advantage to prevent water accumulation and ponding. However, the performance of porous asphalt mixture in low temperature environment has not been thoroughly investigated, and this forms the subject of research in the present study. The mineral aggregate gradation of porous asphalt mixture was designed based on Bailey method, and the low temperature performance of porous asphalt mixture was studied by means of the low temperature bending test. The factors affecting the low temperature performance of porous asphalt mixture were analyzed through the orthogonal experimental design method, and the effects of porosity, modifier content, aging condition, and test temperature on the low temperature performance of porous asphalt mixture were evaluated. The results showed that the modifier content was the most important factor affecting the low temperature performance of porous asphalt mixture, followed by the test temperature, while the porosity and the aging condition were the least. Among the three performance evaluation indicators, namely the flexural tensile strength, maximum bending strain, and bending stiffness modulus, the maximum bending strain had the highest sensitivity to the porosity. It can be seen from the single factor influence test of porosity that there existed an approximately linear relationship between the maximum bending strain and the porosity of porous asphalt mixture, and the maximum bending strain decreased with increasing porosity. Furthermore, in order to ensure the good working performance of porous asphalt mixture in low temperature environment, the porosity should also satisfy the required limits of the maximum bending strain.

Keywords: pavement; porous asphalt mixture; low temperature performance; orthogonal experimental design; porosity

Citation: Wang, J.; Ng, P.-L.; Gong, Y.; Su, H.; Du, J. Experimental Study of Low Temperature Performance of Porous Asphalt Mixture. *Appl. Sci.* **2021**, *11*, 4029. https://doi.org/10.3390/app11094029

Academic Editor: Amir Tabakovic

Received: 27 March 2021
Accepted: 25 April 2021
Published: 28 April 2021

1. Introduction

The existence of water film on road surface in rainy weather affects the effective contact between the tire of vehicle and the road surface. When the water film on road surface reaches a certain thickness, the braking distance of the vehicle will increase, and skid resistance of tires will decrease. These might cause drift of the vehicle and possibly endanger road-users' safety. Moreover, problems such as splashing water, water mist, and night glare caused by the discharge of water from road surface also adversely affect driving safety. The porous asphalt pavement is a surface layer paved with open-grade asphalt with interconnected gaps, and an impervious bonding layer is constructed underneath the surface layer [1]. In rainy weather, the rainwater infiltrates downwards through the porous surface layer, arrives at the surface of the impervious layer and then drains from the side along the edge of the pavement [2]. Moreover, due to the macro-texture of the surface, a higher skid resistance can be obtained for porous asphalt mixture when compared to other surface layer materials [3]. Therefore, the use of porous asphalt pavement is a method to effectively prevent the accumulation and ponding of water on road surface [4,5], which has been conceived and adopted in numerous countries including France, Germany, Spain, United Kingdom, New Zealand, Japan and the United States [6–12].

Porous asphalt is a composite material comprising of mineral materials, binders and additives. From pavement engineering viewpoint, the selection of raw materials is the key to determine the pavement properties of porous asphalt mixture [13,14]. In the process of mix proportioning, the identification of design objectives is of prime importance [15]. To determine the best combination of aggregate and binder so as to achieve good overall performance of the asphalt pavement, the mix proportion design of asphalt mixture must be engineered, which includes the identification of design objectives, the control of gradation and the determination of binder content [16,17]. Adjusting the design goals according to the application requirements is the prerequisite for ensuring the durability and economy of the pavement structure. Due to the diversity of design goals, there is no established universal specification for the mix proportion design, and different design methods for porous asphalt mixtures have been proposed for different design goals. Currently, many researchers have studied the mix proportion design of porous asphalt. Jiang [18] and Xing [19] studied the relationship between porosity and passing rates of series of sieve apertures through laboratory tests, and improved the mix proportions design method of porous asphalt. Zhang [20] applied the Bailey method to the mix proportion design of porous asphalt, and proposed the concept of "reserved porosity" to determine the target porosity. Luo et al. [21] proposed a new open-graded porous asphalt mixture that used epoxy asphalt as the binder to improve the pavement durability, and the results showed superior overall performance of the epoxy modified open-graded porous asphalt mix compared to conventional open-graded porous asphalt mixes. Qian and Lu [22] conducted an experimental study and performance evaluation of small particle porous epoxy asphalt mixture, and the mixture was designed with the aid of a binder drainage test and a raveling test. They found that the proposed porous epoxy asphalt mixture had good mechanical properties while retaining satisfactory frictional resistance and permeability. Xiao and Zhang [23] applied the Coarse Aggregate Void Filling (CAVF) method to the mix proportions design of porous asphalt, and proposed the method to determine the optimum compaction temperature of porous asphalt mixture.

Although the porous asphalt has many advantages, there are also some drawbacks in its application. The low temperature cracking problem of asphalt pavement is reported to be prominent in some cold regions [24–27]. Survey results showed that more than one third of the damages of porous asphalt in cold regions were attributed to its inferior low temperature cracking resistance performance [28]. Asphalt mixture has the characteristics of stress relaxation. When the ambient temperature decreases slowly, the tensile stresses in asphalt mixture caused by temperature drop that accumulate gradually with time would be relieved by relaxation and would not reach the ultimate tensile stress level to cause cracking. On the other hand, the thermal contraction cracks of asphalt mixture often occur in extreme weather where the temperature descends rapidly, such as that caused by an acute cold current or cold wave. In such weather condition, the tensile stresses induced in asphalt mixture accumulate at a faster rate than the relaxation, so that cracks would appear when the ultimate tensile stress level is attained. It should be noted that for porous asphalt pavement in cold regions, in addition to the cracks caused by the ambient temperature reduction, the fluctuation of ambient temperature around the freezing point will cause the water trapped inside internal voids to undergo freeze and thaw and induce damages [29–31].

Porous asphalt pavement is more susceptible to freeze-thaw cycles in winter compared to ordinary dense gradation asphalt pavement [32]. Therefore, porous asphalt pavement used in cold regions should be designed to have excellent low temperature performance.

At present, research on porous asphalt has focused on various aspects including moisture susceptibility [2,22], fatigue resistance and healing effect [33], material optimization [34–36], filtration characteristics [37], rutting resistance [22,38], ravelling resistance [39], and noise reduction characteristics upon siltation [40]. However, research on low temperature performance of porous asphalt mixture is relatively in lack [41]. Fortier and Vinson [42] studied the relationship between low temperature cracking and aging performance of asphalt mixture and highlighted the importance of the appropriate use of modifier. Cheng et al. [43] found that the incorporation of ethylene bis stearic acid amide (EBS) and stearic acid amide (SA) could improve the low temperature deformability, reduce the failure stiffness modulus, and improve the low temperature crack resistance of warm-mixed porous asphalt mixture. Liu et al. [44] explored the feasibility of the application of porous asphalt mixture with steel slag for seasonal frozen regions, and ascertained the optimal replacement percentage of natural aggregate. In their test, steel slag coarse aggregate was used to replace basalt coarse aggregate at four levels (25%, 50%, 75%, 100%) by equal volume, and the results indicated that the low temperature cracking resistance of the mixture was significantly enhanced and acoustic emission energy was uniformly released by the incorporation of steel slag. Ma et al. [45] analyzed the effect of aging on the low temperature performance of porous asphalt mixture through the Thermal Stress Restrained Specimen Test (TSRST), and the results showed that the low temperature performance of porous asphalt mixture decreased after aging, while it was not much different from that of dense graded asphalt mixture. However, other factors such as porosity, modifier content, and test temperature also had effects on the low temperature performance of porous asphalt mixture. From literature review, the effects of these factors had not been adequately investigated, thus further research is needed.

In this paper, the low temperature performance of porous asphalt mixture is studied by means of the low temperature bending test. The mineral aggregate gradation of porous asphalt mixture is designed based on Bailey method [46,47], and the effects of porosity, modifier content, aging condition, and test temperature on the low temperature performance of porous asphalt mixture are assessed through orthogonal experimental design. The works have enabled the development of porous asphalt mixture with desirable engineering properties and resistance to low temperature.

The porous asphalt mixture developed in this research has been successfully applied to the viaduct bridge pavement of Liangdang County to Hui County section of G316 National Expressway in China. G316 National Expressway is a major highway from southeast part to northwest part of China, connecting Fuzhou City in Fujian Province and Lanzhou City in Gansu Province. The total length of the expressway is 2915 km. The section from Liangdang County to Hui County is located in Gansu Province in northern part of China and has a length of 53.4 km. Its major construction works took place from year 2016 to 2018. The porous asphalt surface layer has a thickness of 40 mm and has been testified to have good drainage performance as well as good overall performance in the cold environment of northern China.

2. Low Temperature Performance Test

The commonly used low temperature performance test methods include indirect tensile test, thermal stress restrained specimen test [24], low temperature bending test, creep test, and shrinkage coefficient test. Gong [48] conducted a comprehensive comparative analysis of the commonly used low temperature performance test methods from the aspects of test accuracy and reliability, correlation between test objective and low temperature performance, test practicality and operability, and found that indirect tensile test and low temperature bending test were better test methods for evaluating low temperature performance of porous asphalt mixture. In this paper, low temperature bending test was used for studying the low temperature performance of porous asphalt mixture.

2.1. Raw Materials

The aggregate included S10, S12 and S16 diabase gravel produced in Heyuan, Guangdong Province of China. S10 and S12 were coarse aggregates with a maximum size of 13.2 mm and S16 was fine aggregates. The filler was limestone powder, and the asphalt used in the test was 70# matrix asphalt initially developed in a refinery laboratory in Nanjing City of China (the notion 70# refers to a needle penetration range of 70 ± 10 at 25 °C in 5 s using a 100 g needle). The modifier was SINOTPS modifier manufactured by Shenzhen Oceanpower New Material Technology Company (Opmaterial) Ltd., China. It was a high-viscosity modifier and was a proprietary mixture of thermoplastic elastomer, thermoplastic resin, rubber-type polymers, and additives. Figure 1 depicts the SINOTPS modifier. The mean particle size of the SINOTPS modifier was 2.5 mm.

Figure 1. SINOTPS modifier.

The specifications and technical indicators of coarse aggregate are respectively summarized in Tables 1 and 2, and the specifications and technical indicators of fine aggregate are respectively tabulated in Tables 3 and 4. According to the test method described in Chinese Standard JTG E20-2011 [49], screening tests for aggregate and limestone powder were conducted. Figure 2 illustrates the test process, and the results are listed in Table 5. The technical requirements of limestone powder and the corresponding test results are given in Table 6.

Table 1. Specifications of coarse aggregate.

Designation	Nominal Size Range (mm)	Percentage Passing through the Following Sieve Apertures (%)				
		19.0 mm	13.2 mm	9.5 mm	4.75 mm	2.36 mm
S10	10~15	100	90~100	0~15	0~5	—
S12	5~10	—	100	90~100	0~15	0~5

Figure 2. Aggregate and limestone powder screening.

Table 2. Test results of technical indicators of coarse aggregate.

Test Item	Unit	Test Result	
		5~10 mm	10~15 mm
Crushed value	%	6.8	
Los Angeles abrasion loss	%	9.7	
Apparent relative density	—	2.930	2.962
Water absorption	%	0.74	1.11
Soundness (mass loss with sodium sulfate)	%	5.2	
Needle flake granule content	%	10.2	8.5
Silt content (<0.075 mm particle content by water washing method)	%	0.5	0.9
Soft mineral content	%	1.3	0.7
Polished stone value	—	45	
Adhesion to asphalt	—	5	

Note: Tests were carried out according to Chinese Standard JTG E20-2011 [49].

Table 3. Specifications of fine aggregate.

Designation	Nominal Size Range (mm)	Percentage Passing through the Following Sieve Apertures (%)						
		4.75 mm	2.36 mm	1.18 mm	0.6 mm	0.3 mm	0.15 mm	0.075 mm
S16	0~3	100	80~100	50~80	25~60	8~45	0~25	0~15

Table 4. Technical indicator requirements and test results of fine aggregate.

Test Item	Unit	Test Result	Indicator Requirement
Apparent density	g/cm^3	2.929	≥2.60
Silt content (<0.075 mm particle content by water washing method)	%	1.2	≤3

Table 5. Screening results of aggregates and filler material.

Materials		Percentage Passing through the Following Sieve Apertures (%)									
		16 mm	13.2 mm	9.5 mm	4.75 mm	2.36 mm	1.18 mm	0.6 mm	0.3 mm	0.15 mm	0.075 mm
Coarse aggregate	S10	100	91.3	35.9	0.8	0.4	0.4	0.4	0.4	0.4	0.4
	S12	100	100	99.1	3.3	0.6	0.6	0.5	0.5	0.5	0.5
Fine aggregate	S16	100	100	100	97.0	68.3	47.4	33.5	22.9	17.4	11.9
Limestone powder		100	100	100	100	100	100	100	99.2	95.7	77.6

The properties of the matrix asphalt affect the properties of high-viscosity modified asphalt. The matrix asphalt used for preparing high-viscosity modified asphalt should meet the technical requirements of road petroleum asphalt in Chinese Standard JTG F40-2004 [50]. In this study, 70# matrix asphalt developed in Nanjing was adopted, and the test results of technical indicators per JTG E20-2011 [49] are shown in Table 7. It can be seen from the table that the matrix asphalt was classified per JTG F40-2004 [50] as a grade A asphalt, which is suitable for all grades of highways. Therefore, it is suitable to employ the matrix asphalt for preparing high viscosity modified asphalt.

Table 6. Technical indicator requirements and test results of limestone powder.

Test Item		Unit	Test Result	Indicator Requirement
Apparent density		g/cm^3	2.699	≥2.50
Water content		%	0.43	≤1
Particle size range (by washing method)	<0.6 mm	%	100	100
	<0.15 mm	%	95.7	90~100
	<0.075 mm	%	77.6	75~100
Appearance		—	No agglomeration	No agglomeration
Hydrophilic coefficient		—	0.46	<1
Plasticity index		%	2.9	<4

Table 7. Technical indicator requirements and test results of 70# matrix asphalt.

Test Item		Unit	Test Result	Indicator Requirement
Needle penetration (25 °C, 5 s, 100 g)		0.1 mm	70.5	60~80
Softening point (ring and ball method)		°C	48.5	≥46
Ductility (10 °C)		cm	24.1	≥20
Density		g/cm^3	1.031	Measured value
Relative density		—	1.035	Measured value
Rolling thin film oven heated (RTFOT)	Quality change	%	−0.6	±0.8
	Residual needle penetration ratio	%	69.1	≥61
	Residual ductility (10 °C)	cm	7.8	≥6

Note: Tests were carried out according to Chinese Standard JTG E20-2011 [49].

Currently, the high-viscosity modifiers commonly used in porous asphalt mixture include synthetic polymers such as SBS and proprietary systems such as SINOTPS. Gong [48] found that SINOTPS modifier had certain superiority. For instance, the ductility of SINOTPS modifier at low temperature was significantly better than that of SBS modifier. In view of the desirable performance of SINOTPS modifier, it was adopted in the present work. As confirmed by the experimental trial in this study, SINOTPS had good compatibility with asphalt and could melt to integrate with the asphalt quickly.

Gong [48] selected 70# asphalt as matrix asphalt, and tested the properties of modified asphalt with SINOTPS modifier content of 6%, 9%, 12%, and 15% respectively. By comparing the indicators of modified asphalt with different contents of SINOTPS modifier, it was found that at a modifier content of 12% or 15%, all the indicators of modified asphalt could meet the requirements of Chinese Standard JTG F40-2004 [50]. At 15% content of SINOTPS modifier, the viscosity of modified asphalt at 60 °C and 135 °C was much higher than that of modified asphalt with 12% content of SINOTPS. However, it should be noted that excessive viscosity of modified asphalt would adversely affect the field construction works. Considering the economic factors and aversion of negative effect on production and construction, the optimum content of SINOTPS in modified asphalt was set to be 12% in this study, and the technical indicators of the modified asphalt per JTG E20-2011 [49] are shown in Table 8.

Table 8. Technical indicators of modified asphalt (12% SINOTPS).

Test Item		Unit	Matrix asphalt/SINOTPS = 88/12
Needle penetration (25 °C, 100 g, 5 s)		0.1 mm	49.5
Softening point (ring and ball method)		°C	93.9
Ductility (5 °C)		cm	45.3
Elastic restitution		%	96.8
Dynamic viscosity (60 °C)		Pa·s	69,618
Viscosity (135 °C)		Pa·s	2.883
Viscous toughness		N·m	23.47
Toughness		N·m	15.57
Rolling thin film oven heated (RTFOT)	Quality change	%	0.07
	Residual needle penetration ratio	%	76.2
	Residual ductility (5 °C)	cm	29.1

Note: Tests were carried out according to Chinese Standard JTG E20-2011 [49].

2.2. Design of Mix Proportion

In the current experimental programme, the gradation design method based on Bailey method was adopted [51,52], and the gradation design process is illustrated in Figure 3. At the outset, to conduct the gradation design of porous asphalt mixture using Bailey method, the parameters including the 0.075 mm sieve passing percentage, design density, volume ratio of coarse aggregate, and reserved porosity need to be determined. The reasonable gradation obtained based on Bailey method is listed in Table 9. The bitumen/aggregate ratio was 5.36%, the porosity was 20.55%, and the Marshall stability was 4920 N. In the gradation design, two types of coarse aggregates (denoted as CA-1 and CA-2), one type of fine aggregate (denoted as FA) and one type of filler (denoted as MF) were used. The percentage by mass passing through 0.075 mm sieve aperture was 5% and the design density was 103% of the loose density. The volume ratio of coarse aggregate (CA-1:CA-2) was determined to be 55:45 according to the requirements in JTG F40-2004 [50]. The reserved porosity was determined to be 25% according to the requirement of the target porosity.

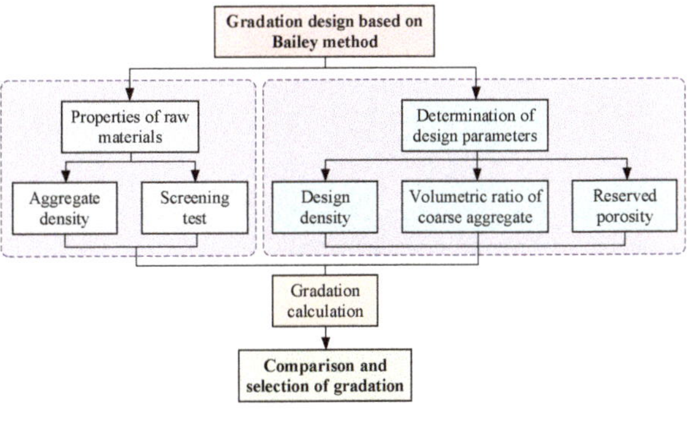

Figure 3. Gradation design of porous asphalt mixture based on Bailey method.

Table 9. Mineral aggregate composition of gradation scheme.

Bitumen/Aggregate Ratio (%)	Percentage of Mineral Aggregate (%)			
	S10	S12	S16	Limestone powder
5.36	46.68	37.02	12.83	3.47

Because of porous asphalt is an open-graded mixture, its optimal bulk density, Marshall stability and flow are not directly comparable with dense graded asphalt. Therefore, it is not viable to determine the optimum asphalt content of porous asphalt mixture directly by using Marshall test method. Instead, the desirable range of asphalt content for porous asphalt mixture should be first determined by employing the binder drainage test and Cantabro test, and then the optimum asphalt content can be determined with respect to the functional requirements of porous pavement. The binder drainage test measures the mass of binder leakage from asphalt mixture through the perforated metal basket [53], and it determines the upper bound of bitumen/aggregate ratio. The Cantabro test measures the mass of disintegration from asphalt mixture under the effect of simulated traffic load in the Los Angeles machine [54], and it determines the lower bound of bitumen/aggregate ratio.

Reference is made to Table 9 of the mineral aggregate composition of porous asphalt mixture as the starting point. Assuming the initial bitumen/aggregate ratio as 5.36%, five different bitumen/aggregate ratios were taken in the proximity of the initial bitumen/aggregate ratio as tabulated in Table 10, where the bitumen/aggregate ratio ranged from 3.6% to 5.6% with equal increment of 0.5%. For each mix in the table, after Marshall test specimens were fabricated and demoulded as shown in Figures 4 and 5 respectively, binder drainage test was conducted to obtain the binder leakage loss rate and Cantabro test was conducted to obtain the cohesion loss rate. The test results are summarised in Table 10. The relationships between binder leakage loss rate, cohesion loss rate and bitumen/aggregate ratio are respectively plotted in Figure 6. From the inflection points of the two curves in Figure 6, the lower bound of bitumen/aggregate ratio determined by the Cantabro test was 4.15%, and the upper bound of bitumen/aggregate ratio determined by the binder drainage test was 5.05%.

Table 10. Test results of binder drainage test and Cantabro test.

Test Item	Bitumen/Aggregate Ratio (%)				
	3.6	4.1	4.6	5.1	5.6
Binder leakage loss rate (%)	0.03	0.05	0.11	0.20	0.65
Cohesion loss rate (%)	6.25	4.25	4.43	4.48	3.19

Figure 4. Marshall test specimens.

Figure 5. Demoulding process of Marshall test specimens.

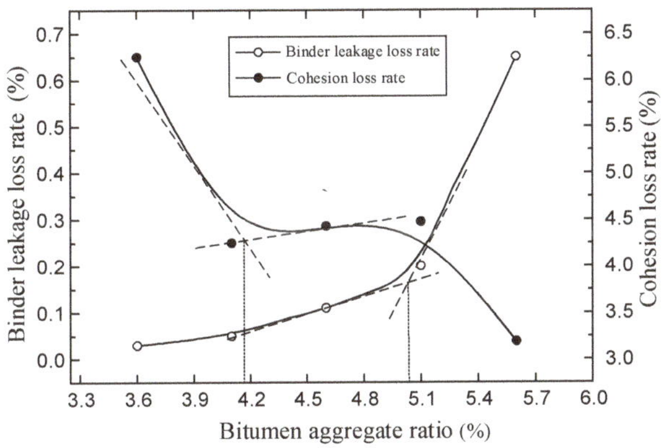

Figure 6. Relationships between binder leakage loss rate, cohesion loss rate and bitumen/aggregate ratio.

Porous asphalt mixture has the characteristic of high porosity. In order to ensure the durability of pavement structure, the thickness of porous asphalt film should be thicker than other types of asphalt mixtures. Therefore, the optimum asphalt content tends to have a larger value. Equation (1) can be used for computing the optimum bitumen/aggregate ratio [50].

$$OAC = 0.25OAC_{min} + 0.75OAC_{max} \quad (1)$$

where OAC is the optimum bitumen/aggregate ratio (in %); OAC_{min} is the lower bound of bitumen/aggregate ratio determined from the Cantabro test (in %); OAC_{max} is the upper bound of bitumen/aggregate ratio determined from the binder drainage test (in %).

The optimum bitumen/aggregate ratio calculated from Equation (1) is 4.8%, and the optimum amount of asphalt is determined to be 4.58%. Upon obtaining the reasonable gradation and the optimum bitumen/aggregate ratio of porous asphalt mixture, these parameters need to be validated by examining whether the performance of the resulting porous asphalt mixture could meet the requirements stipulated in JTG F40-2004 [50]. The verification results are demonstrated in Table 11. It can be seen from the table that the properties of the porous asphalt mixture so obtained can fulfill the application requirements.

Table 11. Performance verification of porous asphalt mixture.

	Test Item	Unit	Test Result	Indicator Requirement
Marshall test	Porosity	%	20.7	18~22
	Marshall stability (double-sided compaction 50 times)	N	5480	≥5000
	Flow value	mm	2.38	2~4
Water permeability	Water permeability coefficient	mL/15 s	1602	>800
High temperature stability	Dynamic stability (60 °C, 0.7 MPa)	times/mm	5458	≥5000
Water immersion stability	Residual stability of water immersion	%	94.0	≥85
	Water immersion loss	%	2.49	≤15
	Freeze-thaw splitting strength ratio	%	91.4	≥80

2.3. Orthogonal Experimental Design

Orthogonal experimental design [55] is a commonly used scientific method for designing multi-factor experiments. It uses a set of standardized orthogonal tables to arrange experiments, and the technique has been widely adopted in the planning and design of experimental trials in production engineering and quality control [56]. In the implementation, the combinations of different experimental factors and levels are evenly matched and rationally divided. The experimental results are processed by statistical methods and as a result scientific conclusions can be obtained. The main advantage of orthogonal experimental design is to enable the selection of representative combinations of experimental factors instead of exhausting all possible combinations. Through the analysis of some representative experimental results, the influence degree of different factors on each evaluation indicator can be obtained, and the best combination of multiple factors can be obtained to achieve the experimentation purpose.

In this study, the choice of factors in the orthogonal experimental design was based on the environmental conditions and the representative mix designs. Four factors, namely the porosity (in %), modifier content (in %), aging, and test temperature (in °C), were included. For each factor, three levels were considered. The resulting matrix can cover a wide range of design scenarios of porous asphalt mixture in low temperature conditions.

(1) Orthogonal table design

The mineral aggregate gradation and optimum asphalt content of porous asphalt mixture obtained in Section 2.2 were used in the test. The types and properties of raw materials have been described in detail in Section 2.1. Based on the above discussions, the selected levels of modifier content and porosity are presented in Table 12, henceforth, a four-factor three-level orthogonal table as tabulated in Table 13 is set up.

Table 12. Values of levels of each factor.

Level Number	Factor			
	A Porosity	B Modifier content	C Aging	D Test temperature
1	16.2%	9%	No aging	0 °C
2	20.7%	12%	Short-term aging	−10 °C
3	23.8%	15%	Long-term aging	−20 °C

Table 13. Orthogonal test table.

Group Number	Porosity (%)	Modifier Content (%)	Aging	Test Temperature (°C)
1	16.2	9	No aging	0
2	16.2	12	Short-term	−10
3	16.2	15	Long-term	−20
4	20.7	9	Short-term	−20
5	20.7	12	Long-term	0
6	20.7	15	No aging	−10
7	23.8	9	Long-term	−10
8	23.8	12	No aging	−20
9	23.8	15	Short-term	0

To simulate the aging of porous asphalt mixture from construction process to service stage, the aging of porous asphalt mixture was divided into short-term and long-term aging levels. Short-term aging refers to the aging of porous asphalt mixture caused by temperature during the paving construction process from mixing plant to pavement compaction. Long-term aging refers to the aging caused by external environmental factors including air, sunlight, precipitation, and traffic loads during the service of porous asphalt pavement. The simulation methods for aging in this test were as follows:

I) Short-term aging

The fresh porous asphalt mixture was evenly spreaded in an enamel pan, with the mass per unit area in loose state of approximately 21 to 22 kg/m^2. It was placed in an oven and heated at 135 °C under forced ventilation for 4 h, then was taken out for test.

II) Long-term aging

The specimens were prepared from porous asphalt mixture after short-term aging according to the required time, size, and moulding method. The specimens were allowed to cool down for no less than 16 h and then were demoulded. Subsequently, the specimens were placed in the oven at 85 °C, and were continuously heated under forced ventilation for 120 h. Afterwards, the oven door was opened so that the specimens were natural cooled for no less than 16 h to room temperature. Lastly, the specimens were taken out for test.

(2) Evaluation indicator

I) Flexural tensile strength

The flexural tensile strength directly reflects the ability of porous asphalt mixture to resist bending failure under low temperature condition. The higher the flexural tensile strength of porous asphalt mixture, the better the low temperature performance of porous asphalt mixture.

II) Maximum bending strain

The bending strain reflects the deformation ability of porous asphalt mixture and is a main indicator for evaluating the low temperature performance. The larger the maximum bending strain, the higher the energy absorbing capacity of porous asphalt mixture before failure, and the better the low temperature crack resistance of porous asphalt mixture.

III) Bending stiffness modulus

The bending stiffness modulus is the ratio of flexural tensile strength to maximum bending strain. It reflects the flexibility of porous asphalt mixture under low temperature condition. A smaller bending stiffness modulus means a higher flexibility and is generally accompanied by a better low temperature crack resistance of the porous asphalt mixture.

3. Experimental Method and Results

3.1. Range of Evaluation Indicator Values

The rutting panel specimens were prepared and were cut into small prismatic trabecular specimens. The low temperature bending test was conducted at a loading rate of 50 mm/min and under the preset temperatures. The fabrication and loading process of prismatic specimens are depicted in Figures 7 and 8, respectively.

(a) (b)

Figure 7. (**a**) Fabrication of prismatic specimen; (**b**) prismatic specimen prior to testing.

Figure 8. Bending test of specimen.

The flexural tensile strength, maximum bending strain, and bending stiffness modulus of specimens under the corresponding conditions were obtained as presented in Table 14. Based on the test results, to obtain the influence degree of various factors on the evaluation indicators, intuitive analysis was performed on the test data. By performing range analysis of the data listed in Table 14, the range analysis table as given in Table 15 can be obtained. In Table 15, k_i represents average response value of each evaluation indicator at each level of factors, where i means the factor level number, i = 1, 2, 3. According to the value of k_i at each factor level, the ranges of different evaluation indicators corresponding to the factor levels can be obtained, and the order of the primary and secondary factors can be judged according to the value of the range.

Table 14. Results of low temperature bending test.

Group Number	Flexural Tensile Strength (MPa)	Maximum Bending Strain (με)	Bending Stiffness Modulus (MPa)
1	5.60	4996	1120
2	5.26	5388	976
3	5.15	5667	908
4	5.57	4832	1153
5	5.16	5491	939
6	5.05	6257	807
7	5.17	5220	991
8	5.56	5063	1098
9	5.22	5915	882

Table 15. Range analysis of results of low temperature bending test.

Evaluation Indicator	Mean Value	Factor Type			
		A Porosity	B Modifier content	C Aging	D Test temperature
Flexural tensile strength (MPa)	k_1	5.33	5.45	5.40	5.20
	k_2	5.26	5.32	5.35	5.36
	k_3	5.32	5.14	5.16	5.46
	Range	0.07	0.31	0.24	0.26
Maximum bending strain (με)	k_1	5550.33	5016.00	5438.67	5621.67
	k_2	5526.67	5314.00	5378.33	5467.33
	k_3	5399.33	5946.33	5459.33	5187.33
	Range	176.33	930.33	81.00	434.33
Bending stiffness modulus (MPa)	k_1	1001.33	1088.00	1008.33	924.67
	k_2	966.33	1004.33	1003.67	980.33
	k_3	990.33	865.67	946.00	1053.00
	Range	35.00	222.33	62.33	128.33

3.2. Influence Analysis of Factors

By analyzing the values of range in Table 15, the order of influence of factors on each of the evaluation indicators can be obtained. The order of influence of factors on the flexural tensile strength is as follows: modifier content > test temperature > aging > porosity. The order of influence of factors on the maximum bending strain is as follows: modifier content > test temperature > porosity > aging. The order of influence of factors on the bending stiffness modulus is as follows: modifier content > test temperature > aging > porosity.

The influence of three levels of each factor on evaluation indicators is graphically represented in Figure 9. In the figure, A to D indicate the factor type as shown in Table 15, and 1 to 3 is the level number of each factor.

(1) Porosity

It can be seen from Figure 9 that the porosity has effect on the three low temperature performance evaluation indicators, but the effect is limited to within 4% over the range of porosity studied. When the porosity increased from 16.2% to 20.7%, the flexural tensile strength, maximum bending strain, and bending stiffness modulus of porous asphalt mixture decreased by 1.3%, 0.4%, and 3.5%, respectively. When the porosity further increased from 20.7% to 23.8%, the flexural tensile strength and bending stiffness modulus increased by 1.1% and 2.5% respectively, while the maximum bending strain decreased by 2.3%. Theoretically, an increase in porosity of asphalt mixture will reduce the flexural tensile strength. On the other hand, when the porosity is higher, the connected porosity also increases. The existence of interconnected pores could relieve the expansion pressure generated by the freezing action, and this could partially offset the adverse effect of porosity on flexural tensile strength at low temperature. Through the analysis of orthogonal test results, it is found that among the three evaluation indicators, the maximum bending strain has the highest sensitivity to porosity. Therefore, it is advisable to design a single factor influence test to study the influence of porosity on the maximum bending strain, as described later in this paper.

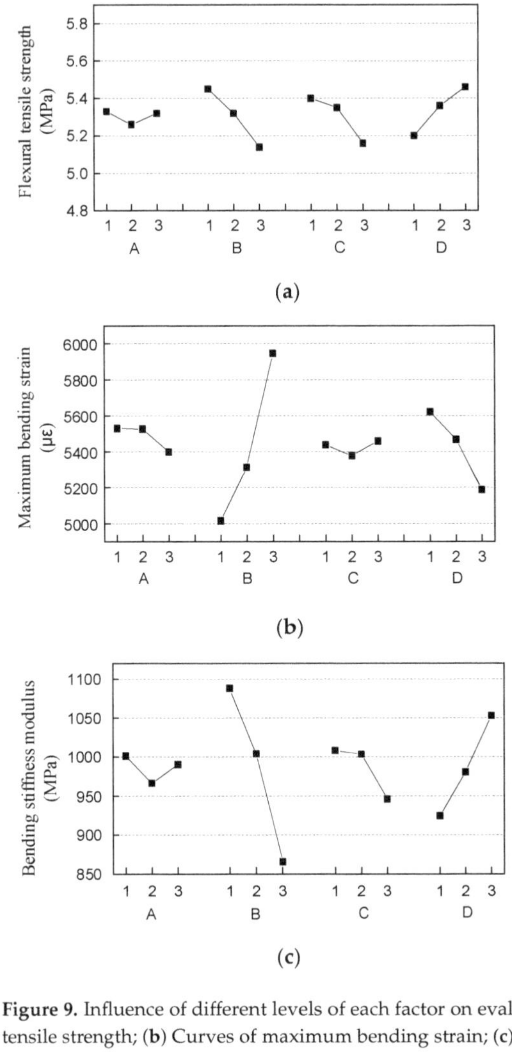

Figure 9. Influence of different levels of each factor on evaluation indicators. (**a**) Curves of flexural tensile strength; (**b**) Curves of maximum bending strain; (**c**) Curves of bending stiffness modulus.

(2) Modifier content

The modifier content is the most important factor affecting the low temperature performance of porous asphalt mixture. As can be seen from Figure 9, by increasing the modifier content from 9% to 15%, the flexural tensile strength decreased by 5.7%, the maximum bending strain increased by 18.5%, and the bending stiffness modulus decreased by 20.4%. The results demonstrate that SINOTPS modifier can significantly improve the deformation capacity and energy absorption capability of porous asphalt mixture, so that the porous asphalt mixture exhibits better low temperature performance.

(3) Aging

Aging has relatively little effect on the low temperature performance of porous asphalt mixture. The factors of porosity and aging are together the least and the second-least influential among the four factors. From Figure 9, it can be seen that the flexural tensile strength, maximum bending strain, and bending stiffness modulus of porous asphalt mixture with short-term aging are respectively 0.9%, 1.1%, and 0.5% lower compared to those of porous asphalt mixture with no aging. The flexural tensile strength and bending stiffness modulus of porous asphalt mixture with long-term aging are respectively 4.4% and 6.2% lower whereas the maximum bending strain is 0.4% higher, compared to those of porous asphalt mixture with no aging. The modest decrease in low temperature performance of porous asphalt mixture through aging was in line with the findings by Ma et al. [45]. The slight effect of aging should be due to the effectiveness of the high viscosity SINOTPS modifier in improving the antiaging performance of porous asphalt mixture.

(4) Test temperature

The influence of test temperature on low temperature bending test results is less significant than that of modifier content. At this juncture, it is noteworthy that the brittle point is a threshold temperature of asphalt mixture at which its behaviour is transitioned from plastic to brittle. In low temperature bending test, in order to prevent possible sudden brittle fracture from affecting the accuracy of the results, the test temperature is generally set at higher than the brittle point temperature. For porous asphalt, the brittle point is lower than that of ordinary dense graded asphalt [57], and the test temperature can generally be set at −10 °C to −20 °C. Consequently, in order to investigate the effect of temperature changes on the low temperature performance of porous asphalt mixture, three levels of test temperature, namely 0 °C, −10 °C, and −20 °C, were adopted in the test.

From Figure 9, it can be seen that when the test temperature was decreased from 0 °C to −20 °C, the flexural tensile strength increased by 5.0%, the maximum bending strain decreased by 7.7%, and the bending stiffness modulus increased by 13.9%. The results indicated that in this temperature range, the porous asphalt mixture still exhibited plastic deformation, and the brittle point temperature was not reached [58]. The maximum bending strain decreased with the test temperature, indicating that the lower the test temperature, the less capable the porous asphalt mixture in resisting deformation. At the same time, the bending stiffness modulus of porous asphalt mixture increased with decreasing test temperature, indicating a reduction in the low temperature crack resistance of porous asphalt mixture.

3.3. Influence of Porosity on Maximum Bending Strain

The porosity is amongst the most concerned indicator in the design of porous asphalt mixture. To further analyze the influence of porosity on the low temperature performance of porous asphalt mixture, the single factor influence test of porosity was conducted. The test selected the maximum bending strain as the evaluation indicator.

The experiment adopted the same mineral aggregate gradation and raw materials as those in the orthogonal test, the modifier content was 12%, and the bitumen/aggregate ratio was 4.8%. The mixing and moulding process of porous asphalt mixture was consistent with that in the orthogonal test, and the porous asphalt mixture was not subjected to aging. According to the results analysis of orthogonal test, as the porosity increased from 16.2% to 23.8%, the maximum bending strain decreased by 3.3% only. To attain higher statistical confidence, in the single factor influence test of porosity, on the basis of the orthogonal test results, another two groups of specimens with porosity of 18.4% and 26.3% were added. The five groups of specimens were subjected to low temperature bending test at −10 °C. With reference to the test results, the relationship between the maximum bending strain ε_B (in με) and the porosity n_0 (in %) was obtained, as depicted in Figure 10.

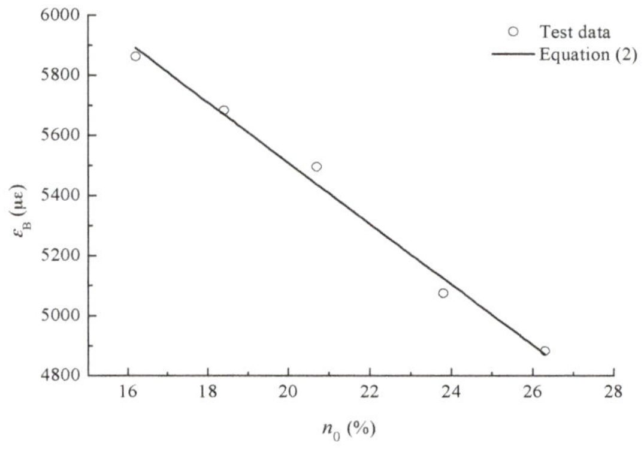

Figure 10. Relationship between maximum bending strain and porosity.

It is evident from Figure 10 that there exists an approximately linear correlation between the maximum bending strain and the porosity of porous asphalt mixture, and the maximum bending strain decreases as the porosity increases. The relationship between the maximum bending strain and the porosity is expressed in Equation (2) with coefficient of determination $R^2 = 0.986$.

$$\varepsilon_B = 7527.7 - 100.9n_0 \quad (16.2 \leq n_0 \leq 26.3) \tag{2}$$

when designing the mix proportion of porous asphalt mixture, in order to ensure good working performance of porous asphalt mixture in low temperature environment, the porosity should also conform to the required limits of the maximum bending strain in relevant technical standards. In the cold regions of northern China, when applying porous asphalt pavement, the maximum bending strain of porous asphalt mixture should not be less than 5000 με in accordance with JTG F40-2004 [50]. From Equation (2), it is noted that when the porosity is controlled within 25%, thus the limit requirement on maximum bending strain is satisfied.

Finally, the possible use of recycled asphalt pavement to enhance the sustainability of highway infrastructure has attracted vast research interest in recent years [59–61]. Recycled asphalt mixtures, which are prepared by blending reclaimed asphalt pavement, virgin bitumen, and mineral additives, offer various advantages including resources recycling, cost saving, and reduced environmental impact [62]. Further research on the utilization of reclaimed asphalt in porous asphalt pavements and their low temperature performance is recommended.

4. Conclusions

In this paper, the low temperature performance of porous asphalt mixture has been studied through the low temperature bending test. The following major conclusions can be drawn:

(1) Through orthogonal experimental design, four main factors affecting the low temperature performance of porous asphalt mixture have been analyzed. The results revealed that the descending order of influence of factors on flexural tensile strength of porous asphalt mixture was: modifier content, test temperature, aging condition, and porosity. The descending order of influence of factors on maximum bending strain was: modifier content, test temperature, porosity, and aging condition; and the descending order of influence of factors on bending stiffness modulus was: modifier content, test temperature, aging condition, and porosity.

(2) The modifier content was the most important factor affecting the low temperature performance of porous asphalt mixture, followed by test temperature. By increasing the modifier content from 9% to 15%, the flexural tensile strength decreased by 5.7%,

the maximum bending strain increased by 18.5%, and the bending stiffness modulus decreased by 20.4%. The results demonstrated that SINOTPS modifier could significantly improve the deformation capacity and energy absorption capability of porous asphalt mixture. By decreasing the test temperature from 0 °C to −20 °C, the flexural tensile strength increased by 5.0%, the maximum bending strain decreased by 7.7%, and the bending stiffness modulus increased by 13.9%. The results indicated that the lower the test temperature, the less capable the porous asphalt mixture in resisting deformation, and the poorer the low temperature crack resistance of porous asphalt mixture.

(3) The factors of porosity and aging were the least and the second-least influential among the four factors. The porosity affected the three low temperature performance evaluation indicators, but the effect was limited to within 4% when the porosity increased from 16.2% to 23.8%. Compared with porous asphalt mixture with no aging, the flexural tensile strength and bending stiffness modulus of porous asphalt mixture with long-term aging were respectively 4.4% and 6.2% lower, whereas the maximum bending strain was 0.4% higher. The slight effect of aging should be due to the effectiveness of the SINOTPS modifier in improving the anti-aging performance of porous asphalt mixture.

(4) The orthogonal test results revealed that the maximum bending strain had the highest sensitivity to porosity among the three performance evaluation indicators, namely the flexural tensile strength, maximum bending strain, and bending stiffness modulus. The effect of porosity on low temperature performance of porous asphalt mixture has been further examined through the single factor influence test. The results showed that there existed an approximately linear correlation between the maximum bending strain and the porosity of porous asphalt mixture, and the maximum bending strain decreased with increasing porosity. Furthermore, when designing the mix proportions of porous asphalt mixture, to ensure good working performance of porous asphalt mixture in low temperature environment, the porosity should also satisfy the limit requirements of the maximum bending strain in relevant technical standards. Further research on the low temperature performance of the porous asphalt mixture containing recycled asphalt materials is recommended.

Author Contributions: Conceptualization, J.W. and Y.G.; methodology, Y.G.; formal analysis, J.W. and Y.G.; investigation, J.W., P.-L.N. and Y.G.; resources, P.-L.N. and Y.G.; data curation, Y.G.; visualization, H.S.; validation, H.S. and J.D.; writing—original draft preparation, J.W., P.-L.N. and Y.G.; writing—review and editing, J.W., P.-L.N. and H.S.; project administration, J.D.; supervision, J.D.; funding acquisition, P.-L.N. and J.D. All authors have read and agreed to the published version of the manuscript.

Funding: This research was funded by the National Natural Science Foundation of China under the National Project for Research and Development of Major Scientific Instruments (Grant No. 51727813) and the European Regional Development Fund (Project No. 01.2.2-LMT-K-718-03-0010) under grant agreement with the Research Council of Lithuania (LMTLT).

Institutional Review Board Statement: Not applicable.

Informed Consent Statement: Not applicable.

Data Availability Statement: The data presented in this study are available on request from the corresponding authors.

Conflicts of Interest: The authors declare no conflict of interest.

References

1. Speight, J.G. *Asphalt Materials Science and Technology*; Butterworth-Heinemann: Oxford, UK, 2015.
2. Poulikakos, L.D.; Partl, M.N. Evaluation of moisture susceptibility of porous asphalt concrete using water submersion fatigue tests. *Constr. Build. Mater.* **2009**, *23*, 3475–3484. [CrossRef]
3. Pérez-Acebo, H.; Gonzalo-Orden, H.; Findley, D.J.; Rojí, E. A skid resistance prediction model for an entire road network. *Constr. Build. Mater.* **2020**, *262*, 120041. [CrossRef]
4. Poulikakos, L.D.; Gilani, M.S.; Derome, D.; Jerjen, I.; Vontobel, P. Time resolved analysis of water drainage in porous asphalt concrete using neutron radiography. *Appl. Radiat. Isotopes* **2013**, *77*, 5–13. [CrossRef] [PubMed]

5. Kiselev, A.; Zhang, H.; Liu, Z. The effect of two-phase mixing on the functional and mechanical properties of TPS/SBS-modified porous asphalt concrete. *Constr. Build. Mater.* **2021**, *270*, 121841. [CrossRef]
6. Afonso, M.L.; Dinis-Almeida, M.; Fael, C.S. Study of the porous asphalt performance with cellulosic fibres. *Constr. Build. Mater.* **2017**, *135*, 104–111. [CrossRef]
7. Zhang, Y.; Verwaal, W.; van de Ven, M.F.C.; Molenaar, A.A.A.; Wu, S.P. Using high-resolution industrial CT scan to detect the distribution of rejuvenation products in porous asphalt concrete. *Constr. Build. Mater.* **2015**, *100*, 1–10. [CrossRef]
8. Isenring, T.; Koster, H.; Scazziga, I. Experiences with porous asphalt in Switzerland. *Transport. Res. Rec.* **1990**, *1265*, 41–53.
9. Ruiz, A.; Alberola, R.; Pérez, F.; Sánchez, B. Porous asphalt mixtures in Spain. *Transport. Res. Rec.* **1990**, *1265*, 87–94.
10. Nicholls, J.C. *Review of UK Porous Asphalt Trials*; TRL Report 264; Transport Research Laboratory: London, UK, 1997.
11. Fletcher, E.; Theron, A.J. *Performance of Open Graded Porous Asphalt in New Zealand*; NZ Transport Agency Research Report 455; New Zealand Transport Agency: Wellington, New Zealand, 2011.
12. Chen, J.S.; Yang, C.H. Porous asphalt concrete: A review of design, construction, performance and maintenance. *Int. J. Pavement Res. Technol.* **2020**, *13*, 601–612. [CrossRef]
13. Ni, F.J.; Xu, H.; Leng, Z.; Liu, Q. Influence of asphalt properties on porous asphalt mixture performance. *J. Traffic Transp. Eng.* **2003**, *3*, 1–4.
14. Tan, Y.Q.; Song, X.; Ji, L.; Chen, G.; Wu, X. Influence of coarse aggregate performance on high temperature performance of asphalt mixture. *China J. Highw. Transp.* **2009**, *22*, 29–33.
15. Nicholls, J.C.; Carswell, I.G. *The Design of Porous Asphalt Mixtures to Performance-Related Criteria*; TRL Report 497; Transportation Research Laboratory: Wokingham, Berkshire, UK, 2001.
16. Hunter, R.N.; Self, A.; Read, J. *The Shell Bitumen Handbook*, 6th ed.; ICE Publishing: Westminster, London, UK, 2015.
17. Kett, I. *Asphalt Materials and Mix Design Manual*; Noyes Publications: Westwood, NJ, USA, 1998.
18. Jiang, W. Study on Mix Proportion Design Method and Road Performance of Porous Asphalt Pavement Mixture. Master's Thesis, Chang'an University, Xi'an, China, 2008.
19. Xing, M.L. Study on Composition Design and Performance of Porous Asphalt Mixture. Master's Thesis, Chang'an University, Xi'an, China, 2007.
20. Zhang, M.M. Study on Mix Proportion Design and Technical Performance of Porous Asphalt Mixture. Master's Thesis, Chang'an University, Xi'an, China, 2009.
21. Luo, S.; Lu, Q.; Qian, Z. Performance evaluation of epoxy modified open-graded porous asphalt concrete. *Constr. Build. Mater.* **2015**, *76*, 97–102. [CrossRef]
22. Qian, Z.; Lu, Q. Design and laboratory evaluation of small particle porous epoxy asphalt surface mixture for roadway pavements. *Constr. Build. Mater.* **2015**, *77*, 110–116. [CrossRef]
23. Xiao, X.; Zhang, X.N. Design of porous asphalt mixture based on CAVF method. *J. Highway Transp. Res. Dev.* **2016**, *33*, 7–12.
24. Gražulytė, J.; Vaitkus, A.; Andrejevas, V.; Gribulis, G. Methods and criteria for evaluation of asphalt mixture resistance to low temperature cracking. *Balt. J. Road Bridge Eng.* **2017**, *12*, 135–144. [CrossRef]
25. Kliewer, J.E.; Zeng, H.; Vinson, T.S. Aging and low-temperature cracking of asphalt concrete mixture. *J. Cold Reg. Eng.* **1996**, *10*, 134–148. [CrossRef]
26. Sebaaly, P.E.; Lake, A.; Epps, J. Evaluation of low-temperature properties of HMA mixtures. *J. Transp. Eng.* **2002**, *128*, 578–586. [CrossRef]
27. Yalghouzaghaj, M.N.; Sarkar, A.; Hamedi, G.H.; Hayati, P. Application of the surface free energy method on the mechanism of low-temperature cracking of asphalt mixtures. *Constr. Build. Mater.* **2021**, *268*, 121194. [CrossRef]
28. Zhang, Y. Study on Low Temperature Anti-Cracking Performance and Mix Proportion Design Method of Asphalt Mixture in Permafrost Regions. Master's Thesis, Chang'an University, Xi'an, China, 2004.
29. Bäckström, M. Ground temperature in porous pavement during freezing and thawing. *J. Transp. Eng.* **2000**, *126*, 375–381. [CrossRef]
30. Lamothe, S.; Perraton, D.; Di Benedetto, H. Contraction and expansion of partially saturated hot mix asphalt samples exposed to freeze-thaw cycles. *Road Mater. Pavement Des.* **2015**, *16*, 277–299. [CrossRef]
31. Vu, V.T.; Chupin, O.; Piau, J.M.; Hammoum, F.; Bouron, S. Experimental study and modeling of the behavior of partially saturated asphalt concrete under freezing condition. *Constr. Build. Mater.* **2018**, *163*, 169–178. [CrossRef]
32. Lebens, M. *Porous Asphalt Pavement Performance in Cold Regions*; Office of Materials and Road Research, Minnesota Department of Transportation: St Paul, MN, USA, 2012.
33. Liu, Q.; Schlangen, E.; van de Ven, M.; van Bochove, G.; van Montfort, J. Evaluation of the induction healing effect of porous asphalt concrete through four point bending fatigue test. *Constr. Build. Mater.* **2012**, *29*, 403–409. [CrossRef]
34. García, Á.; Schlangen, E.; van de Ven, M.; van Bochove, G. Optimization of composition and mixing process of a self-healing porous asphalt. *Constr. Build. Mater.* **2012**, *30*, 59–65. [CrossRef]
35. Mo, L.T.; Huurman, M.; Woldekidan, M.F.; Wu, S.P.; Molenaar, A.A.A. Investigation into material optimization and development for improved ravelling resistant porous asphalt concrete. *Mater. Design* **2010**, *31*, 3194–3206. [CrossRef]
36. Vila-Cortavitarte, M.; Lastra-González, P.; Calzada-Pérez, M.Á.; Indacoechea-Vega, I. Analysis of the influence of using recycled polystyrene as a substitute for bitumen in the behaviour of asphalt concrete mixtures. *J. Clean. Prod.* **2018**, *170*, 1279–1287. [CrossRef]

37. Hu, X.; Dai, K.; Pan, P. Investigation of engineering properties and filtration characteristics of porous asphalt concrete containing activated carbon. *J. Clean. Prod.* **2019**, *209*, 1484–1493. [CrossRef]
38. Wang, X.; Gu, X.; Dong, Q.; Wu, J.; Jiang, J. Evaluation of permanent deformation of multilayer porous asphalt courses using an advanced multiply-repeated load test. *Constr. Build. Mater.* **2018**, *160*, 19–29. [CrossRef]
39. Mo, L.T.; Huurman, M.; Wu, S.P.; Molenaar, A.A.A. 2D and 3D meso-scale finite element models for ravelling analysis of porous asphalt concrete. *Finite Elem. Anal. Des.* **2008**, *44*, 186–196. [CrossRef]
40. Chen, S.; Lin, X.C.; Zheng, C.F.; Guo, X.D.; Chen, W.X. Evaluation of siltation degree of permeable asphalt pavement and detection of noise reduction degree. *Appl. Sci.* **2021**, *11*, 349. [CrossRef]
41. Zhang, K.; Kevern, J. Review of porous asphalt pavements in cold regions: The state of practice and case study repository in design, construction, and maintenance. *J. Infra. Preserv Resil.* **2021**, *2*, 4. [CrossRef]
42. Fortier, R.; Vinson, T. Low-temperature cracking and aging performance of modified asphalt concrete specimens. *Transport. Res. Rec. J. Transport. Res. Board* **1998**, *1630*, 77–86. [CrossRef]
43. Cheng, Y.C.; Chai, C.; Liang, C.Y.; Chen, Y. Mechanical performance of warm-mixed porous asphalt mixture with steel slag and crumb-rubber–SBS modified bitumen for seasonal frozen regions. *Materials* **2019**, *12*, 857. [CrossRef] [PubMed]
44. Liu, H.B.; Zhu, B.; Wei, H.B.; Chai, C.; Chen, Y. Laboratory evaluation on the performance of porous asphalt mixture with steel slag for seasonal frozen regions. *Sustainability* **2019**, *11*, 6924. [CrossRef]
45. Ma, X.; Ni, F.J.; Chen, R.S.; Wang, Y. Evaluation of low-temperature performance of porous asphalt mixture. *J. Highway Transp. Res. Dev. (English Ed.)* **2010**, *4*, 33–37. [CrossRef]
46. Vavrik, W.R.; Pine, W.J.; Huber, G.; Carpenter, S.H.; Bailey, R. The Bailey method of gradation evaluation: The influence of aggregate gradation and packing characteristics on voids in the mineral aggregate. *J. Assoc. Asphalt Paving Technol.* **2001**, *70*, 132–175.
47. Vavrik, W.R.; Pine, W.J.; Carpenter, S.H. Aggregate blending for asphalt mix design: Bailey method. *Transp. Res. Rec. J. Transp. Res. Board* **2002**, *1789*, 146–153. [CrossRef]
48. Gong, Y.H. Study on the Composition Design and Structural Performance of Porous Asphalt Concrete Bridge Deck Pavement. Master's Thesis, Beijing Jiaotong University, Beijing, China, 2018.
49. JTG E20-2011. *Standard Test Methods of Bitumen and Bituminous Mixtures for Highway Engineering*; China Communications Press: Beijing, China, 2011.
50. JTG F40-2004. *Technical Specification for Construction of Highway Asphalt Pavements*; China Communications Press: Beijing, China, 2004.
51. Haryanto, I.; Takahashi, O. Effect of aggregate gradation on workability of hot mix asphalt mixtures. *Balt. J. Road Bridge Eng.* **2007**, *2*, 21–28.
52. Vavrik, W.R. *Bailey Method for Gradation Selection in HMA Mixture Design*; Transportation Research Board Circular E-C044; Transportation Research Board: Washington, DC, USA, 2002.
53. Hardiman, H.; Hamzah, M.O.; Mohammed, A.A. Binder drainage test for porous mixtures made by varying the maximum aggregate sizes. *Civ. Eng. Dimen.* **2004**, *6*, 26–31.
54. Cox, B.C.; Smith, B.T.; Howard, I.L.; James, R.S. State of knowledge for Cantabro testing of dense graded asphalt. *J. Mater. Civ. Eng.* **2017**, *29*, 04017174. [CrossRef]
55. Taguchi, G.; Chowdhury, S.; Wu, Y. *Taguchi's Quality Engineering Handbook*; Wiley: Hoboken, NJ, USA, 2005.
56. Taguchi, G. Quality engineering (Taguchi methods) for the development of electronic circuit technology. *IEEE Trans. Reliab.* **1995**, *44*, 225–229. [CrossRef]
57. Wang, T.; Xiao, F.; Amirkhanian, S.; Huang, W.; Zheng, M. A review on low temperature performances of rubberized asphalt materials. *Constr. Build. Mater.* **2017**, *145*, 483–505. [CrossRef]
58. Nejad, F.M.; Aghajani, P.; Modarres, A.; Firoozifar, H. Investigating the properties of crumb rubber modified bitumen using classic and SHRP testing methods. *Constr. Build. Mater.* **2012**, *26*, 481–489. [CrossRef]
59. Arshad, M.; Ahmed, M.F. Potential use of reclaimed asphalt pavement and recycled concrete aggregate in base/subbase layers of flexible pavements. *Constr. Build. Mater.* **2017**, *151*, 83–97. [CrossRef]
60. Pi, Y.H.; Li, Z.; Pi, Y.X.; Huang, Z.; Li, G.C. Performance evaluation of recycled asphalt pavement materials and cold recycling mixtures designed with vibratory compaction method. *Appl. Sci.* **2019**, *9*, 3167. [CrossRef]
61. Moon, K.H.; Cannone Falchetto, A. Double-recycled reclaimed asphalt pavement: A laboratory investigation at low temperatures based on different mathematical approaches. *Materials* **2020**, *13*, 3032. [CrossRef]
62. Yang, Z.; Zhuang, G.Y.; Wei, X.S.; Wei, J.T.; Yu, H.Y.; Xu, W. Quantitative analysis of the blending degree of virgin and RAP binders in recycled asphalt mixtures with a high RAP content. *Appl. Sci.* **2018**, *8*, 2668. [CrossRef]

applied sciences

MDPI

Article

Correlation of Pavement Distress and Roughness Measurement

Abdullah I. Al-Mansour * and Amr A. Shokri *

College of Engineering, King Saud University, P.O. Box 800, Riyadh 11421, Saudi Arabia
* Correspondence: amansour@ksu.edu.sa (A.I.A.-M.); eng.amralsobihi86@gmail.com (A.A.S.)

Abstract: Riyadh City established and implemented a Pavement Maintenance Management System (PMMS) through the General Directorate of Maintenance and operation. The system was created to address the difficulties that come with maintaining and reserving the pavement network. To evaluate pavement conditions, Riyadh (PMMS) uses visual checks, structural capacity roughness, and skid resistance. An Urban Distress Index (UDI) is calculated during the visual assessment process. Distressed pavement types, severity, and quantity are taken into account when calculating UDI values. As a result, the procedure gathers extensive data on the pavement's condition. However, the procedure is time-consuming and very costly. The Automatic Road Analyzer car provides data on road roughness in accordance with the International Roughness Index (IRI). The IRI data are often generated quite quickly and at a cheap cost as compared to the distress survey. This study's aim is to examine whether a sample of Riyadh city pavement sections can be connected to the IRI depending on the distress type. The research develops statistical models that correlate IRI values with several distress-types associated with roadway classes. Correlating the International Roughness Index values to distress type will eliminate the necessity to implement the manual inspection at a network- level. This saves money and time for PMMS employees when preparing annual maintenance requirements and setting priorities. The finding of the study, of the relationship between the pavement distresses and the International Roughness Index showed a statistically significant relationship between pavement roughness and some ride-quality distresses, like depression and patching, as well as some non-ride quality distresses like potholes and rutting. In addition, for both main and secondary streets, an analysis of variance shows the existence of a correlation between the two variables.

Keywords: International Roughness Index; pavement condition evaluation; correlation; Saudi Arabia

Citation: Al-Mansour, A.I.; Shokri, A.A. Correlation of Pavement Distress and Roughness Measurement. *Appl. Sci.* **2022**, *12*, 3748. https://doi.org/10.3390/app12083748

Academic Editors: Amir Tabakovic, Jan Valentin and Liang He

Received: 5 March 2022
Accepted: 2 April 2022
Published: 8 April 2022

1. Introduction

Evaluation of pavement condition is one of the major elements of any pavement maintenance management system (PMMS). Through pavement condition, maintenance needs, budget, and priority are determined. Pavement condition usually involves distress surveys, roughness and structural capacities, and skid resistance. Most, if not all highway agencies incorporate at least one of these components [1].

There are two levels of pavement evaluation: network level and project level. Network level evaluation gives us a comprehensive overview of the network's condition. In project level evaluation, detailed information on pavement conditions are determined.

Riyadh city and through Riyadh Municipality, and the General Directorate for Maintenance and Operation, have been implementing PMMS utilizing visual condition surveys, roughness measurements, structural capacities, and skid resistance to evaluate pavement condition. Visual checks and pavement roughness are usually performed annually on the street network. Pavement skid resistance is conducted annually on only main streets. Evaluation of structure is only performed at a project-level and is based on manual and roughness evaluation results [2].

In general, visual condition surveys include analyzing the distress type, as well as its severity level, extent, and site. As there are various authorities involved, there are many techniques for conducting pavement condition surveys. The city of Riyadh has its pavement inspection procedure. As part of the procedure, a local pavement condition evaluation called the Urban Distress Index (UDI) should be established. The index is scaled from 0 to 100, with 100 representing excellent pavement condition. The UDI values are determined by visual assessment of the type, severity, and extent of pavement distress. Typically, an investigator is sent to a specific section of pavement in order to gather the necessary distress data to report the current types, severity levels, and quantities of distress. The data is gathered by physically walking through the segment. The process delivers specific information regarding the state of each segment of pavement. Furthermore, due to the large size of the city's pavement network, the process is both time-consuming and expensive.

Road pavement surface roughness is generally used to assess the riding condition of a road pavement surface. It is closely linked to the serviceability of pavement, which is a measurement of the pavement's physical characteristics. Profilometric, vehicle response, and subjective evaluation are the three types of roughness measurement techniques currently in use. The most accurate and well-suited for detailed analysis are profilometric techniques.

Riyadh PMMS uses the profilometric method to measure pavement roughness. The International Roughness Index (IRI) is used to calculate roughness measures. The IRI and the road profile are computed at regular intervals and are both shown on the computer. The horizontal and vertical profile data obtained from the profilometric method are converted into a response of the vehicle's motion by utilizing the technique of IRI calculation, which is predicated by a mathematical model. The IRI value is calculated by dividing the displacement units by the length units. At speeds of (40–50) kilometers per hour, roughness measurements are taken. Due to this, roughness measurements were acquired in a reasonably short period of time and at a cheap cost.

The major goal of this study is to determine the relationship between pavement distress assessment and roughness measures in the Riyadh roadway network. The pavement distresses were ride quality-related distresses such as cracking, patching, potholes, depression, rutting, and raveling.

Roughness measurements are indicated by the International Roughness Index (IRI). The analysis included establishing correlation factors between the distress types of riding quality and IRI and developing models to forecast the pavement distress type values based on IRI and highway class.

2. Literature Review

2.1. Background

Any pavement management system (PMS) is required to contain data of the pavement condition. The main goal of collecting and evaluating the condition of the pavement is to evaluate the current state of pavement condition at the time of inspection in order to determine urgent pavement maintenance requirements and plan for future objectives [2].

Pavement management systems (PMS) are increasingly being employed at all levels of government for effective highway management. Most PMS techniques may provide support at two levels: the network-level and the project-level. The data collected at the network level gives management comprehensive information on the whole system. The network-level level frequently provides information for planning reasons and financial analyses. Project-level information, on the other hand, contains construction, engineering design, and cost accounting details. The amount of data necessary for each level varies greatly. The general goals of network-level pavement condition evaluation are to analyze current pavement conditions, determine immediate pavement repair needs, and make a strategy for future requirements. Distress survey, roughness, structural capacity, and skid resistance are four common approaches for evaluating pavement conditions.

2.2. The Visual Inspection and Roughness Relationship

Research by Mubaraki was performed to find out the relationship between IRI and specific types of distress. Raveling, rutting, and cracking were among the distresses that occurred. Three relationships linking IRI and raveling, rutting, and cracking have been observed. According to the findings of this investigation, raveling and cracking are related to IRI and it might be categorized as "ride quality distress." Furthermore, IRI was shown to have no statistically significant correlation with rutting [3].

Abdelaziz and his team built an artificial neural network (ANNs) and regression models to predict IRI as a function of pavement distresses. The LTPP database was used to create these models. The models that were constructed included prediction of IRI as an age function, IRI initial, alligator fatigue cracking (all severities), transverse cracking (all severities), and rut depth standard deviation (RUTSD). The correlation coefficient (R2) of the regression analysis was 0.57, but the correlation coefficient (R2) of the ANNs model was 0.75 [4,5].

In the same way [6], examined data from California's locations in the Bay area of San Francisco city to see if there was a correlation between IRI and pavement distress. To do this, the researchers set out with the objective of developing a surface roughness model that could be used to predict car operating expenses on highways in the Bay region of San Francisco city. In this study, 39 measurements were taken at 15 m intervals over a 152.4 m test section, the researchers developed a link between IRI and pavement distress. The outcome was the development of a linear link between IRI and a composite Pavement Condition Index (PCI. It was computed by taking into account the type and level of distress observed for each segment of the roadway. The PCI scale ranges from zero to one hundred, with one hundred representing ideal road conditions. The model obtained an adjusted R2 of 0.52, In other words, the aggregate pavement quality index could predict slightly more than half of the variation in IRI.

In the United States, researchers investigated the association between pavement surface roughness and various types of distresses on the pavement [7]. The research included 462 pavement portions from 37 projects throughout Michigan. The sections of road pavement were studied to explore the connection between pavement surface roughness and distress. Assumption number one in this research was that an excess roughness results in increased dynamic loads on axles, which can then result in a tangible acceleration of pavement degradation. If this relationship is established, then it will be possible to plan preventive maintenance (PM) action to smooth the pavement surface.

The transportation department in the State of Connecticut utilized data in order to determine a relationship between the pavement distresses and IRI. Rutting and total quantity of cracking were the only distress types used. They called the associations they discovered "relatively weak." R2 values for various lengths of pavement sections were 0.177 for 10 m, 0.242 for 30 m, and 0.299 for 90 m. They also found that utilizing neural networks did not enhance the findings, thus, they concluded that the IRI could not be utilized to forecast the general condition of pavements. Their study concentrated on pavement segments less than 90 m in length, and a simple summation was used to aggregate the value of cracks [8].

Other researchers have discovered that when the IRI is compared to other performance measures, the R2 value is greater. For example [9], investigated the correlation between IRI and cracks, depth of rutting, potholes, patchwork, and raveling, and discovered that a R2 value of 0.77 was the best they could come up with. They checked asphalt roads in India for guidance. Another study [10] investigated the relationship between the IRI and the PSI in Dubai by using a nonlinear regression model. They discovered that the R2 for differing types of asphalt roads was 0.67 and 0.44, respectively.

Pavement roughness is a key indicator of pavement performance since it directly reflects pavement serviceability to road users. Recently, certain governments and provinces in Canada and the US are utilizing IRI in terms of an objective indicator of their pavement network conditions in their business plans [11]. For example, The Federal Highway Administration (FHWA) in the United State utilizes IRI as a performance metric to describe and monitor its National Highway System's pavement performance. There are two states in America, Kansas and Washington, that utilize IRI rating percentages to represent the status of their highway network. In recent years, it has been a widespread practice among several highway organizations to measure IRI at the network level. Many road agencies now collect roughness on a routine basis since the IRI is a scale that may be transferred, is repeatable, and stable. Roughness is measured at the network level once a year or twice a year as a component of the evaluation of pavements, which is critical in setting maintenance and rehabilitation priorities.

3. Methodology and Data Collection

3.1. Study Methodology

There are two primary phases in the study's methodology. Collecting the information required for a sample group of streets in Riyadh City is the first phase in the process. The data includes information on the type, size, verity, and density for each distress and the IRI for each section of pavement that is included. The collected data of main streets and secondary streets were treated separately to avoid the possible effect of highway class and traffic categories. The second phase in the methodology is to investigate the relationship between IRI and all other types of distress.

3.2. Data Collection

The city of Riyadh is divided into smaller components for management and maintenance purposes. The city is divided into branch municipalities, districts, and regions. The region is a subdivision of the district and it's surrounded by four main streets and contains no main street within its entire area. The two main categories of the Riyadh street network are main streets and secondary streets. The main street is defined as the street that connects two main streets and is more than twenty-five meters wide or has an island. Main streets account for about 27 percent of the overall network area in Riyadh City. Secondary streets are all the categories of streets outside the definition of main streets. Secondary streets represent about 73 percent of the overall network area. Main and secondary streets are divided into sections. The segment of the main street is defined as distance between two intersections in the main street. The segment for the secondary street is defined as a region within a district. The segments for main and secondary streets are divided into several sample units; a representative unit of sample for the main streets was 100 m in length per lane. The sample unit of the secondary street is a street in a region.

The mean source of data was the General Directorate for Maintenance and Operation, Riyadh Municipality. A representative sample of pavement sections from both main and secondary streets was randomly selected. A section may have two to three lanes. In secondary streets, each data point represents the section. A total of 1619 main street sections were included in the study, with a total distance of more than 898.258 km. For secondary street sections a total of 140 were sections covered in the study. These represent an area of about 1814 square km.

The data for both main and secondary streets include distress data, pavement, condition, and International Roughness Index for each observation period. Time between the pavement condition survey and roughness measurements were less than 6 and 3 months for main and secondary streets, respectively. Table 1 presents the types and codes for distress that appear in both main and secondary streets according to the Riyadh Pavement Maintenance Management System (PMMS).

Table 1. Distress Types and Codes in Riyadh PMMS.

Main Streets		Secondary Streets	
Type of Distress	**Distress Code**	**Type of Distress**	**Distress Code**
Fatigue Cracking	D1	Block Cracking	D2
Block Cracking	D2	Long. and Trans. Cracking	D3
Long. and Trans. Cracking	D3	Patching	D4
Patching	D4	Potholes	D5
Potholes	D5	Depressing	D6
Depressing	D6	Raveling	D11
Rutting	D7		
Shoving	D8		
Bleeding	D9		
Polished Aggregate	D10		
Raveling	D11		
Patching Cracks	D12		
Patching Depression	D13		

3.3. Relationship between Types of Pavement Distress and IRI

Statistical correlation analysis was carried out to analyze the interaction between IRI and pavement distresses. Riding quality distresses are expected to show a considerable effect on IRI values. That means that distress is characterized as riding quality distress if a high degree of linkage exists between the value of IRI and the distress density for a pavement section. In contrast, distress is characterized not affecting riding quality if the correlation between distress density and IRI is not significant.

The analysis of IRI and distress types involved investigating the linkage between the value of IRI and each distress type. It is expected that a section of pavement will have more than one distress. Therefore, pavement sections are grouped based on the distress that is dominating in each section. The IRI can then be related to riding quality distresses through statistical models. Figure 1 describes the suggested procedure of analysis.

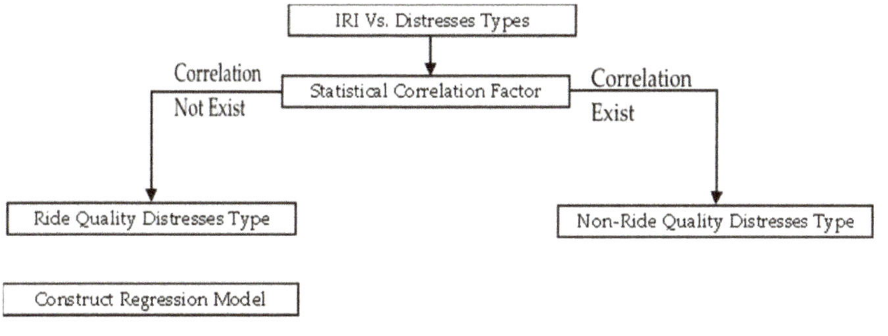

Figure 1. Analysis of IRI and type of distress.

4. Data Analysis and Results

4.1. Statistical Information

The modeling approach applied in the analysis is based on statistical regression. The fundamental form of a regression equation is as follows [12]:

$$Y = b_0 + b_1 X \tag{1}$$

The regression equation can be solved to estimate the predicted values of the regression coefficients (b_0 and b_1).

To perform the analysis procedure, two types of computer software were used to examine and study the data. This software was a spreadsheet program and a statistical program. The data was entered into an Excel spreadsheet, which then generated the graphs (charts) that show the overall trend of the data. The "Minitab" statistical software was used for the following analysis [13]:

- Conduct descriptive and general statistics for the IRI and various distress types.
- Perform correlation tests between IRI and various distress types.
- Model the correlation between IRI and distress types.

In the preliminary analysis, the statistical program (Minitab) was used to select the general statistics of the IRI and various distress type data for both main and secondary streets.

4.2. Distresses Related to Ride Quality

Pavement roughness measures are generally used to determine the riding quality surface of a roadway's pavement. Pavement roughness is usually connected to pavement serviceability. The qualities of the pavement's physical characteristics are measured by pavement serviceability. So, roughness is connected to the opinions of persons who drive on the road. It has a major influence on comfort, safety, vehicle operating costs, and travel speed.

This section attempts to research pavement distresses to generate the types of distresses that are related to riding quality for both main and secondary streets. All pavement distresses in the study region were studied for their impact on the IRI measurement.

- Main Street sections

For main street sections, ten types of distress appear in the study sample. The cracking distresses fatigue cracking (code D1), block cracking (code D2), longitudinal and transverse cracking (code D3) and patching cracks (code D12) have been aggregated into one group called cracking distress. In addition, the distress types of depression (code 6) and depression patching have been assembled into one group called depression distress. After grouping, the distress types were reduced to six. The distress groups in the main sections are as follows:

Cracking (codes D1, D2, D3 and D12); Patching (code D4); Potholes (code D5); Depression (codes D6 and D13); Rutting (code D7); Raveling (code D11).

The relation between IRI and each of the distress types in main street samples is investigated and discussed in the subsequent paragraphs.

1. Cracking vs. IRI

Figure 2a presents crack distress as dominant in the total sample units. It is clear from the graph that a visible scattering exists in the roughness and density of distress data. The chart doesn't show a strong link between the IRI value and cracking density. Furthermore, the IRI and the crack density correlation coefficient is 0.167. The low correlation factor means that the two variables have a poor linear relationship. The p-value is equal to zero. This means that there is enough relation between IRI and distress density. The statistical test result does not correspond to previous experiences indicating that cracking distress is a non-ride quality form of distress at all levels of severity.

2. Patching vs. IRI

Figure 2b shows that the patching and IRI values have a positive relationship. The graph, on the other hand, does not demonstrate any clear-cut conclusions regarding the relationship's linearity. IRI and patching density have a correlation coefficient of 0.471 and a p-value of zero. The coefficient has a positive sign which mean that patching density and IRI have a positive relationship. The linear relationship distinguished around 47 percent of the IRI and patching density observations, according to the correlation factor value. The hypothesis test obtained a p-value of 0, indicating that there is sufficient evidence of a link between IRI and patching density at a 95% confidence level.

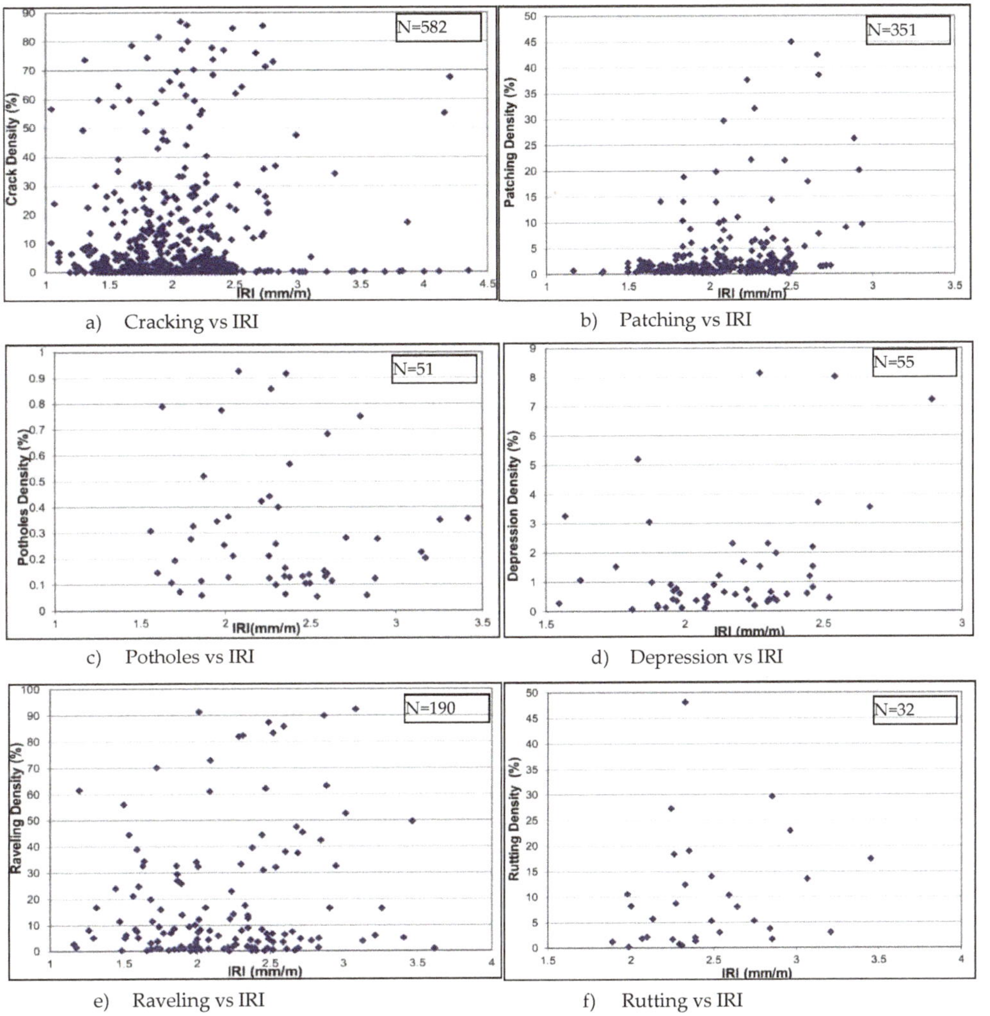

a) Cracking vs IRI b) Patching vs IRI

c) Potholes vs IRI d) Depression vs IRI

e) Raveling vs IRI f) Rutting vs IRI

Figure 2. Distress types vs. IRI for Main Road, N = number of sections included.

3. Potholes vs. IRI

Figure 2c represents the relation between pothole density and IRI. The pothole distress was the dominant distress. The graph shows that the relationship's linearity is not well defined. The coefficient of correlation is 0.123. The *p*-value is 0.390, which is larger than the level of significance of 0.05. As a result, there is not enough evidence of a link between IRI and the number of potholes.

4. Depression vs. IRI

The graph in Figure 2d shows the relationship between IRI and depression density. The graph shows that depression and IRI values have a positive relationship. IRI and depression density have a correlation coefficient of 0.342 and a *p*-value of 0.011. The positive sign indicates that depression density and IRI have a positive relation. When the correlation coefficient equals 0.34, there is no clear choice about the relationship's linearity. The *p*-value for the hypothesis test is 0.011, which is less than the 0.05 level of significance.

The result is that we have sufficient evidence to understand that there is a statistically significant correlation between IRI and depression density with 95 percent confidence.

5. Raveling vs. IRI

Figure 2e shows the relationship between IRI and distress of raveling density. The correlation coefficient is 0.399 and the *p*-value is zero. There is no clear-cut decision about the linearity of the relation. The positive sign indicates a positive relation between raveling density and IRI. In the hypothesis test the *p*-value= 0, therefore there is sufficient evidence of the relationship between IRI and raveling density.

6. Rutting vs. IRI

Figure 2f shows the correlation between IRI and rutting density. The graph shows that rutting density and IRI values have a positive relationship. there is no clear choice about the relationship's linearity. The correlation coefficient is 0.357, meaning that the linear relationship only affected 35.7 percent of the IRI and rutting density observations. There is insufficient evidence of a link between IRI and rutting density since the *p*-value is 0.045, which is extremely near to the 0.05 level of significance.

- Secondary Street Sections

There are six distress types included in secondary pavements of the Riyadh street network. The cracking distresses Block Cracking (code D2) and Longitudinal and Transverse Cracking (code D3) have been aggregated to one group called cracking distress. After grouping the distress types were reduced to five. The total distress groups in the second section are Cracking (codes D2 and D3) Patching (code D4); Potholes (code D5); Depressing (code D6) and Raveling (code D11).

The relation between IRI and each group of distress types in secondary Streets is discussed in the subsequent paragraphs.

1. Cracks vs. IRI

The graph in Figure 3a shows the relationship between IRI and crack density. The correlation coefficient is 0.28, which indicates that the linear relationship only affected 28 percent of the IRI and crack density observations. There is insufficient evidence of a link between IRI and crack density since the *p*-Value of the hypothesis test is zero. The positive sign means a positive correlation between IRI and carack density, the low relationship factor indicates insignificant linearity relation between crack density and IRI.

2. Patching vs. IRI

The relation between IRI and patching distress density is seen in Figure 3b. The graphic shows that patching density and IRI have a positive relationship. The coefficient of correlation is 0.514. The positive sign indicates that patching density and IRI have a positive relationship. Furthermore, the correlation factor indicates that the linear relationship distinguished around 51.4 percent of the IRI and patching density data. The p-value for the hypothesis test is zero. As a result, the relation between IRI and patching density has sufficient evidence at a 95 percent confidence level.

3. Potholes vs. IRI

Figure 3c shows the relationship between IRI and pothole distress density. The correlation factor is 0.322 indicating low linearity of the relationship. The *p*-value is 0.208 which is greater than the 0.05 level of significance; so, there is insufficient evidence of the relation between the value of IRI and pothole density. This result could be because potholes are usually avoided during roughness measurements.

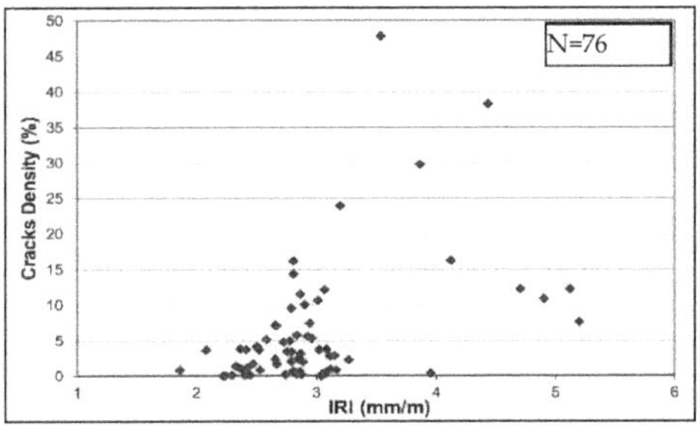

a) Crack vs IRI

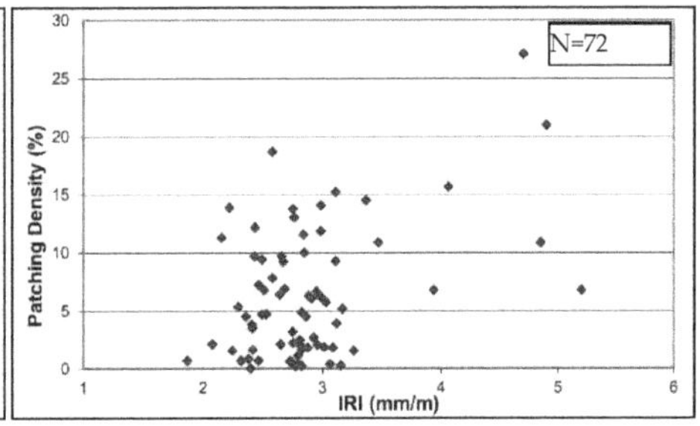

b) Patching vs IRI

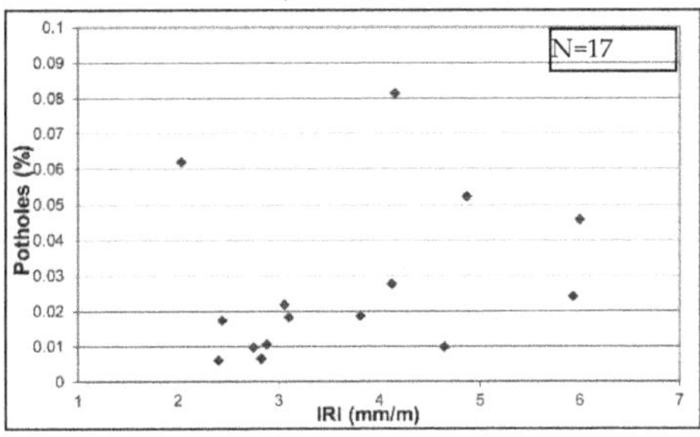

c) Potholes vs IRI

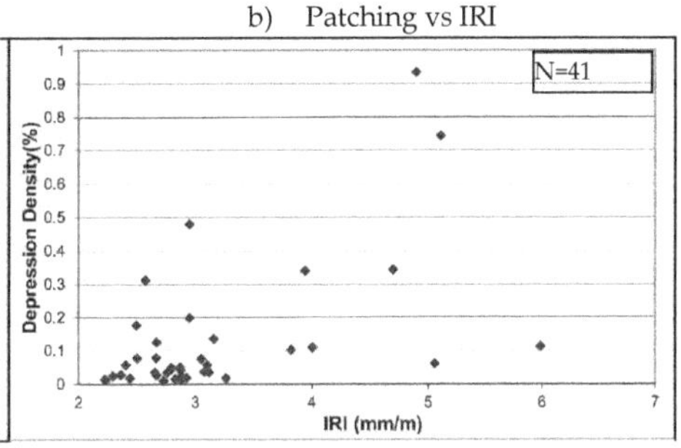

d) Depression vs IRI

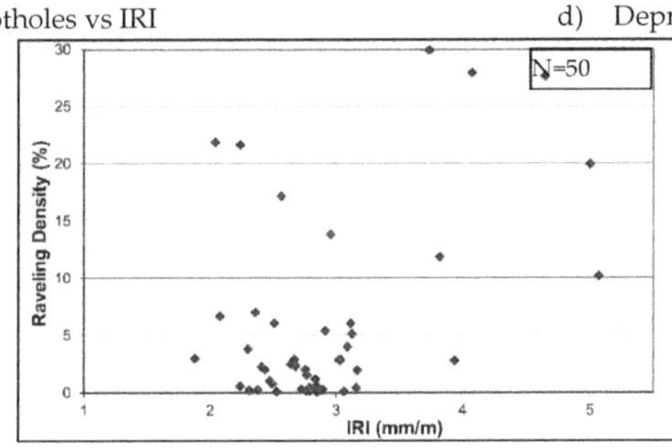

e) Raveling vs IRI

Figure 3. Distress types vs. IRI for secondary roads, N = number of sections included.

4. Depression vs. IRI

The correlation between IRI and depression distress density is presented in Figure 3d. The graph shows that patching density and IRI have a positive relationship. When the density of depressions in the pavement sample is high, the IRI value increases. The linear relationship distinguished about 43.6 percent of the IRI and depression density data, on the basis of the correlation factor value. The hypothesis test gave a p-value of 0.004, which is less than the 0.05 level of significance, showing that there is sufficient relation between IRI and depression density at the 95 percent confidence level.

5. Raveling vs. IRI

The graph in Figure 3e shows the relationship between the value of IRI and the raveling distress density. The graph shows a relationship between raveling and IRI. The linear relationship distinguished about 38.7% of the IRI value and raveling density data, according to the correlation factor value. The p-value in the hypothesis test is 0.006. As a result, there is sufficient evidence of the association between the value of IRI and the raveling density at the 95 percent confidence level.

4.3. Distresses Density and the IRI Model

In this section, the regression method was chosen to create models that related IRI with the density of different distresses. The linear regression model would be in the form:

$$IRI = b_0 + b_1 \text{ distress}(1) + b_2 \text{ distress}(2) + b_3 \text{ distress}(3) \ldots + b_n \text{ distress}(n); \quad (2)$$

where; $b_0, b_1, \ldots, b_n$: regression coefficients

For each street category, two regression models were generated. The first model looked at how different types of distress impacted pavement roughness (IRI). The previous results in detecting ride quality types of distress were then confirmed using this model. The second model was created to find the best match between IRI and the different types of ride quality distress. The two models were checked and analyzed using a variety of statistical tests. The *t*-test was used to evaluate the relationship between the two regression coefficients (b_0, b_1) and the dependent variable at the 95% confidence level. Analysis of variance was used to determine the regression line's quality, or if the variation in IRI is dependent on the distress data. Finally, the correlation coefficient (R2) was used to show how well the regression model was fitted to the data.

- Main Street Sections
- The influence of pavement roughness on six different types of distress in the main street sections in Riyadh city was studied and analyzed. The study developed and tested two regression models; the first model attempted to link the IRI index to all types of distress in main street samples. The second regression model was developed to connect the IRI index to different types of distress that have been identified as ride quality distresses. The linear regression looked like this:

$$IRI = b_0 + b_1 \, Cr + b_2 \, Pa + b_3 \, Po + b_4 \, Dr + b_5 \, Ru + b_6 \, Ra; \quad (3)$$

where; b_0, b_1, b_2, b_3, b_4, b_5, b_6 are regression coefficients; Cr = crack distress density; Pa = patching distress density; Po = potholes distress density; Dr = depression distress density; Ru = rutting distress density; Ra = raveling distress density.

1. Model-1—connects IRI to all observable distresses:

$$IRI = 2.09 + 0.00265 \, Cr + 0.00554 \, Pa + 0.00906 \, Po + 0.0193 \, Dr + 0.00531 \, Ru + 0.00317 \, Ra \quad (4)$$

Summary of the model parameters is presented in Table 2.

Table 2. Summary of *t*-tests for main streets (Model-1).

Predictor	Coef.	St.Dev.	T	*p*
Constant	2.09494	0.01593	131.54	0
Cr	0.0026529	0.0002124	12.49	0
Pa	0.0055395	0.000327	16.94	0
Po	0.009056	0.00501	1.81	0.071
Dr	0.019317	0.002978	6.49	0
Ru	0.005306	0.00176	3.01	0.063
Ra	0.0031703	0.0003066	10.34	0

The *p*-values for most regression coefficients are zero, meaning that there is a significant correlation between the types of distress and the estimated value of IRI. The *p*-values of the potholes (Po) and rutting (Ra) coefficients, on the other hand, are more than the 0.05 level of significance. As a result, there is sufficient evidence to exclude potholes and rutting in estimating the roughness index.

2. Analysis of Variance

Source	DF	SS	MS	F	p
Regression	6	203.339	33.890	132.26	0.000
Residual Error	1610	412.547	0.256		
Total	616	615.886			

The *p*-values for the ANOVA-tests for the model are zero, meaning that there is no significant variation in the IRI predicted values caused by the regression model. The results show a strong connection between IRI and various types of distress.

3. Squared multiple correlation coefficient (R2): R-Sq = 32.5% R-Sq(adj) = 32.3%

The squared multiple correlation coefficient (R2) is 32.3 percent. This means that 32.3 percent of the IRI values can be reflected and described by the regression model. It also shows that while there are statistically significant relationships between IRI and some types of distress, the relationship isn't strong enough to use distress types as an accurate measure of roughness conditions.

(a) Model-2—connects IRI to ride quality distresses:

In this case, the types of distress that had a strong relationship with the IRI index are called ride quality types of distress. This distress includes cracks distress (Cr), patching distress (Pa), depression distress (Dr), and raveling distress (Ra).

1. The regression model developed is as follows:

$$IRI = 2.10 + 0.00267\ Cr + 0.00553\ Pa + 0.0195\ Dr + 0.00316\ Ra \tag{5}$$

A summary of the regression model is shown in Table 3.

Table 3. Summary of *t*-tests (Model-2).

Predictor	Coef.	St.Dev.	T	p
Constant	2.09875	0.01594	131.67	0
Cr	0.0026746	0.0002129	12.56	0
Pa	0.0055299	0.0003278	16.87	0
Dr	0.019547	0.002987	6.55	0
Ra	0.003156	0.0003075	10.26	0

Since all the regression coefficients have *p*-values equal to zero, this indicates that there is a statistically significant link between the different types of ride quality distress and the expected levels of IRI.

2. Analysis of Variance

Source	DF	SS	MS	F	p
Regression	4	200.181	50.045	194.06	0.000
Residual Error	1612	415.705	0.258		
Total	1616	615.886			

The *p*-values for the ANOVA-tests for the model are zero, meaning that there is no significant variation in the IRI predicted values caused by the regression model. The results show a strong connection between IRI and various types of distress. The F-value for model-2 is greater than that in model-1 meaning improvement of model-2 over model-1.

3. Squared multiple correlation coefficient (R2): R-Sq = 33.0% R-Sq(adj) = 32.8%

The squared multiple correlation coefficient (R2) is 32.8 percent. This means that 32.8 percent of the IRI values can be reflected and described by the regression model. It also shows that while there are statistically significant relationships between IRI and some types of distress, the relationship isn't strong enough to use distress types as an accurate measure of roughness conditions.

- Secondary street sections

The effect of the pavement roughness on five distress types in the section of the secondary streets in the Riyadh city network has been studied and analyzed. The study developed and tested two regression models; the first one was to relate the IRI index to all distresses in the secondary street samples, whereas the second model was to relate IRI index to ride quality distress types. The distresses included in the regression evaluations were, cracks distress (Cr), patching distress (Pa), potholes distress (Po), depression distress (Dr), and raveling distress (Ra). The linear regression model would be in the form:

$$\text{IRI} = b_0 + b_1\,\text{Cr} + b_2\,\text{Pa} + b_3\,\text{Po} + b_4\,\text{Dr} + b_5\,\text{Ra}; \qquad (6)$$

where the number of regression coefficients equal 6.

The distress in the secondary streets class chosen to be ride quality type distresses that showed a statistically significant association with the IRI were cracks distress (Cr), patching distress (Pa), depression distress (Dr), and raveling distress (Ra). The linear regression model would be in the form:

$$\text{IRI} = b_0 + b_1\,\text{Cr} + b_2\,\text{Pa} + b_3\,\text{Dr} + b_4\,\text{Ra}; \qquad (7)$$

where the number of regression coefficients equals 5.

(a) Model-1 the linear regression equation for the first model that relates IRI against all apparent distresses:

$$\text{IRI} = 2.59 + 0.0422\,\text{Cr} + 0.0411\,\text{Pa} + 5.75\,\text{Po} + 1.32\,\text{Dr} + 0.00996\,\text{Ra} \qquad (8)$$

The estimated regression coefficients and the overall model statistics were shown in Table 4.

Table 4. Summary of t-test for secondary streets (Model-1).

Predictor	**Coef.**	**St.Dev.**	**T**	***p***
Constant	2.594	0.1317	19.7	0
Cr	0.042227	0.00892	4.73	0
Pa	0.04107	0.01259	3.26	0.001
Po	5.752	4.146	1.39	0.168
Dr	1.3185	0.4064	3.24	0.001
Ra	0.009962	0.005005	1.99	0.039

The *p*-values for most of the regression coefficients are zero, clearly showing a significant linear correlation between the estimated values of IRI and the different types of distress. The *p*-value of the potholes (Po) coefficient is equal to 0.168, which is greater than the 0.05 level of significance therefore there is significant evidence to exclude the potholes coefficient which demonstrates no linear relationship between IRI and potholes distress.

1. Analysis of Variance

Source	DF	SS	MS	F	*p*
Regression	5	70.405	14.081	17.56	0.000
Residual Error	134	107.463	0.802		
Total	139	177.868			

Based on the ANOVA-test results, the *p*-values for the values of IRI response variables are equal to zero, meaning that the regression model does not account for any statistically significant variance in the IRI response variables. Furthermore, the data give sufficient evidence of the presence of a dependence relationship between the value of IRI and distresses.

2. Squared multiple correlation coefficient (R2): R-Sq = 38.7% R-Sq(adj) = 36.9%

The value of the squared multiple correlation coefficient (R2) is 36.9%. This means that 36.9% of the IRI values can be described by the regression model. This shows adequate logic of the regression model relating IRI and distress data.

(b) Model-2 the linear regression equation for the second model that relates IRI vs. ride quality distresses:

$$\text{IRI} = 2.5889 + 0.043121\ \text{Cr} + 0.04366\ \text{Pa} + 1.3193\ \text{Dr} + 0.010589\ \text{Ra} \quad (9)$$

The estimated regression coefficients and the overall model statistics were shown in Table 5.

Table 5. Summary of t-tests for main streets (Modal-2).

Predictor	Coef.	St.Dev	T	p
Constant	2.5889	0.1321	19.6	0
Cr	0.043121	0.008927	4.83	0
Pa	0.04366	0.01249	3.49	0.001
Dr	1.3193	0.4077	3.24	0.002
Ra	0.010589	0.005002	2.12	0.036

In this case, the p-values for all ride quality regression coefficients are equal to zero, meaning that there is strong evidence to accept the regression coefficient of all distress densities in the regression model with 95 percent confidence. This indicates a statistically significant linear association between the different types of ride quality distresses and IRI ratings.

1. Analysis of Variance

Source	DF	SS	MS	F	p
Regression	4	68.861	17.215	21.32	0.000
Residual Error	135	109.007	0.807		
Total	139	177.868			

According to the ANOVA-test results, the p-values for the response variable is equal to zero, meaning that there is no significant amount of variation within the response variable in the regression model. The data reflect sufficient evidence of a dependent relationship between IRI and distressed values. Note that the F-value for model-2 = 21.32, which is greater than the F-value from model-1 which = 17.56, demonstrating an improvement in model-2 over model-1.

2. Squared multiple correlation coefficient (R2): R-Sq = 39.6% R-Sq(adj) = 37.3%

The value of the squared multiple correlation coefficient (R2) equals 37.3% which means that 37.3 percent of the IRI values are described and evaluated by the regression model showing adequate logic of the regression model relating IRI and distress data. Note that R2 in the second model (37.3) is greater than the first one (36.9) indicating an improvement of model-2 over model-1. It also shows that while there are statistically significant correlations between IRI and some types of distress, these relationships aren't strong enough for distress types to be used as a predictable measure for roughness conditions.

5. Findings

The influence of all of the distress types that are present in the study area, on the IRI measurement was analyzed in order to study the riding quality-related types of distress. From the main street sections, six distress groups exist in the Riyadh network samples. While the secondary streets involve five distresses groups. Tables 6 and 7 conclude the final analysis of the relationship between the roughness evaluation index (IRI) and the pavement distress density for both main and secondary streets, sequentially. The correlation between the value of IRI and distress types has been investigated by two important statistical tests:

1. Testing the individual correlation hypothesis between the value of IRI and each type of distress.
2. The regression coefficients were used to evaluate the overall regression model that links the IRI value with the different types of distress.

Table 6. Summary of correlation factors and regression models in main street sections.

Distress Type	**Correlation Factor**	**Correlation Hypothesis Test**
Cracks	r = 0.167	$p = 0.000$
Patching	r = 0.471	$p = 0.000$
Potholes	r = 0.123	$p = 0.390$
Depression	r = 0.342	$p = 0.011$
Rutting	r = 0.357	$p = 0.045$
Raveling	r = 0.399	$p = 0.000$
Regression Model Equation: IRI = 2.10 + 0.00267 Cr + 0.00553 Pa + 0.0195 Dr + 0.00316 Ra		
Regression Coefficients Test (*t*-test)		
Distress Type	Hypothesis Test	
Cracks	$p = 0.000$	
Patching	$p = 0.000$	
Potholes	$p = 0.071$	
Depression	$p = 0.000$	
Rutting	$p = 0.003$	
Raveling	$p = 0.000$	
Analysis Of Variance $p = 0$		
Regression Model Adequacy R2 = 32.8%		

Table 7. Summary of correlation factors and regression models in secondary street sections.

Distress Type	**Correlation Factor**	**Correlation Hypothesis Test**
Cracks	r = 0.280	$p = 0.000$
Patching	r = 0.541	$p = 0.000$
Potholes	r = 0.322	$p = 0.208$
Depression	r = 0.436	$p = 0.004$
Raveling	r = 0.327	$p = 0.006$
Regression Model Equation: IRI = 2.10 + 0.00267 Cr + 0.00553 Pa + 0.0195 Dr + 0.00316 Ra		
Regression Coefficients Test (T-test)		
Cracks	0.280	$p = 0.000$
Patching	0.514	$p = 0.000$
Potholes	0.322	$p = 0.168$
Depression	0.436	$p = 0.001$
Raveling	0.387	$p = 0.036$
Analysis Of Variance $p = 0$		
Regression Model Adequacy R2 = 37.3%		

Tables 6 and 7 show that p-values for cracking, depression, patching, and raveling in both main and secondary streets are near zero, indicating a significant relationship between these distress types and IRI. As a result, the findings of the research in both the main and secondary street categories suggests that there is a significant relationship between IRI and cracking, patching, depression, and raveling with a 95% confidence level. Statistical testing on the overall model between IRI and distress types confirmed these findings. The p-values for each regression coefficient were obtained in the regression coefficients test. In both the major and subsidiary streets, the regression coefficients for cracking, depression, patching, and raveling are close to zero (less than 0.05), indicating a substantially linear relationship between IRI and each regression coefficient. IRI is related to different types of distress in different ways. Patching and depression had the highest correlation to IRI, according to the values of the correlation factors. While raveling has a lower relationship with IRI, cracking has the lowest. It was also found that the correlation factors for all the distresses versus IRI did not reach 50%, meaning that the relationship describes less than half of these relationships. The relationship is not strong enough to consider pavement conditions as an indicator of IRI. As a result, the correlation and regression coefficients testing showed that cracking, patching, depression, and raveling are ride quality types of distress with a 95% confidence level. However, it is unrealistic to claim that IRI can completely describe pavement distress conditions.

It's also shown in Tables 6 and 7, that potholes and rutting, point toward a relationship that is not significant with IRI values in both the main and secondary streets. The p values (which are larger than 0.05 level of significance) of both potholes and rutting demonstrates that these distress types have no strong evidence of a significant relationship with IRI. These results were proven through the statistical tests on the overall model between IRI and distress types. The p-values for each regression coefficient were obtained in the regression coefficients test. Potholes and rutting have regression coefficients larger than 0.05, meaning that there is no significant linear relationship between IRI and each distress type. As a result, the correlation and regression coefficient tests show that potholes and rutting are non-ride quality distress types and that IRI could not describe potholes and rutting distress conditions with 95 percent confidence.

6. Conclusions

The relationship between IRI and pavement distresses was investigated using statistical correlation analysis. The purpose of the research was to measure the level of distress related to riding quality. The relationship between distress density and IRI values was determined using a statistical correlation test. In the pavement portions, the model related IRI and distress types were calculated. The correlation between the value of IRI and distress types has been investigated by two important statistical tests:

1. Testing the individual correlation hypothesis between the value of IRI and each type of distress.
2. The regression coefficients were used to evaluate the overall regression model that links the IRI value with the different types of distress.

The results of the study indicate that there is a significant correlation between IRI and cracking, patching, depression and raveling in both main and secondary street groups, with a 95% confidence level. It has also shown that potholes and rutting distress types did not have a significant relationship to IRI values in both the main and secondary streets. This has led to a conclusion that is based on the statistical investigation of the relationship between IRI and distress types: cracking, patching, depression, and raveling could possibly be characterized as ride-quality type distresses. While potholes and rutting distress, could probably be described as non-ride quality type distresses.

Roughness and the extent of correlation can be helpful in detecting the type of distress. This analysis concludes that IRI can either be used to evaluate pavement quality or to monitor pavement deterioration. Rapid measurements of IRI using the Automatic Road Analyzer (ARAN) can ease the process of traditional pavement visual inspection. Using sample visual inspection ratings from pavement distress images can also be used to determine an approximate IRI value without having a Road Analyzer and allow subsequent planning and evaluation to proceed. However, not all surface distresses can be detected by roughness, especially some types of distress that are of very low severity.

Author Contributions: Conceptualization; Formal analysis; Funding acquisition; Investigation; Methodology; Project administration; Resources; A.I.A.-M., Visualization, A.I.A.-M. and A.A.S.; Writing—original draft, A.I.A.-M.; Writing—review & editing, A.I.A.-M. and A.A.S. All authors have read and agreed to the published version of the manuscript.

Funding: The authors would like to acknowledge the support provided by Researchers Supporting Project Number (RSP2022R424), King Saud University, Riyadh, Saudi Arabia.

Institutional Review Board Statement: Not applicable.

Informed Consent Statement: Not applicable.

Data Availability Statement: Available on request and with regulations.

Acknowledgments: Authors would like to thank Riyadh Municipality, General Administration for Operation and Maintenance for providing the data. The authors acknowledge King Saud University for providing the funding of this research paper.

Conflicts of Interest: The authors declare no conflict of interest.

References

1. Al-Mansour, A.; Al-Swailmi, S.; Al-Swailem, S. Development of pavement performance models for Riyadh street network. *Transp. Res. Rec.* **1999**, *1655*, 25–30. [CrossRef]
2. Al-Mansour, A.I.; Al-Swailem, S.S. Pavement condition data collection and evaluation of riyadh main street network. *J. King Saud Univ. Sci.* **1999**, *11*, 1–17. [CrossRef]
3. Mubaraki, M. Highway subsurface assessment using pavement surface distress and roughness data. *Int. J. Pavement Res. Technol.* **2016**, *9*, 393–402. [CrossRef]
4. El-Hakim, A.; El-Aziz, A.; Nader, E.; El-Badawy, S.M.; Afify, H.A. Validation and Improvement of Pavement ME Flexible Pavement Roughness Prediction Model Using Extended LTPP Database. 2017. No. 17-02203. Available online: https://trid.trb.org/view/1437930 (accessed on 4 March 2022).
5. Abdelaziz, N.; El-Hakim, R.T.A.; El-Badawy, S.M.; Afify, H.A. International Roughness Index prediction model for flexible pavements. *Int. J. Pavement Eng.* **2018**, *21*, 88–99. [CrossRef]
6. Dewan, S.A.; Smith, R.E. Estimating IRI from pavement distresses to calculate vehicle operating costs for the cities and counties of San Francisco Bay area. *Transp. Res. Rec.* **2002**, *1816*, 65–72. [CrossRef]
7. Chatti, K.; Lee, D. Development of a preventive maintenance strategy for minimizing roughness-related pavement damage. *Transp. Res. Board* **2001**, *1769*, 39–45. [CrossRef]
8. Aultman-Hall, L.; Jackson, E.; Dougan, C.E.; Choi, S.-N. Models relating pavement quality measures. *Transp. Res. Rec.* **2004**, *1869*, 119–125. [CrossRef]
9. Chandra, S.; Sekhar, C.R.; Bharti, A.K.; Kangadurai, B. Relationship between pavement roughness and distress parameters for Indian highways. *J. Transp. Eng.* **2013**, *139*, 467–475. [CrossRef]
10. Al-Suleiman, T.I.; Shiyab, A.M.S. Prediction of pavement remaining service life using roughness data—Case study in Dubai. *Int. J. Pavement Eng.* **2003**, *4*, 121–129. [CrossRef]
11. Li, N.; Kazmierowski, T.; Sharma, B. Verification of network-level pavement roughness measurements. *Transp. Res. Rec.* **2001**, *1764*, 128–138.
12. Hayter, A.J. *Probability and Statistics for Engineering and Scientists*; Thomson Brooks/Cole: Belmont, CA, USA, 2007.
13. Minitab Statistical Software Release v.19 for Windows. Available online: https://www.minitab.com/en-us/products/minitab/ (accessed on 4 March 2022).

applied sciences

MDPI

Review

Influential Factors and Evaluation Methods of the Performance of Grouted Semi-Flexible Pavement (GSP)—A Review

Xiaogang Guo and Peiwen Hao *

School of Highway, Chang'an University, Xi'an 710064, China; gangxo@outlook.com
* Correspondence: pwhao@chd.edu.cn

Featured Application: This review provides guidance for the future studies of this new pavement technology, which may be of concern in many fields of road construction, such as heavy-duty pavement, perpetual pavement, reclaimed asphalt pavement, and new airport runway.

Abstract: Grouted Semi-flexible Pavement (GSP) is a novel pavement composed of open-graded asphalt concrete grouted with high-fluidity cement mortar. Due to its excellent load-bearing and anti-rutting performance, it has great potential as anti-rutting overlay and surface in road construction. However, the understanding of GSP performance remains limited and pertinent findings are inconsistent. This article aims to provide a systematic literature review for the articles which were published between 2000 and 2020 on GSP, explore the problems in the recent research, identify knowledge gaps, and deliver recommendations for future research. The influential factors and the relative evaluation methods of GSP performance are summarized and discussed in this article.

Keywords: Grouted Semi-flexible Pavement; influential factor; evaluation method; performance; laboratory test

Citation: Guo, X.; Hao, P. Influential Factors and Evaluation Methods of the Performance of Grouted Semi-Flexible Pavement (GSP)—A Review. *Appl. Sci.* **2021**, *11*, 6700. https://doi.org/10.3390/app11156700

Academic Editor: Amir Tabakovic

Received: 22 June 2021
Accepted: 19 July 2021
Published: 21 July 2021

1. Introduction

Conventional flexible pavement, namely asphalt concrete (AC) pavement, suffers from many distresses including rutting, moisture damage, and fatigue failure after long-term operation, because of weather wearing and traffic volume growing worldwide [1,2]. With soaring demand of transportation, the costs of highway construction and pavement maintenance have increased dramatically over time. As a solution for these problems, Grouted Semi-flexible Pavement (GSP), as known as Resin-Modified Pavement, is a type of high-performance pavement that was derived in France in the 1950s. GSP consists of an open-graded matrix asphalt (OMA) mattress with a 20–35% void rate, grouted by high-fluidity cement mortar (HCM) [3–8]. GSP has great resistance to rutting, top-down cracking, oil corrosion, and fatigue damage, combining advantages of both asphalt and concrete [9–11]. It is cost-effective and has been used in many fields, which may be under unfavorable or complex traffic environments that are slow-speed, heavy-duty, or high-temperature, such as airport runways, factory field pavement, bus rapid transit lanes, and high-performance reclaimed asphalt pavement [12–14].

Construction of GSP is normally a two-phase operation. First, the OMA mattress is prepared and paved by equipment lighter than or equal to that used for AC. After the asphalt is cooled, HCM can be spread on the surface. Due to the good connectivity of voids in OMA, HCM penetrates the whole layer to obtain a very low residential void rate for GSP by rubber scrapers and light vibratory rollers. The GSP eventually forms the required strength for traffic through curing for a few days. The beneficial properties of this material rely on careful construction process control, and recent related challenges are problems for mechanical mechanism investigation and distress prevention, such as cracking.

In Europe, GSP was originally invented as a patent of heat-resistant pavement, namely "*Salviacim*", by French company Jean Lefebvre at Cognac airport in 1954 [3,15]. In 1987, two companies, Densit and Phoenix, jointly developed and promoted the second generation of GSP, called "*Densiphalt*", of which the residual void rate was reduced after being grouted to improve in-field performance [15–17]. In 1989, Roffe, the researcher belonging to the Jean Lefebvre Company in France, published a monograph on the "*Salviacim*", which was still considered a special construction method by the public [3,18].

The U.S. Army issued a series of evaluation reports of GSP and practiced this technology on airport aprons and tank runways from 1976 [19–21]. The Departments of Transportation of states and the U.S. were also concerned for the performance of GSP and conducted many projects to prove its performance [22–24]. Consequently, the specification of GSP by the U.S. Army was established and applied in many fields, and were updated several times [25–28].

Wang et al., conducted a trial road of GSP on Huishen pavement in the Guangdong province in China in 1986 [29]. In 1995, Xu et al., introduced laboratory tests for GSP and presented the bending tensile strength as an indicator to control its quality [30]. Subsequently, Zhang and Pan adapted the Main Aggregate Filling (MAF) method and the orthogonal tests to design GSP in 2000 [31,32]. Further, Hao et al., proposed that the optimal asphalt content of OMA should be determined by the Cantabro test and the Schellenberg Binder Drainage test, while the low-temperature cracking of GSP was found to be the main problem affecting its operational life [4–6].

Although GSP was just used originally as a construction technique for special pavements such as airport aprons, it is now considered as a rather new technology because it is found to be strongly different from rigid pavement and flexible pavement for their working behaviors. With the theory of perpetual pavement arousing and traffic load increasing in the recent years, GSP has been welcomed into a rapidly developing period and has become a potential high-performance and cost-effective pavement which has a higher complex shear modulus than AC. However, standards of this technology are seldom published in the world. There are still many unsolved problems in GSP design, such as low-temperature cracking, unreasonable testing parameters, and undeveloped laboratory methods, etc., which need intensive research [6,33]. The poor understanding of these problems limited further development and application of GSP.

The objectives of this study are to summarize the findings in GSP over the past two decades and analyze the relationships of various factors and the methods of laboratory tests on the performance to identify knowledge gaps, promote this technology and deliver recommendations for future research. 143 papers have been selected and reviewed in total on three principles: first, the study should illustrate a qualitative and quantitative evaluation for GSP; second, the research should come from a new perspective for factors and properties of GSP; third, new findings and different conclusions are included for more discussion. Finally, relevant recommendations and future studies are proposed.

2. Influential Factors on the Performance of GSP

2.1. Microstructure

Raw materials and structures of GSP are the essential elements affecting laboratory performance and also yielding technical problems. In a microscopic view, the relationships among asphalt, aggregate, and cement particles build the macro-strength of GSP. Indeed, the micro-analysis method is most commonly used in laboratory tests to demonstrate the intuitive influential factors of micromorphology on the behavior of GSP.

In recent years, resolution of micromorphology techniques, especially the Scanning Electron Microscope (SEM), has reached to the nanometer level, which can capture the microstructure views between different particles to analyze their relationships [34,35]. The GSP samples are observed in various instruments, resolutions, and observing ranges. The methods and results are shown in Table 1 and Figure 1.

Table 1. Methods of microscopic observation for GSP.

Authors	Observation Methods	Zoom Ratios	Observing Ranges
Ling et al. [36]	SEM	500 times	Interfaces
Huo et al. [37]	SEM (200SXVFT-IR), Fourier Infrared Spectrum Analyzer	5000 times	Interface transition zone, Chemical or asphalt aging reaction
Hou et al. [38]	SEM (Model JSM-6390LV, JEOL)	10,000 times	Microcracking, Fiber-like network

Figure 1. The GSP microscopic images observed by SEM: (**a**) Microstructure at 90-day age; (**b**) Interface and fibrous structures; (**c**) Wrapping phenomenon of asphalt film; (**d**) Generation of micro-cracking and bridging effect of the fiber-like network.

GSF has higher density and solidity than traditional AC, which contributes to great complex shear modulus and compressive strength [36]. The strength of GSP is reinforced with increase of age (at ages of 7 d, 28 d, and 90 d), and the inner hydrated cement forms a fibrous-like network that penetrates through asphalt films, as shown in Figure 1 [39]. This exclusive phenomenon, refered to as three-dimensional lattices or a mosaic spatial network, bridges asphalt and cement, which increases the thickness of interface transition zones and enhances the bonding force between asphalt and cement [36]. Meanwhile, the un-hydrated cement plays a role as mineral powder to strengthen the adhesion between asphalt and aggregate.

The propagation of micro-cracks in GSP can be also investigated by SEM (Figure 1). The observation results indicate cracking emerges from the bottoms of samples and expands along with the interface between asphalt and cement [38,40]. This reveals that the asphalt-cement interface is the weak interface existing through the whole depth of GSP structure, approved by the Heavy Vehicle Simulator test and the Dynamic Cone Penetrometer test [41,42]. Additionally, the reaction of internal among raw materials is mainly physical, because no chemical or asphalt aging processes are identified at the interfaces by the Fourier Infrared Spectrum Analyzer (FISA) (200SXVFT-IR) [37].

In general, these characteristics of micro-structure determine the macro properties of GSP, which has high rutting resistance and low cracking resistance. Moreover, the micro-analysis method can effectively and intuitively gain the relationships of each of these components and raw materials. On the other hand, this method is also limited, owing to the subjectivity of corresponding researchers and the complexity of GSP's structure. For example, some different fibrous structures can be observed from GSP samples, but which has a better coherent strength between materials is unknown because it is hard to get quantitative analysis from SEM images.

2.2. *Raw Materials and Admixtures*

GSP is a complex composite, combining multiple materials, such as asphalt binder, aggregate, cement, and admixtures. Generally, the effects of them can be separated into two independent steps: intermediate mixes (OMA and HCM) and final grouted or combined mix (GSP). In other words, the properties and the formation of raw materials and admixtures can directly determine the performance of intermediate mixes, which further affects the quality of the following GSP.

OAC is an asphalt mattress with high void rate similar to the open grade friction course (OGFC). According to design principles of OGFC, aggregate gradation can contribute to the volume of void and binder is able to provide inner cohesive strength in OAC. Likewise, strength and fluidity are two main properties of HCM, and are controlled by two factors: formation and types of raw materials.

2.2.1. Aggregate Gradation

Open-grade aggregate is designed primarily as a skeleton of OAC for GSP. It mainly maintains compressive pressures from traffic loads and forms inter-air voids of OAC. Therefore, a good gradation can not only provide a good compressive strength but also construct an even and interconnective space of void.

An optimal gradation mostly can be realized by two design methods: an experiential method and a volumetric method called the main aggregate filling (MAF) method. The latter assumes coarse aggregate is supporting the main structure of OAC and can achieve the required void volume by filling fine aggregate and binder into the compacted coarse aggregate. Due to back-calculation of the void volume, this method is concise to control the final void rate of OMA [43,44]. However, interconnectivity of void is overlooked in this method, which is another important factor for permeability of OAC equally to void volume [45]. Therefore, factors affecting particle sizes and consecutiveness of aggregate gradation are considered to establish the relationships between the given characters above, illustrated in Figure 2, Tables 2–4.

Table 2. Effects of gradations on permeability of OAC [46,47].

Gradation	Density [g/cm^3]	Relative Density	Void [%]	Permeability Rate
Gradation A	1.83	2.46	25.57	1.34(10^{-3} [m/s])
Gradation B	1.85	2.46	24.85	1.11(10^{-3} [m/s])
Gradation C	1.87	2.46	23.78	0.94(10^{-3} [m/s])
BSI-4%	N/A	N/A	33.05	362.9(m/day)
Densiphalt12-4.5%	N/A	N/A	32.08	247.68(m/day)

Table 3. Effects of gradations on physical properties of OAC [47].

Gradation	Cantabro Loss [%]	Indirect Tensile Strength Test [kPa]
BSI-4%	33.2	113.35
Densiphalt12-4.5%	34.91	136.8

Table 4. Effects of gradations on physical properties of GSP [48,49].

Gradation	Compressive Rebound Modulus Test		Low Temperature Bending Test			
	Compressive Strength [Mpa]	Rebound Modulus [Mpa]	Maximum Load [N]	Flexural Strength [Mpa]	Maximum Strain [με]	Stiffness Modules [Mpa]
Continuous	4.66	1913.48	802.30	6.26	1140	5483.77
Uniform	5.23	2812.60	866.48	6.73	1040	6486.53
Composite gradation	N/A	N/A	N/A	6.95	1838	3780.6
Upper limit	N/A	N/A	N/A	6.73	1943	3467
Lower limit	N/A	N/A	N/A	7.57	1523	4970.3

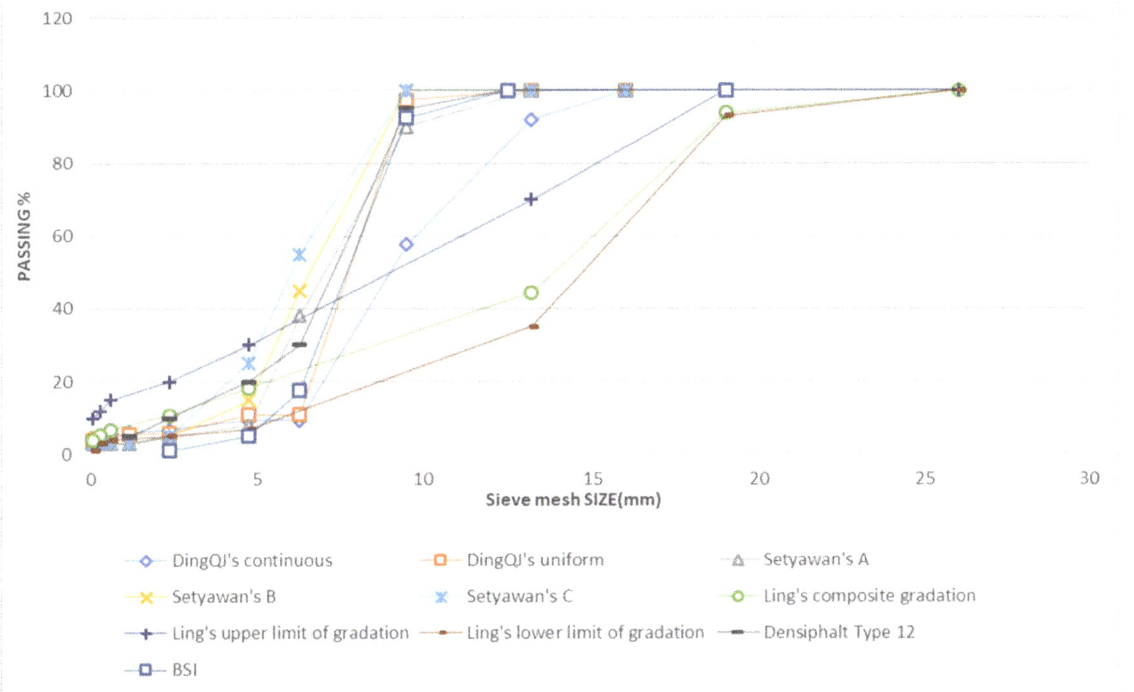

Figure 2. Some used gradations with different particle sizes and proportions.

The main coarse aggregate used in GSP normally falls in a range of 4.75 mm–13.2 mm particle size. Several combinations of different tiers in this range can form OMA with the requested void rate. Under the similar void rate, gradation with a large particle size will benefit the whole structure. Moreover, the water permeability of samples which have a large proportion in larger particles (14~10 mm) is significantly higher than that of average samples (Table 2) [46]. A lower residual void rate of GSP will be obtained after grouting [50,51]. Overall, a coarser aggregate gradation can be more conducive to forming a larger void space, which has better interconnection, to make the grouted cement mortar more easily able to penetrate the OMA mat [46,52]. Subsequently, flexural tensile strength and compressive strength of the following GSP are also be improved, which are positively linked with low-temperature performance and high-temperature performance, respectively [49]. On the contrary, some other properties may be slightly harmed with the increase of particle size. Cantabro loss rises to a certain value, which indicates that the cohesion between aggregate in GSP is weakened.

From a different perspective, Ding et al., proposed that uniform gradation could conduct smoother void spaces and less stress concentration than continuous gradation. He compared two types of gradations: uniform gradation (one tier: 10~5 mm) and continuous gradation (three tiers: 13~10 mm, 10~5 mm, and 5~3 mm) [53]. It was found that GSP with uniform gradation was superior to that with continuous gradation in the properties of Marshall stability, splitting strength, compressive resilience, and low-temperature bending capacity [53].

Saboo et al., evaluated seven types of gradations using the hierarchical ranking strategy considering parameters as void content, permeability, abrasion resistance, and tensile strength of OAC; BSI-4% and Densiphalt12-4.5% were selected as the optimal gradation which had a main tier of 4.75 mm~12.5 mm [47]. In general, gradation with the main particle sizes in the scope of 4.75 mm~12.5 mm is appropriate to improve the performance of GSP to form a better structure (Figure 2).

However, a conflict exists in previous hypotheses that mainly concerns coarse particles. For example, a given continuous gradation may also contain larger-tier aggregate. The reason may be that the influence of gravel morphology and fine aggregate is overlooked which needs further research [52]. In our opinion, the two viewpoints are all correct, though only the applying scope of these principles is different. The definitions for describing gradation types are difficult to quantify, due to the existence of few samples with untested properties, such as residual void rate and water permeability. Large particle size and uniform gradation can both contribute to the formation of evener and larger void space.

2.2.2. Asphalt Binder

The types of asphalt binder in OMA include base asphalt, rubber asphalt, SBS-modified asphalt, and high-viscosity modified asphalt. Due to various characteristics at different temperatures, the optimal type and content of asphalt binder as two factors varies in OMA design, as shown in Table 5. Orthogonal experiment uncovers the effects of binder on the relative OMA or GSP sample, as illustrated in Tables 5 and 6.

Table 5. Optimal binder and high-temperature performance of OMA.

Author	Binder Types	Void Rate [%]	Optimum Asphalt Content [%]	Unconfined Compressive Strength [kPa]	Marshall Stability [kN]	Cantabro Loss [%]
Setyawan et al. [46]	50-pen straight run asphalt	27.44	4	1691.59	N/A	18.7
	7%SBS Asphalt	27.20	4	906.56	N/A	10.1
	100-pen straight run asphalt	27.74	4	760.64	N/A	34.3
Ling et al. [48]	Rubber-asphalt	22.1	3.6	N/A	6.80	23.8
Zhao et al. [54]	High-viscosity asphalt	26.0	3	N/A	3.6	19.5
	Rubber-asphalt	26.2	3	N/A	2.7	38.4
	SBS asphalt	25.9	3	N/A	3.0	32.5
	Ordinary asphalt	25.8	3	N/A	2.4	48
Wang et al. [55]	4.4%SBS asphalt	25	4.4	N/A	4.6	N/A
Luo et al. [56]	SBS asphalt	22.47	3.4	N/A	4.15	14.7

Table 6. Optimal binder and low temperature performance of OMA.

Author	Asphalt Types	Test	Test Condition	Strength [Mpa]	Modulus [Mpa]
Setyawan [46]	50-pen straight run asphalt	Indirect Tensile Strength	20 °C	259.65	953.71
	7%SBS-Modified Asphalt			155.82	331.44
	100-pen straight run asphalt			125.26	548.68

Asphalt binder with high viscosity and low penetration, such as 50-pen straight-run asphalt, SBS asphalt, rubber-asphalt, and high-viscosity asphalt, can increase strength and reduce Cantabro loss for OMA which will result in good high-temperature performance of GSP. However, binder types have little impact on void space structure in OMA, as well as strength and moisture resistance of GSP [46,54].

OMA samples using SBS asphalt as binder can obtain favorable water permeability. Setyawan et al., explained that modified asphalt protected OMA from drainage that caused void-blocking [46]. Therefore, the performance of following GSP samples was also enhanced, especially in low-temperature cracking resistance [46,54,55].

Rubber asphalt also promotes high-temperature performance of GSP. It is considered as a potential candidate due to its excellent cost-effectiveness and environment-friendliness [48].,High-viscosity asphalt also shows advantages of flexural tensile strain and stiffness modulus at a low temperature, which is associated with better anti-cracking performance [54].

Optimal binder content of OMA can be determined by three main parameters in orthogonal testing: compressive strength, tensile strength, and Cantabro loss. The value of it mostly falls in a range of 3%~5%, due to different test methods. However, it can be argued that a high content rate of binder in OMA can increase toughness of OMA, leading to good anti-cracking performance of GSP [57], because the Marshall stability of OMA continues to rise with the increase of the binder content rate, even up to 9% [58].

Besides these two factors: binder type and content in OMA, bonding force for the asphalt-cement interface is overlooked in steps of influential factors analysis, which can contribute to low-temperature performance of GSP. To achieve a criterion for AC (the flexural tensile strain in small beam bending test should exceed 2000 $\mu\varepsilon$ at -10 °C [59]), the factors for anti-cracking characteristics of GSP should be considered of both OMA and HCM to bridge two materials in whole structure in design.

2.2.3. High-Fluidity Cement Mortar

High-fluidity cement mortar (HCM) is a specialized grouting material which can penetrate OMA due to its high fluidity to construct a fibrous-like network structure inside the asphalt mattress. Strength and shrinkage rate are the two main factors of HCM controlling the performance of GSP. Therefore, to investigate their effects, various formulas of raw materials and different additives are compared, as shown in Table 7.

Table 7. Formulas of raw materials and performance of HCM.

Author	Water-binder Ratio	Sand-cement Ratio	Water Reducing Agent	Expansion Agent	Others	Fluidity (0.5 h) [s]	Compressive Strength (7 d) [Mpa]	Flexural Strength (7 d) [Mpa]	Drying Shrinkage Rate (60 d) [0.001]
Hu et al. [60,61]	0.45	0.25	0.3% polycarboxylic acid	8% UEA	N/A	12.98	19.4	4.92	3.188
Pei et al. [62,63]	0.56	N/A	1.0% TH-928 polycarboxylic acid	10% UEA	0.008% Saponin	≈10.5	≈25	≈6	≈1.34
	0.56~0.58	N/A	N/A	N/A	10% Fly ash, 10% Mineral powder	11~14	15~20	2.2~5.7	1.4~2.7
	0.61~0.63	15	N/A	N/A	10% Fly ash	12~17	9~15	2.7~5.7	1.0~1.8
Koting et al. [43,64]	0.32	N/A	2% super plasticizers	N/A	5% of Silica powder	14.2	57.5	5.8	N/A
Saboo et al. [45]	0.57~0.59	0.3	2% naphthalene	N/A	N/A	10~14	20~24.5	N/A	N/A
	0.48~0.50	0.3	1% polycarboxylate	N/A	N/A	10~14	21~25	N/A	N/A
Ling et al. [49]	0.65	0.14	N/A	N/A	6% Fly ash, 10% Mineral powder	11.4	17.2	4.4	N/A
	0.65	0.2	N/A	N/A	10% Mineral powder, 10% Polymer	11.1	12.6	5.9	N/A

The water-cement ratio is the critical factor affecting both fluidity and strength of cement mortar. The fluidity of HCM is positively associated with the water-cement ratio and negatively related to the sand-cement ratio [65]. However, an excess water-cement ratio or a short sand-cement ratio will lead to high dry shrinkage [60,61]. In addition, with the increase of water-cement ratio, the bleeding rate of HCM shows an upward trend, which may cause void-blocking and slurry leakage [66,67]. Therefore, Cheng et al., recommends that the water-cement ratio should be less than or equal to 0.55 for HCM, while Saboo et al., advises a ratio scope of 0.4~0.6 [45,66]. Additionally, 10~14 s is a recommended range for HCM fluidity [45,49,67].

The additive type is another factor in determining the performance of HCM. Superplasticizer and fly ash are commonly used to increase fluidity of HCM and minimize residual void after grouting. Some studies indicates that silica fume, mineral powder, and ultrafine sand also have good effects on fluidity [43,64,68,69]. Expansion agent UEA can reduce the shrinkage of HCM and furtherly enhance the anti-cracking ability of following GSP [60–63,70]. However, the strength of cement mortar is inevitably underestimated by fluidity improving in the methods above. Therefore, to achieve the balance design between fluidity and strength, a range of strength value is proposed as 10~30 Mpa, and furtherly narrowed to 20~25 Mpa for HCM [45,49]. Although other factors such as the shrinkage and flexural strength of HCM have important effects on the low-temperature performance of GSP, their values are not considered as parameters by recent studies comprehensively.

Cai et al., used ABAQUS software to simulate the shrinkage and expansion of cement mortar in GSP to estimate the inner stresses [71]. Index values of the deformation were calculated and determined to protect GSP from cracking that is caused by stress concentration of shrinkage, as shown in Table 8.

Table 8. Recommended deformation indexes of cement mortar in GSP [71].

Void Ratio of OMA [%]	Maximum Shrinkage Deformation [%]	Maximum Expansion Deformation [%]
20	0.13	0.39
25	0.14	0.38
30	0.15	0.34

2.3. Admixtures

Admixtures for GSP refer to specific materials adding to OMA or HCM to improve workability or strength of following GSP. Resin is the admixture early used in GSP which can be traced back to 1976 in the U.S. Amy [72], and polymers have been used to enhance the anti-cracking performance of GSP from the 1980s on the Huishen highway, China [73]. Over the past 40 years, more and more types of admixtures have been developed into a big family for GSP.

According to different ingredients, GSP admixtures can be roughly classified into five categories: polymers, fibers, interface modifiers, emulsified asphalt, and other new functional admixtures. New functional admixtures have been recently explored for extension in road functions such as emission-reducing, water conservation, and weather cooling. In addition, two blending methods are usually used for admixtures of GSP: firstly, admixture is blended into cement mortar to reduce rigidity and dry shrinkage of HCM, which furtherly improve cracking resistance for following GSP; secondly, it is mixed within asphalt to enhance the strength of OMA, which achieves a high rutting resistance for the whole structure. Admixtures and their effects on GSP are illustrated in Table 9.

Table 9. Admixtures and their effects on GSP.

Categories	Citations	Ingredients	Blending Method	Dosage [1] [%]	Rutting Test [times /mm]	Low-Temperature Strength		
						Methods	Tensile Strength −10 °C [Mpa]	Tensile Strain −10 °C [με]
Polymer	Ling et al. [74]	DL latex	Into cement	P/C = 1.0	20,138	Splitting test	1.548	4590
		YH resin		P/C = 5.0	21,457		1.955	2740
		BD emulsion		P/C = 4.7	19,079		1.222	5170
	Luo et al. [56]	Latex	Into cement	P/C = 3.0	≈13,500	Small beam bending test	≈10.5	≈2400
	Ling et al. [49]	Styrene-butadiene emulsion	Into cement	P/C = 10	15,750	Small beam bending test	6.95	1838
	Huang et al. [75]	Waste rubber powder	Into cement	P/C < 20	Increase of 47.5%	Small beam bending test	N/A	1132
	Ling et al. [48]	Rubber-asphalt	Into asphalt	P/A = 21	15,000	Small beam bending test	2.18	5140
	Xu et al. [76]	Carboxylated styrene-butadiene latex	Into cement	P/C = 10	14,318.18	Small beam bending test	6.47	1807.23
	Wang et al. [77]	Carboxyl Latex	Into cement	P/C = 8	21,724	Small beam bending test	7.15	1527

Table 9. *Cont.*

Categories	Citations	Ingredients	Blending Method	Dosage [1] [%]	Rutting Test [times /mm]	Low-Temperature Strength		
						Methods	Tensile Strength −10 °C [Mpa]	Tensile Strain −10 °C [με]
Fibre	Cheng et al. [78]	PVA fiber solution	Into cement	F/C = 5	N/A	Small beam bending test	9.16	3656.82
Interface modifier	Yang et al. [79]	Silane Coupling agent	Into cement	I/C = 0.4	N/A	Small beam bending test	6.97	≈1025
	Xu et al. [76]	Silane coupling agent	Into cement	I/C = 0.5	16,578.95	Small beam bending test	7.17	1826.72
Emulsified asphalt	Zhang et al. [80]	Cationic emulsified asphalt	Into cement	A/C = 31.3	N/A	Small beam bending test	11.05	3159.26
	Huang et al. [81]	Emulsified asphalt	Into cement	A/C = 30	N/A	Small beam bending test	6.56	1446
	Xu et al. [76]	Cationic emulsified asphalt	Into cement	A/C = 5	42,000	Small beam bending test	6.98	2133.29
Combination	Sun et al. [82]	Waterborne epoxy-emulsified asphalt	Into cement	E/A = 60; C/A = 55	N/A	Small beam bending test	6.18	2662
	Gong et al. [83]	Modified Agent-100	Into asphalt	M/A = 14	N/A	Semi-circular bending test	12.91	N/A
		Modified Agent-100 incorporated fiber	Into asphalt	M/A = 14; F/A = 0.2			10.18	

[1] A = Asphalt; C = Cement; P = Polymer; E = Epoxy; F = Fibre; I = Interface modifier; M = Modifier.

2.3.1. Polymers

Polymers are the most traditional admixtures used in GSP, including latex, resin, SBS, styrene-butadiene rubber (SBR), and waste rubber [74,84]. Latex and resin are efficient to smooth the high rigidity of HCM and reinforce the tensile strength of the whole structure because of their hydrophilicity, which results in bonding at interfaces between materials [56,76]. However, they have little or no influence on the high-temperature performance of GSP. That is because adding polymers into cement mortar will impair the compressive strength of HCM.

In contrast, GSP with rubber as an admixture has little promotion to the low-temperature cracking resistance, due to the weak interface between rubber and cement [48,75], although the fatigue life of rubber-modified GSP is prolonged with the increase of rubber content [85]. In this category, waste rubber is thought to be a cost-effective and environmental-friendly admixture as a substitute to replace raw materials with a high blending ratio up to 20% [75]. Its weakness of insufficient strength can be enhanced by adding an interface modifier, which will be introduced later in this paper [86].

2.3.2. Fibers

Fiber plays a role in reinforcing tensile strength in GSP [87]. It can be blended into GSF in two ways: adding into OMA alone or into both OMA and HCM. However, the effect of fiber mainly works in OMA structure, as there is no obvious advantage showed by the latter method [78].

Nevertheless, it is possible that fiber may harm GSP. GSP samples with 0.3% loose cellulose fiber show a high abrasion loss, which is difficult to meet the requirement as a surface course [46]. Moreover, the tensile strength of fiber-modified GSP may decrease sometimes, even though the fatigue life was increased [83]. The reason is unknown.

2.3.3. Interface Modifiers

Interface modifier commonly refers to silane coupling agent (SCA) or other additives, which can enhance the cohesive strength of interfaces between asphalt and cement to protect weak interfaces from cracking [79]. With the increase of the dosage of SCA, all properties of GSP are improved, especially shear resistance [88,89].

However, SCA needs to be combined with other additives such as rubber or latex to loosen the inner stress in HCM, due to its high stiffness after reaction. It is found that the workability and fluidity of HCM modified by SCA are improved, while the dry shrinkage rate and bleeding rate are greatly alleviated. Therefore, the following GSP samples are strengthened in the cracking and rutting resistance [90].

2.3.4. Emulsified Asphalt

Anionic emulsified asphalt and cationic emulsified asphalt are two types of emulsified asphalt which are always blended with cement to produce cement asphalt (CA) as grouting material for GSP. Cationic CA efficiently increases the flexibility of grouting material, compared to anionic CA and average cement mortar [80]. In addition, CA (especially Cationic CA) has better cohesion with asphalt than average cement mortar, which can also act as a kind of interface modifier [76].

Despite advancement in anti-cracking resistance, CA-modified GSP will be inevitably sabotaged in both the compressive strength of grouting material and the rutting resistance of GSP. Accordingly, Xu et al., presents a method using SBS in OMA to offset the decrease in strength of CA-modified GSP [76].

2.3.5. Complex Admixtures

The purpose of complex admixtures for GSP is to combine advantages of different admixtures. However, some combinations are not always as expected, because of underlying contradictive influences of combined admixtures on GSP. Therefore, in its composition design, more factors and more procedures should be considered in a comprehensive way to achieve desired outcomes, comparing to the individual components [83].

2.3.6. New Functional Admixtures

Some novel admixtures are applied in GSP to extend functions of a road, such as environment conservation and road life extension. For example, water-retaining material is mixed into grouting material to produce water-retaining GSP. This pavement can cool down 8 °C–10 °C for eight hours for the surrounding area by evaporating the water absorbed from rain, while it maintains a high anti-rutting performance [91,92]. Phase-change material (PCM) is also added into GSP to adjust the ambient temperature. The PCM-modified GSP can be protected from the damage of temperature fluctuation by the thermal storage capacity of PCM. It is proved that GSP mixed with 5% PCM can relieve the temperature of pavement by 11.5 °C for 4 h [93,94].

3. Evaluation Methods of the Performance of GSP

3.1. Common Laboratory Testing Methods

Common traditional laboratory tests are used for the evaluation of GSP, including: the Marshall test and the rutting test to assess high-temperature performance, splitting test, semi-circular bending test, small beam bending test at −10 °C to measure low-temperature performance, and Immersion Marshall Residual Stability test to estimate moisture resistance, as shown in Table 10.

Table 10. Traditional laboratory tests and performance of GSP.

Authors	Void Rate of OMA	Marshall Test		Rutting Test	Immersion Marshall Test	Small Beam Bending Test (−10 °C)	Splitting Test (−10 °C, 1 mm/min)
		Stability [kN]	Flow Value [0.1 mm]	Dynamic Stability [times/mm]	Residual Stability [%]	Tensile Strength [Mpa]	Tensile Strength [Mpa]
Xu et al. [30,95]	25%	>25	10~25	>12,000	N/A	4.82	25%
Hao et al. [6]	25%	17.12	19.8	10,242	85.2	6.408	25%
Wu et al. [96,97]	25%	33.1	51.4	22,096	95.9	N/A	25%
Dong [98]	24.9%	17.7	32.4	15,750	78.4	7.57	24.9%
Zhang [99]	25%	7.38	27.1	21,000	100.4	6.71	25%
Pang et al. [100,101]	22%	N/A	N/A	>6000	N/A	N/A	22%
Hou et al. [38]	29.5%	N/A	N/A	15,750	110%	5.2	29.5%
Li et al. [102]	25%	>20	N/A	>10,000	>75	N/A	25%
Hu et al. [103]	30%	N/A	N/A	21,725	N/A	N/A	30%
Huo et al. [104]	25%	29.52	N/A	22,096	95.9	N/A	25%

3.1.1. High-Temperature Performance

The Marshall test is most commonly employed to evaluate high-temperature performance of GSP and then results can be easily compared by this index from different projects. GSP has greater Marshall stability and lower flow value than traditional AC, due to its enhanced strength and fewer residual voids by grouting cement mortar. Additionally, the compressive strength of GSP continues to improve with growth of the OMA void [6]. GSP containing reclaimed asphalt pavement (RAP) is also evaluated by the Marshall test, which shows a promising stability value of 53.9 kN [105].

The Rutting test is another popular method, and shows the great anti-rutting resistance of GSP. Average dynamic stability of GSP can be up to 15,000 times/mm, compared to the value of AC-16 at only 2000 times/mm [49,106]. GSP samples can keep their excellent function even at a high test-temperature of 70 °C [103]. Additionally, the Static Indentation Creep test is also adopted to illustrate these characteristics of GSP [107].

3.1.2. Low-Temperature Performance

Evaluation methods for the low-temperature performance of GSP include the small beam bending (SBB) test, splitting test, and semi-circular bending (SCB) test. Additionally, the creep test and indirect tensile strength (ITS) test are also employed at −10 °C to inspect the anti-cracking resistance of GSP [6,108].

Tensile strength of GSP can approximated to AC in the SBB test, while tensile strain is found lower than AC. This is because the excess stiffness of GSP limits the deformation of whole structure. However, admixtures can be efficient to accomplish a significant enhancement in tensile strength of GSP [38]. That is, the modified GSP shows improved low -temperature performance in the SCB test, creep test, and ITS test at −10 °C [108].

However, the laboratory behavior of GSP in splitting tests is far different from that in SBB tests, and both tensile strength and strain are significantly lower than AC's [70]. Modifiers also had a mild effect on GSP in this test. Therefore, these methods need further investigation and selection to uncover properties of GSP concisely. Further, Ding et al., investigated recycled Semi-flexible Pavement material by splitting test at 20 °C, which indicated its anti-cracking resistance was positively linked with viscoelasticity of the binder [109].

3.1.3. Moisture Resistance

GSP is considered as a promising anti-moisture pavement as it prevents water permeation by its high density and low residual void. This property can be assessed by freeze-thaw test and Immersion Marshall test. The freeze-thaw test is divided into the freeze-thaw splitting test (ST) and freeze-thaw indirect tensile strength (ITS) test. Thus, the results of retained Marshall stability (RMS) and tensile strength ratios (TSR) are summarized and illustrated in Figure 3 [6,38,49,70,77,103,104,110,111].

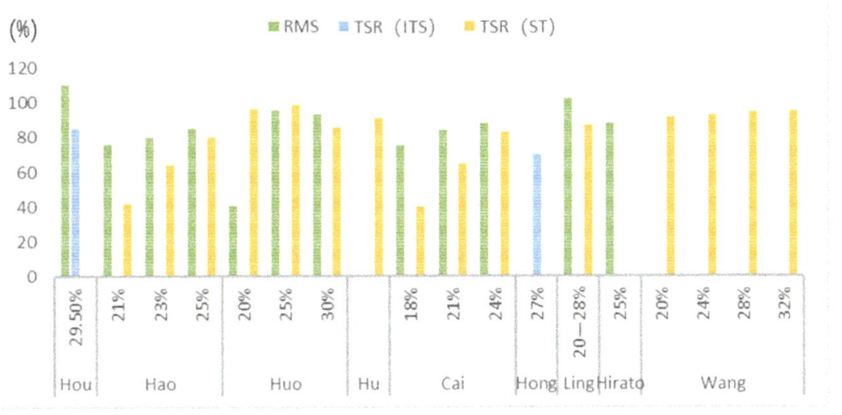

Figure 3. Moisture resistance of GSP.

Moisture resistance of GSP is significantly enhanced with high void rate OMA. The reason is that a high void rate can lead to void interconnection and saturated grouting [112]. From the figure, results of Immersion Marshall testing are commonly higher and easier to reach design requirement than those in freeze-thaw tests, because of the more rigorous experimental environment of the latter. There is another interesting phenomenon where some RMS results exceed 100%. This phenomenon may be triggered by further hydration of cement in GSP. Therefore, from this aspect, the freeze-thaw test may be more accurate than the RMS test in the evaluation of moisture resistance for GSP.

3.1.4. Oil Corrosion Resistance

GSP is made to possess an excellent oil corrosion resistance by its good density. Hao et al., adopted oil corrosion resistance test to assess GSP samples. Marshall samples were soaked in #90 gasoline for 24 h to test its retained Marshall stability (RMS), which was 88.5% twice bigger than AC [6]. Hirato et al., also immersed Marshall samples in kerosene for 48 h to attain RMS values, which exceeded 80% [111].

3.1.5. Impact Resistance

Impact resistance of GSP was just tested for some airport lanes by Split Hopkinson Pressure Bar equipment, which was raised by Kolsky to measure stress pulse propagation [113]. The peak stress and failure modes of samples were collected and analyzed by Dong-Hua Test Real-Time Data Measurement Analysis Software System under various air pressures and different OMA void rates. It was found that GSP with 25% void rate had the best impact resistance in the low-pressure areas (0.25 Mpa and 0.30 Mpa), and the 27%-GSP sample could resist the peak stress value of 19.67 Mpa in the high-pressure areas (0.35 Mpa and 0.4 Mpa) [113].

3.1.6. Anti-Weather-Exposure Ability

Anti-weather-exposure ability of GSP was tested by conserving samples in an exposing environment for 7 days, 90 days, 180 days, and 240 days. Its strength and fatigue life were shown no decline in Marshall stability test and cyclic wheel load test [114].

In addition, thermal cracking equipment was adopted to evaluate long-necked specimens at −5 °C. It was found the thermal resistance of GSP was positively related to the content and viscosity of binders [115].

3.2. Fatigue Life Performance

The fatigue life of GSP has raised more concern in recent years, because of its different fatigue life behavior from asphalt and concrete [116]. The results of fatigue life of GSP are summarized in characteristics and correction factors, which are attained by the Indirect Tensile Fatigue test (ITFT), Fatigue Bending test, Cyclic Wheel Load test, Proportion-scale Accelerated Road test, and Full-scale Accelerated Road test. Test methods and fatigue equations of these studies are also shown in Table 11.

Table 11. Fatigue life behavior of GSP.

Author	Test	Loading Mode	Void Rate of OMA	Fatigue Equation	Failure Standard	Fatigue Correction Factor
Oliveira et al. [117,118]	Two-point bending test	Stress mode	25%	$N_f = 1.5659 \times 10^3 \left(\frac{1}{\varepsilon}\right)^{0.23}$ R2 = 0.82	10% residual stiffness	45
		Strain mode	25%	$N_f = 2.2034 \times 10^3 \left(\frac{1}{\varepsilon}\right)^{0.2474}$ R2 = 0.87	50% initial stiffness	data
	ITFT	Strain mode	25%	$N_f = 3.4809 \times 10^2 \left(\frac{1}{\varepsilon}\right)^{0.1323}$ R2 = 0.9134	9 mm vertical displacement or cracking	45
Ling et al. [119]	Four-point bending test	Stress mode	23%	$N_f = 1.0624 \times 10^5 \left(\frac{1}{\sigma}\right)^{4.6636}$ R2 = 0.9526	Fracture	128.5
Huang et al. [120]	Three-point bending test	Stress mode	30%	$N_f = 2.4177 \times 10^5 \left(\frac{1}{\sigma}\right)^{0.6166}$ R2 = 0.9134	Fracture	N/A
Wang et al. [55]	Four-point bending test	Strain mode	25%	$N_f = 8.3 \times 10^{20} \left(\frac{1}{\varepsilon}\right)^{7.0972}$ R2 = 0.9915	Fracture	N/A
Ding et al. [109]	Dynamic splitting tensile test on mechanics testing system (MTS)	Stress mode	N/A	$N_f = 5.5 \times 10^7 \left(\frac{1}{\sigma}\right)^{1.2903}$	Fracture	1703.3
Hou et al. [38]	Four-point bending test	Stress mode	28.9%	$N_f = 1.7056 \times 10^6 \left(\frac{1}{\sigma}\right)^{4.8951}$ R2 = 0.9956	Fracture	N/A
Wang et al. [77]	ITFT	Stress mode	28%	$N_f = 6.1589 \times 10^8 \left(\frac{1}{\sigma}\right)^{2.0113}$ R2 = 0.9684	Fracture	N/A
Gong et al. [83]	Semi-circular bending	Stress mode	22%	$N_f = 3.0199 \times 10^4 \left(\frac{1}{\sigma}\right)^{4.44}$ R2 = 0.92	Fracture	N/A
Yang et al. [121,122]	Cyclic wheel loading test	Times	26%	$N_e = 4.8551 \times \left(\frac{1}{\varepsilon}\right)^{0.3483}$ R2 = 0.9311	20 mm cracking	N/A

3.2.1. Characteristics of Fatigue Life of GSP

Raw materials have effects on the fatigue characteristics of GSP. To assess the fatigue life of GSP, the indirect tensile fatigue test (ITFT) and two-point bending test are commonly used under stress or strain control mode. Fatigue curves illustrate that modified asphalt (polymer asphalt or rubber asphalt) and low-shrinkage cement mortar dominate the fatigue life behavior, while binder content has a slight influence [123]. Moreover, cement may be the most effective material for fatigue life in GSP, compared with asphalt and aggregate [120]. The fatigue life also increases with the growth of the void rate in OMA, which means more cement mortar will be grouted in OMA [77]. Additionally, from the aspect of test conditions, the fatigue life decreases with the rising test temperature and cannot be changed with different loading frequencies [119,124].

The fatigue life of GSP has a linear relationship with stiffness modulus, of which the equation is expressed as below [125]:

$$\log N_f = \log \alpha + \beta \cdot \log\sigma + \gamma \cdot \log E_{100} \tag{1}$$

α, β, γ are the regression parameters. E_{100} is the stiffness parameter on the 100th cycle of the fatigue test according to the European standard, and R^2 could reach 0.79.

In general, GSP shows a better fatigue life than traditional AC, especially in a low-stress level [120,125]. However, the fatigue life of GSP decreases more rapidly than that of AC and more slowly than that of semi-rigid material at the high-stress level. Due to this reason, GSP may be inferior to AC in anti-fatigue performance with a higher stress [38]. The cumulative fatigue life of GSP will be also shorter than that of AC based on MINER's Linear Fatigue Damage Accumulation theory, according to the traditional fatigue failure standard (50% of initial stiffness) [126].

However, far different from the pattern of fatigue curves of AC, GSP does not have a sudden drop after loss of the 50% initial stiffness, and thus can still maintain its work capacity instead of failing with loading time [117]. Consequently, 10% residual stiffness is presented to be a new failure standard for GSP, and then the fatigue life of GSP is found to be greatly underestimated [127]. Therefore, failure standards are important and can lead to different results for the evaluation of fatigue life. Additionally, the strain control mode is recommended for fatigue life test of GSP due to its thin course structure applied in field, which shows a better fatigue life than the stress-control mode [123,128].

In 2015, Yang et al., adopted the Cyclic Wheel Loading test on GSP with different OMA void ratios (20%, 23%, 26%, and 30%) [121,122]. The failure standard was defined as a 20 mm crack on the surface. Finally, the repeated wheel loading times N_e was derived, which was close to the actual fatigue life in the field [121].

3.2.2. Fatigue Correction Factor

The correction factor or shift factor of fatigue life is an experiential effective coefficient expressing the relationship between the fatigue life in test and that in field. According to complex test conditions including material types, test methods, loading mode, and testing temperatures, the shift factor is difficult to be unified and determined through different studies.

Ling et al., calculated the fatigue correction coefficient of GSP as follows [119]:

$$N_f = \frac{1}{5} \times \frac{1}{3} \times 0.50 \times \frac{60}{365} \cdot N_e = 5.48 \times 10^{-3} N_e \tag{2}$$

Specifically, the intermittent time coefficient was selected as 5; the stress-reduction multiple of the fatigue life was 3 times; the transverse distribution coefficient was adopted as 0.5; and the unfavorable season days was 60 [119].

Similarly, in another test, Ding et al., considered different factors from Ling: the intermittent time coefficient was 7; the crack propagation coefficient was 20; the Days number of the unfavorable season was 60; and the transverse distribution coefficient was 0.5. The result was expressed as follows [109]:

$$N_e = \frac{1}{0.5} \times 7 \times 20 \times \frac{365}{60} \cdot N_f = 1.703 \times 10^3 N_f \tag{3}$$

Olivia et al., indicated the fatigue life correction factor of GSP should be calculated based on ITFT data, considering the intermittent time and loading mode, as shown in Table 12 [117,118].

Table 12. Fatigue Correction Coefficient of GSP [117,118].

Loading Mode	Fatigue Life	Correction Factor
Continuous loading	17,739	1.0
1000 cycles Loading and Equivalent Intermittent Time	40,980	2.3
2 cycles loading + 1 cycle intermission	50,823	2.9
1 cycle loading + 1 cycle intermission	140,050	7.9
1 cycle loading + 2 cycles intermission	734,395	41.4

Due to an additional factor of 1.1 for the lateral load distribution, 45 was finally determined as the correction factor [117,118]. However, this factor was still thought to be a conservative value, because the loading intermission in field was much longer than that in experiment.

3.3. Computationand Simulation Method

3.3.1. Finite Element Method under Various Contact Models

The Finite Element Method (FEM) is widely utilized in study of material engineering. FEM software such as ANSYS, ABAQUS, and BISAR are practiced in computation and simulation for GSP. Physical properties of GSP are calculated under various contact models according to different computational hypothesis and parameters from experience and tests, as shown in Tables 13 and 14.

Table 13. Computational parameters of GSP in FEM.

Author	Thickness H [cm]	Elastic Modulus E [Mpa]	Poisson's Ratio [μ]
Yu et al. [129]	15	1200	0.25
Huang et al. [130]	10	2000	0.25
Chen et al. [131]	15	2000	0.25
Xu et al. [132]	20	2500	0.25
Wang et al. [133]	5	2000	0.25
Yang et al. [122]	30	3600	0.15
Setyawan et al. [40]	6	7600	0.24

Table 14. Results of FEM Calculation.

Author	Maximum Shear Stress (Strain)	Maximum Tensile Stress (Strain)	Maximum Deflection Value [mm]
Huang et al. [130]	0.235 Mpa	0.155 Mpa	0.572
Chen et al. [131]	N/A	0.1224 Mpa	0.480
Xu et al. [132]	444.5με	266.1με	3.502
Wang et al. [133]	621.1 Kpa	≈800 Kpa	0.213
Yang et al. [122]	≈1.28 Mpa	≈1.25 Mpa	N/A
Setyawan et al. [40]	0.576 Mpa	3.80 Mpa	N/A

FEM can calculate load status of GSP structures using the Elastic Layered System model in three dimensions (3D) and two dimensions (2D). This model easily describes the relationship between the traffic load and the whole structure of GSP. It is found that the bonding of interlayers can reduce the shear stress, bottom tensile stress, and rebound deflection in GSP [133,134]. Thus, the structure shows a better anti-shear capability than AC, which results in a high rutting resistance [120–131]. This method also fits well and gains consistent conclusion in many circumstances, such as airports, which have different structures and requirements, [122,132].

The toughness of GSP is calculated using the Visco-Elastoplastic model as the contact model by the ABAQUS software. It is found that the toughness is affected by the OMA void: when it was 25%, GSP has the highest toughness of 11.358 kJ/m^2, showing the best anti-cracking capacity [135].

A 2D sectional image model is established by a CCD digital camera and CAD software to analyze the position of cracking for GSP. OMA is assumed to be visco-elastoplastic, and cement mortar is defined as elastic in the computational process. Then, expansion force and contraction force in GSP are calculated by ABAQUS. The results show that expansion of cement mortar has little or no effect on GSP; but contraction can dramatically increase stress on asphalt-cement interfaces to lead to cracking [71]. In short, low-temperature shrinkage of cement mortar is the main reason causing cracking and asphalt-cement interfaces are the weak interfaces.

3.3.2. Compressive Strength Prediction Model

Compressive strength of GSP is assessed by Cube compressive strength test. The values have a linear relationship with that of grouted hydrated cement mortar [136]. The expression is shown as below:

$$y = 1.3619x^{0.4736},\ R^2 = 0.8525 \tag{4}$$

x is the compressive strength of cement mortar; y is the compressive strength of GSP.

And the relationship can be illustrated in Figure 4:

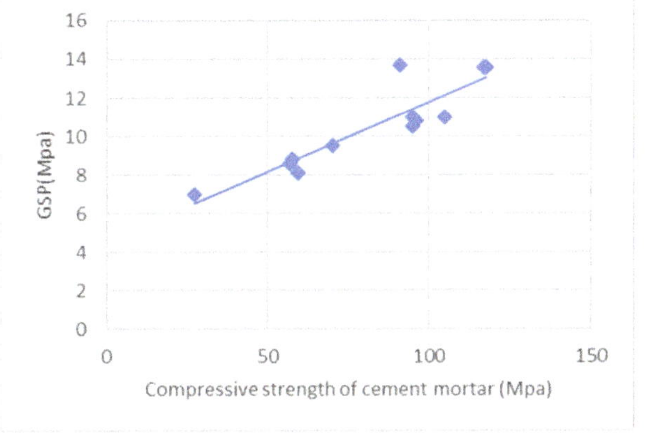

Figure 4. Relationship of compressive strength between cement mortar and GSP [136].

In addition, Wu et al., indicated that void rates of OMA and compressive strength of GSP had an empirical regression relationship as follows [137,138]:

$$y = 8.3x^2 - 319.7x + 5181.4 \tag{5}$$

x was the void rate of OMA (20%~30%); y was the resilient modulus of GSP.

3.3.3. 2S2P1D Model

Cai et al., evaluated the viscoelastic behavior of GSP according to the 2S2P1D model (composed of 2 spring units, 2 parabolic units, and 1 clay pot unit). It was found that the 2S2P1D model had a good correlation with the test results. The equation was expressed as follows [139]:

$$E^*(w) = E_e + \frac{E_g - E_e}{1 + \mu(iw\tau_0)^{-k} + (iw\tau_0)^{-h} + (i\omega\beta\tau_0)^{-1}} \quad (6)$$

$E^*(w)$ was the complex modulus; E_e was the equilibrium modulus; E_g was the glassy modulus; w was the angular frequency; μ was a calibration constant; i was the complex number; τ_0 was the characteristic relaxation time; k and h were constant values defined as $0 < k < h < 1$; and β was a constant that depended on the viscosity of the dashpot.

The reinforcement effect of GSP is positively linked with the OMA void ratio, and the dynamic modulus is associated with the gradation [139]. In other words, cement can play a role to enhance the strength of GSP under high-temperature and low-frequency loads.

3.3.4. Weak Interlayer Model

A cracking model is specially introduced to calculate cohesion strength of interfaces to determine the position of cracking and explain the internal factors through a full-scale Heavy Vehicle Simulator test [42,140]. Then, the strain energy of distortion (SED) is defined as the response parameter to predict weak interlayers. A high SED value presents a higher vulnerability to to cracking at this position. The equation is illustrated as follows:

$$V_0 = \frac{1}{2E}\left(\sigma_x^2 + \sigma_y^2 + \sigma_z^2\right) - \frac{\nu}{E}\left(\sigma_x\sigma_y + \sigma_y\sigma_z + \sigma_x\sigma_z\right) + \frac{1}{2G}\left(\tau_{xy}^2 + \tau_{yz}^2 + \tau_{xz}^2\right), \quad (7)$$

$$SED = V_0 - \frac{1 - 2\nu}{6E}\left(\sigma_x + \sigma_y + \sigma_z\right)^2 \quad (8)$$

SED is the strain energy of distortion ($N{\cdot}m/m^3$); V_0 is the total strain energy per unit volume; E is Young's Modulus (Mpa); ν is Poison's Ratio; G was shear modulus (Mpa); σ is compressive or tensile stress; and τ is shear stress (Mpa).

From the traditional perspective, thicknesses and Poisson's ratios of GSP are selected from experiential values referring to traditional AC pavement, rather than the values from actually measuring. Therefore, though specialized parameters are adopted in tests for GSP, these values vary widely. For instance, the U.S. Air Force recommends that the elastic modulus of GSP should be 12,000 Mpa at 20 °C, following the given modulus-temperature correlation curve (Figure 5), and Poisson's ratio should be 0.27 [27]. However, the design manual in the United Kingdom suggests the elastic modulus should be 8000 Mpa and Poisson's Ratio should be 0.25, according to an Indirect Tensile Stiffness Modulus (ITSM) test at the frequency of 5 Hz and the temperature of 20 °C [118,141]. In contrast, Pożarycki et al., back-calculated the in-situ GSP modulus by a Falling Weight Deflectometer, of which the value reaches to 23,700 Mpa [142]. For this reason, the full-scale pavement test is suggested to obtain the more precise property parameters [140].

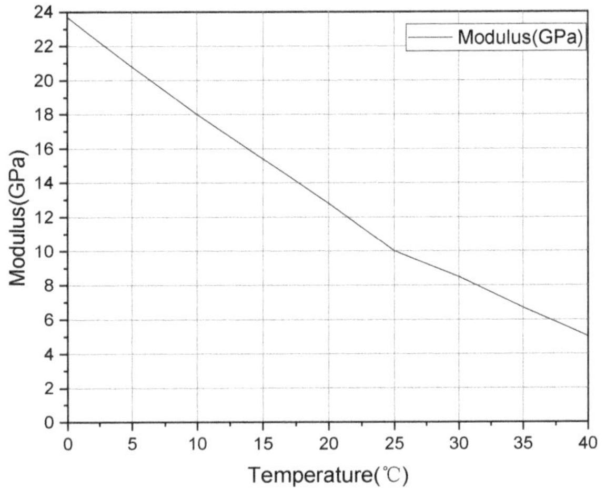

Figure 5. GSP Resilient Modulus Versus Temperature Design Curve.

4. Conclusions

According to the influential factors and the evaluation methods, recent studies of GSP have been investigated over the last twenty years in this review article. It is concluded that:

- Microstructure can be analyzed by microscopic observation method in the study of influential factors. However, the result is generally subjective, because image capturing and judging mainly rely on personal experience. Thus, associating tests should be introduced to acquire results in qualitative and quantitative ways;
- Effects of raw materials such as asphalt content and fine aggregate on GSP are not clear enough, due to assumptions in GSP design. More conditions and parameters need to be revised and developed for the evaluation of GSP;
- Individual admixture cannot completely meet the requirement for the low-temperature performance of GSP, while the combination of admixtures cannot be designed reasonably. More effective modifiers should be developed to improve the cracking resistance of GSP;
- Some other characteristics of GSP including interface strength, skid resistance, and other properties are seldom studied, which should be explored to assess the performance of GSP;
- Models used in GSP simulation are based on assumptions of traditional AC. Therefore, these computations and predictions cannot be accurate, as they ignore the difference in behaviors and parameters between GSP and AC.

5. Future Studies

Based on the current literatures, some recommendations for future studies are provided, as follows:

- Advanced technologies like X-ray CT and AFM technology can be used for characterizing the micro-structure of GSP;
- More factors need to be considered and revised by laboratory tests, including binder content and fine aggregate;
- High-efficiency interface modifiers and combined admixtures need to be developed and evaluated;
- Full-scale accelerated test is recommended to assess the fatigue life and other properties of GSP;
- Models of GSP should be carefully considered according to the influential factors in simulation and computation.

Author Contributions: Conceptualization, X.G. and P.H.; methodology, X.G.; validation, X.G. and P.H.; formal analysis, X.G. and P.H.; investigation, X.G.; resources, X.G.; data curation, X.G.; writing—original draft preparation, X.G.; writing—review and editing, X.G.; visualization, X.G.; supervision, P.H.; project administration, P.H.; funding acquisition, P.H. All authors have read and agreed to the published version of the manuscript.

Funding: This research received no external funding.

Institutional Review Board Statement: Not applicable.

Informed Consent Statement: Not applicable.

Acknowledgments: This work was supported by the China Scholarship Council (Grant No.201906560026).

Conflicts of Interest: The authors declares no conflict of interest.

References

1. Moghaddam, T.B.; Karim, M.R.; Abdelaziz, M. A review on fatigue and rutting performance of asphalt mixes. *Int. J. Sci. Res. Essays* **2011**, *6*, 670–682. [CrossRef]
2. Aalborg. *Densiphalt Handbook*; Densit A/S: Glenview, IL, USA, 2000.
3. Roffe, J.C. *Salviacim–Introducing the Pavement*; Jean Lafebvre Enterprise: Paris, France, 1989.
4. Cheng, L.; Hao, P.-W. Design method of mother-asphalt-mixture with semi-flexible pavement based on the volume method. *J. Xi'an Highw. Univ.* **2002**, *22*, 1–3.
5. Cheng, L.; Hao, P.-W. Mixture of cement slurry with semi-flexible pavement. *J. Xi'an Highw. Univ.* **2002**, *22*, 1–4.
6. Hao, P.-W.; Cheng, L.; Lin, L. Pavement performance of semi-flexible pavement in laboratory. *J. Chang. Univ. (Nat. Sci. Ed.)* **2003**, *23*, 2.
7. Goel, A.; Das, A. Emerging road materials and innovative applications. In Proceedings of the National Conference on Materials and Their Application in Civil Engg, Hamirpur, India, 26–27 August 2004; pp. 1–12.
8. Ahlrich, R.C.; Anderton, G.L. *Construction and Evaluation of Resin Modified Pavement*; Army Engineer Waterways Experiment Station Vicksburg Ms Geotechnical Lab: Vicksburg, MS, USA, 1991.
9. Svenson, K. Estimated lifetimes of road pavements in Sweden using time-to-event analysis. *J. Transp. Eng.* **2014**, *140*, 04014056. [CrossRef]
10. Jacobsen, S. The Effectiveness of Grouted Macadam at Intersections: A Life-Cycle Cost Analysis. Stockholm, KTH School of Architecture and the Built Environment. 2012, p. 101. Available online: http://urn.kb.se/resolve?urn=urn:nbn:se:kth:diva-101813 (accessed on 14 July 2021).
11. Du, Y.; Chen, J.; Han, Z.; Liu, W.J.C.; Materials, B. A review on solutions for improving rutting resistance of asphalt pavement and test methods. *Constr. Build Mater.* **2018**, *168*, 893–905. [CrossRef]
12. Kowalski, K.J.; Król, J.; Radziszewski, P.; Casado, R.; Blanco, V.; Pérez, D.; Viñas, V.M.; Brijsse, Y.; Frosch, M.; Le, D. Eco-friendly materials for a new concept of asphalt pavement. *Transp. Res. Procedia* **2016**, *14*, 3582–3591. [CrossRef]
13. Pratelli, C.; Betti, G.; Giuffre, T.; Marradi, A. Preliminary In-Situ Evaluation of an Innovative, Semi-Flexible Pavement Wearing Course Mixture Using Fast Falling Weight Deflectometer. *Materials* **2018**, *11*, 611. [CrossRef]
14. Zhao, M.; Shen, F.; Qiao, L.; Ding, Q. Recovery and utilization of waste asphalt materials (RAP) on semi-flexible pavement. *Energy Educ. Sci. Technol. Part A Energy Sci. Res.* **2014**, *32*, 3401–3408.
15. Mayer, J.; Thau, M. Jointless pavements for heavy-duty airport application: The Semi-Flexible approach. *Adv. Airfield Pavements* **2001**, 87–100. [CrossRef]
16. Pelland, R.J.; Gould, J.S.; Mallick, R.B. Selecting a rut resistant hot mix asphalt for Boston-Logan International Airport. *Airfield Pavements Chall. New Technol.* **2004**, 390–408. [CrossRef]
17. Collop, A.; Elliott, R. Assessing the mechanical performance of Densiphalt. In *Performance and Durability of Bituminous Materials and Hydraulic Stabilised Composites, Proceedings of the Third European Symposium, Leeds, UK, April 1999*; Springer: Berlin/Heidelberg, Germany, 1999; pp. 343–357.
18. Setyawan, A. *Development of Semi-Flexible Heavy-Duty Pavements*; University of Leeds: Leeds, UK, 2006.
19. Rone, C.L.J.M.P.S. *Evaluation of Salviacim*; US Army Corps of Engineer Waterways Experiment Station: Vicsburg, MS, USA, 1976.
20. Ahlrich, R.C.; Anderton, G.L. *Evaluation of Resin-Modified Paving Process*; Transportation Research Board: Washington, DC, USA, 1991; pp. 32–41. ISBN 030905155X.
21. Emery, J. An evaluation of the performance of concrete blocks on aircraft pavements at Luton airport. In Proceedings of the 3rd International Conference on Concrete Block Paving, Rome, Italy, 17–19 May 1988.
22. Anderton, G.L.; Ahlrich, R.C. *Design, Construction and Performance of Resin Modified Pavement at Fort Campbell Army Airfield*; Army Engineer Waterways Experiment Station Vicksburg Ms Geotechnical Lab: Vicksburg, MS, USA, 1994.
23. Battey, R. *Construction, Testing and Preliminary Performance Report on the Resin Modified Pavement Demonstration Project*; Mississippi Department of Transportation: Grenada, MS, USA, 2004. Available online: https://rosap.ntl.bts.gov/view/dot/24140 (accessed on 14 July 2021).

24. Battey, R.L.; Whittington, J.S. *Construction, Testing and Performance Report on the Resin Modified Pavement Demonstration Project*; Mississippi Department of Transportation: Washington, DC, USA, 2007. Available online: https://rosap.ntl.bts.gov/view/dot/24137 (accessed on 14 July 2021).
25. Anderton, G.L. *User's Guide: Resin Modified Pavement*; Army Engineer Waterways Experiment Station Vicksburg Ms Geotechnical Lab: Vicksburg, MS, USA, 1996.
26. Anderton, G.L. *Engineering Properties of Resin Modified Pavement (RMP) for Mechanistic Design*; Army Engineer Waterways Experiment Station Vicksburg Ms Geotechnical Lab: Vicksburg, MS, USA, 2000.
27. Department of the Air Force. *Engineering Technical Letter (ETL) 01-8: Resin Modified Pavement (RMP) Design and Application Criteria*; Department of the Air Force Headquarters Air Force Civil Engineer Support Agency: Tyndall AFB, FL, USA, 2001.
28. Ahlrich, R.C.; Anderton, G.L. Resin modified pavement in airfield applications. In Proceedings of the Airport Pavement Innovations. Theory to Practice. Proceedings of Conference, Vicksburg, MS, USA, 8–10 September 1993; Available online: http://worldcat.org/isbn/0872629252 (accessed on 14 July 2021).
29. Wang, S.Q.; Lin, X.X.; Lou, H.X. Study on special asphalt mixture, a new type of pavement composite material. *J. East China Highw.* **1989**, *2*, 76–81.
30. Xu, P.H.; Wang, Z.K. Indoor test and research on poured semi-rigid pavement materials. *J. Xi 'an Highw. Jiaotong Univ.* **1995**, *3*, 22–25.
31. Zhang, X.; Pan, D. Design method of Semi-flexible pavement base asphalt mixture. *J. Highw. Eng.* **2000**, *25*, 22–23.
32. Zhang, X.; Pan, D. Preparation of Semi-flexible pavement cement slurry. *J. Highw. Eng.* **2000**, *25*, 6–7.
33. Bonicelli, A.; Preciado, J.; Rueda, A.; Duarte, A. Semi-flexible material: A solution for high-performance pavement infrastructures. In Proceedings of the 3rd World Multidisciplinary Civil Engineering, Architecture, Urban Planning Symposium, WMCAUS 2018, Prague, Czech Republic, 18–22 June 2018. [CrossRef]
34. Loeber, L.; Sutton, O.; Morel, J.; Valleton, J.M.; Muller, G. New direct observations of asphalts and asphalt binders by scanning electron microscopy and atomic force microscopy. *J. Microsc.* **1996**, *182*, 32–39. [CrossRef]
35. Mazumder, M.; Ahmed, R.; Ali, A.W.; Lee, S.-J.J.C.; Materials, B. SEM and ESEM techniques used for analysis of asphalt binder and mixture: A state of the art review. *Constr. Build Mater.* **2018**, *186*, 313–329. [CrossRef]
36. Ling, T.Q.; Tang, Q.Q.; Luo, R. Microstructure analysis and strength theory of semi-flexible pavement materials. In Proceedings of the 4th International Road and Airport Pavement Technology Conference, Kunming, China, 23–25 April 2002.
37. Huo, Y.Z.; Liang, Y.; Huang, B.T. Performance test and mechanism analysis of semi-flexible pavement. *J. Highw.* **2009**, *2*, 52–56.
38. Hou, S.; Xu, T.; Huang, K. Investigation into engineering properties and strength mechanism of grouted macadam composite materials. *Int. J. Pavement Eng.* **2015**, *17*, 878–886. [CrossRef]
39. Cheng, L. *Research on Performance and Design Method of Semi-Flexible Pavement Mixture*; Chang'an University: Xi'an, China, 2002.
40. Setyawan, A.; Zoorob, S.E.; Hasan, K.E. Investigating and Comparing Traffic Induced and Restrained Temperature Stresses in a Conventional Rigid Pavement and Semi-Rigid Layers. *Procedia Eng.* **2013**, *54*, 875–884. [CrossRef]
41. Netterberg, F.; de Beer, M. Weak interlayers in flexible and semi-flexible road pavements: Part 1. *J. S. Afr. Inst. Civ. Eng.* **2012**, *54*, 32–42.
42. De Beer, M. Weak interlayers found in flexible and semi-flexible road pavements. In Proceedings of the 8th RILEM International Conference on Mechanisms of Cracking and Debonding in Pavements, Nantes, France, 7–9 June 2016; pp. 425–430.
43. Koting, S.; Karim, M.R.; Mahmud, H.B.; Abdul Hamid, N.A. Mechanical properties of cement-bitumen composites for semi-flexible pavement surfacing. *Balt. J. Road Bridge Eng.* **2014**, *9*, 191–199. [CrossRef]
44. Yang, G.L.; Wu, K.H.; Wang, W.M.; Meng, F.B.; Li, Y.F.; Bao, X.N. Parent Asphalt Mixture Design of Semi-Flexible Pavement Based on Method of Main Mineral Aggregate Void Filling. *Adv. Mater. Res.* **2012**, *446–449*, 2599–2602. [CrossRef]
45. Saboo, N.; Khalpada, V.; Sahu, P.K.; Radhakrishnan, R.; Gupta, A. Optimal proportioning of grout constituents using mathematical programming for semi flexible pavement. *Int. J. Pavement Res. Technol.* **2019**, *12*, 297–306. [CrossRef]
46. Setiawan, A. Design and properties of hot mixture porous asphalt for semi-flexible pavement applications. *J. Tek. Sipil.* **2009**, *5*, 41–46.
47. Saboo, N.; Ranjeesh, R.; Gupta, A.; Suresh, M. Development of hierarchical ranking strategy for the asphalt skeleton in semi-flexible pavement. *Constr. Build. Mater.* **2019**, *201*, 149–158. [CrossRef]
48. Ling, T.Q.; Dong, Q.; Dong, Y.Y. Application of rubber asphalt in infilled semi-flexible pavement. *J. Chang. Univ. (Nat. Sci. Ed.)* **2009**, *26*, 24–27.
49. Ling, T.-Q.; Zhaoz, Z.-J.; Xiong, C.-H.; Dong, Y.-Y.; Liu, Y.-Y.; Dong, Q. The Application of Semi-Flexible Pavement on Heavy Traffic Roads. *Int. J. Pavement Res. Technol.* **2009**, *2*, 211–217.
50. Husain, N.M.; Mahmud, H.; Karim, M.R.; Hamid, N. Effects of aggregate gradations on properties of grouted Macadam composite pavement. In Proceedings of the 2010 2th International Conference on Chemical, Biological and Environmental Engineering, Cairo, Egypt, 2–4 November 2010; pp. 128–131.
51. Husain, N.M.; Karim, M.R.; Mahmud, H.B.; Koting, S. Effects of Aggregate Gradation on the Physical Properties of Semiflexible Pavement. *Adv. Mater. Sci. Eng.* **2014**, *2014*, 8. [CrossRef]
52. Hou, S.; Xu, T.; Huang, K. Aggregate gradation influence on grouting results and mix design of asphalt mixture skeleton for semi-flexible pavement. *J. Test. Eval.* **2017**, *45*, 591–600. [CrossRef]

53. Ding, Q.J.; Sun, Z.; Shen, F.; Huang, S.L. The performance analysis of semi-flexible pavement by the volume parameter of matrix asphalt mixture. *Adv. Mater. Res.* **2011**, *168–170*, 351–356. [CrossRef]
54. Zhao, G.Q.; Huang, C.; Deng, C. Research on the influence of asphalt type on the performance of semi-flexible pavement materials. *J. Highw. Eng.* **2017**, *42*, 199–201.
55. Wang, W.M.; Gao, D.; Wu, K.H. Study on the properties of semi-flexible pavement materials. *J. Highw. Eng.* **2014**, *1*, 78–82.
56. Luo, S.; Yang, X.; Zhong, K.; Yin, J. Open-graded asphalt concrete grouted by latex modified cement mortar. *Road Mater. Pavement Des.* **2018**, *21*, 1–17. [CrossRef]
57. Plug, C.; de Bondt, A.; van der Woerd, B.; Steensma, G. *Improved Performance Grouted Macadams–High Performance Applications of Grouted Macadam*; Ooms Nederland Holding bv: Avenhorn, The Netherlands, 2006; Available online: http://www.materialedge.co.uk/docs/CP-AdB-Stabifalt(English)[1].pdf (accessed on 14 July 2021).
58. Debbarma, C.; Debbarma, D.; Roy, A.; Zachariah, J.P.; Sarkar, P. A study on the effect of admixture in grouted macadam. In Proceedings of the AMIE Section A & B Summer Examination Commences wef 2nd, Agartala, India, 30 April 2018; p. 62.
59. Liu, B. *Specification for Design of Highway Asphalt Pavement*; Ministry of Transport of the People's Republic of China: Beijing, China, 2017.
60. Hu, S.G.; Zhang, R.G.; Ding, Q.J. Semi-flexible pavement performance of grouting cement mortar research. *J. Highw.* **2009**, *7*, 1–6.
61. Hu, S.; Huang, S.; Ding, Q.; Zhang, R. Study on the cementitious mortar for semi-flexible pavement. In Proceedings of the 1st International Conference on Microstructure Related Durability of Cementitious Composites, Nanjing, China, 13–15 October 2008; pp. 1237–1246.
62. Pei, J.; Cai, J.; Zou, D.; Zhang, J.; Li, R.; Chen, X.; Jin, L. Design and performance validation of high-performance cement paste as a grouting material for semi-flexible pavement. *Constr. Build. Mater.* **2016**, *126*, 206–217. [CrossRef]
63. Zhang, J.; Cai, J.; Pei, J.; Li, R.; Chen, X. Formulation and performance comparison of grouting materials for semi-flexible pavement. *Constr. Build. Mater.* **2016**, *115*, 582–592. [CrossRef]
64. Koting, S.; Karim, M.R.; Mahmud, H.; Mashaan, N.S.; Ibrahim, M.R.; Katman, H.; Husain, N.M. Effects of using silica fume and polycarboxylate-type superplasticizer on physical properties of cementitious grout mixtures for semiflexible pavement surfacing. *Sci. World J.* **2014**, *2014*, 596364. [CrossRef]
65. Vijaya, S.S.A.; Karim, M.R.; Mahmud, H.; Ishak, N.F. Influence of superplasticiser on properties of cementitious grout used in semi-flexible pavement. In Proceedings of the 8th International Conference of Eastern Asia Society for Transportation Studies, Eastern Asia Society for Transportation Studies, Surabaya, Indonesia, 16–19 November 2009; p. 296. [CrossRef]
66. Cheng, Z.Q.; Kong, F.S.; Jia, R.R. Study on bleeding performance of cement-based grout materials for semi-flexible pavement. *J. China Foreign Highw.* **2014**, *36*, 276–279.
67. Gu, X.Y.; Li, L.X.; Cheng, Z.Q. Study on flow performance of cement-based grout materials for semi-flexible pavement. *J. Highw.* **2017**, *62*, 280–285.
68. Wang, Y.J.; Guo, C.Y.; Tian, Y.F.; Wang, J.J. Design of Mix Proportion of Cement Mortar with High-performance Composite Semi-flexible Pavement. *Adv. Mater. Res. Switz.* **2013**, *641–642*. [CrossRef]
69. Hassan, K.; Setyawan, A.; Zoorob, S. *Effect of Cementitious Grouts on the Properties of Semi-Flexible Bituminous Pavements in Performance of Bituminous and Hydraulic Materials in Pavements*; Routledge: London, UK, 2017; pp. 113–120.
70. Cai, J.; Pei, J.; Luo, Q.; Zhang, J.; Li, R.; Chen, X. Comprehensive service properties evaluation of composite grouting materials with high-performance cement paste for semi-flexible pavement. *Constr. Build. Mater.* **2017**, *153*, 544–556. [CrossRef]
71. CAI, X.; Zhang, S.X.; Li, K. Limit stress analysis of weak internal parts of Semi-flexible materials. *J. Taiyuan Univ. Technol.* **2017**, *48*, 652–656.
72. Anderton, G.L. The Resin Modified Pavement System: A New Pavement Surfacing Technology. In *Airport Facilities: Innovations for the Next Century, Proceedings of the 25th International Air Transportation Conference, Austin, TX, USA, 14–17 June 1998*; American Society of Civil Engineers: Austin, TX, USA, 1998.
73. Wang, S.; Lin, X.; Lou, H. Study on the composite material of new pavement-special asphalt mixture. *J. East China Highw.* **1989**, *2*, 15.
74. Ling, T.; Zhou, J.; Zhao, Z.; Li, C.; Dong, Y.; Research, T. Study on Optimization of Polymer Modified Cement Slurry for Poured Semi-flexible Pavement. *J. Highw.* **2009**, *6*, 24–28.
75. Huang, C.; Hong, J.X.; Lin, J.T.; Deng, C.; Li, L. Utilization of Waste Rubber Powder in Semi-Flexible Pavement. *Key Eng. Mater.* **2014**, *599*, 361–367. [CrossRef]
76. Xu, Y.; Jiang, Y.; Xue, J.; Tong, X.; Cheng, Y.J.C. High-Performance Semi-Flexible Pavement Coating Material with the Microscopic Interface Optimization. *Coatings* **2020**, *10*, 268. [CrossRef]
77. Wang, D.; Liang, X.; Jiang, C.; Pan, Y. Impact analysis of Carboxyl Latex on the performance of semi-flexible pavement using warm-mix technology. *Constr. Build. Mater.* **2018**, *179*, 566–575. [CrossRef]
78. Cheng, Z.-Q.; Zhang, X.-Y.; Kong, F.-S. Research on Mechanism and Improving Low Temperature Performance of Semi Flexible Pavement Material by Use of PVA. *J. Wuhan Univ. Technol.* **2016**, *38*, 44–49.
79. Yang, Y.; Huang, S.L.; Ding, Q.J.; Peng, X.Y. The Property Research on Interfacial Modificated Semi-Flexible Pavement Material. *Appl. Mech. Mater.* **2011**, *71–78*, 1090–1098. [CrossRef]
80. Zhang, X.Y.; Cheng, Z.Q.; Kong, F.S. Low Temperature Performance of Semi-flexible Pavement With CA Grouting Material. *J. Beijing Univ. Technol.* **2017**, *43*, 1205–1211. [CrossRef]

81. Huang, C.; Liu, J.; Hong, J.; Liu, Z. Improvement on the crack resistance property of semi-flexible pavement by cement-emulsified asphalt mortar. In Proceedings of the 3rd Mainland, Taiwan and Hong Kong Conference on Green Building Materials, GBM 2011, Wuhan, China, 25–27 November 2011; pp. 26–32.
82. Sun, Z.; Hou, X.X.; He, G.P. Research on the Crack Resistance at Low Temperature and the Mechanism of Semi-Flexible Pavement. *Aer. Adv. Eng. Res.* **2015**, *22*, 533–536.
83. Gong, M.; Xiong, Z.; Chen, H.; Deng, C.; Chen, X.; Yang, J.; Zhu, H.; Hong, J. Evaluation on the cracking resistance of semi-flexible pavement mixture by laboratory research and field validation. *Constr. Build. Mater.* **2019**, *207*, 387–395. [CrossRef]
84. Fang, B.; Xu, T.; Shi, S. Laboratory study on cement slurry formulation and its strength mechanism for semi-flexible pavement. *J. Test. Eval.* **2016**, *44*, 907–913. [CrossRef]
85. Hamzani; Munirwansyah; Hasan, M.; Sugiarto, S. The influence of the using waste tire rubber and natural ziolite as Asphalt and Cement replacements to compressive strength of Semi-Flexible Pavement. In Proceedings of the 8th Annual International Conference 2018 on Science and Engineering, Aceh, Indonesia, 12–14 September 2018; IOP Publishing Ltd: Banda Aceh, Indonesia, 2018.
86. Zhu, J.; Birgisson, B.; Kringos, N. Polymer modification of bitumen: Advances and challenges. *Eur. Polym. J.* **2014**, *54*, 18–38. [CrossRef]
87. Yin, H. *Study on the Performance of Semi-Flexible Pavement Mixed With Fiber*; University Of South China: Hengyang, China, 2013.
88. Hou, Q.Q.; Sun, Z. Study on the properties of semi-flexible pavement modified by water-based epoxy resin. *J. Traffic Stand.* **2013**, *24*, 81–84. [CrossRef]
89. Kong, F.S. *Study on the Performance of HIGH Performance Semi-Flexible Pavement Materials*; Chang'an University: Xi'an, China, 2016.
90. Li, Z.M. *Study on Temperature Stability of Semi-Flexible Pavement Composites Filled with Rubber Powder*; Guangxi University of Science and Technology: Liuzhou, China, 2013.
91. Ling, T.-Q.; Zheng, X.-W.; Ling, M.; Xiong, C.-H.; Dong, Q. Research on performance of water-retention and temperature-fall semi-flexible pavement material. *China J. Highw. Transp.* **2010**, *23*, 7–11.
92. Dong, Q.; Wang, C.; Xiong, C.; Li, X.; Wang, H.; Ling, T. Investigation on the Cooling and Evaporation Behavior of Semi-Flexible Water Retaining Pavement based on Laboratory Test and Thermal-Mass Coupling Analysis. *Materials* **2019**, *12*, 2546. [CrossRef]
93. Liu, G.Y. *Application of Phase Change Materials in Semi-Flexible Pavement*; Chang'an University: Xi'an, China, 2013.
94. Sha, S. Study on Preparation and Application of PEG/SiO2 Compound Phase Change Materials for Semi-flexible Pavement. *Road Mach. Constr. Mech.* **2016**, *33*, 63–66. [CrossRef]
95. Xu, P. Experimental study on composite materials for semi-rigid road surface. *J. Highw. Eng.* **2002**, *15*, 7–10.
96. Wu, G.X.; Zhang, Y.L.; Wang, A.M. Experimental study on high temperature stability of semi-flexible pavement. *J. Chongqing Jiaotong Univ.* **2007**, *26*, 52–55. [CrossRef]
97. Zhang, D.K.; Cheng, C.Q.; Wu, G.X. Analysis of water stability and low temperature crack resistance of semi-flexible pavement. *J. Chongqing Jiaotong Univ.* **2007**, *26*, 55–57. [CrossRef]
98. Dong, Y.Y. *Research on Design Parameters and Construction Technology of High-Performance Semi-Flexible Pavement*; Chongqing Jiaotong University: Chongqing, China, 2008.
99. Zhang, R.K. *Research and Application of High Performance Perfusion Semi-Flexible Pavement Materials*; Wuhan University of Technology: Wuhan, China, 2009.
100. Pang, C.Q.; Yang, Y.L. Research on Properties of Semi-Flexible Mixtures. *J. Highw.* **2004**, *4*, 108–110. [CrossRef]
101. Pang, C.Q. Mixture design for grouting pavements. In Proceedings of the International Workshop on Energy and Environment in the Development of Sustainable Asphalt Pavements, Xi'an, China, 1 June 2010; pp. 262–265.
102. Li, X.; Li, Y. Analysis of road performance of semi-flexible cement asphalt mixture. *J. Shandong Agric. Univ. (Nat. Sci. Ed.)* **2018**, *49*, 815–818.
103. Hu, S.-G.; Huang, S.-L.; Liu, Z.-G.; Ding, Q.-J.; Wu, G.-X. Design and construction of the semi-flexible based permeable anti-sliding noise-reducing pavement. *J. Wuhan Univ. Technol.* **2008**, *30*, 27–31.
104. Huo, Y.-Z.; Liang, Y.; Huang, B.-T.; Huang, F. Laboratory Test and Mechanism Analysis of Road Performance of SemiFlexible Pavement. *Highway* **2009**, *11*, 52–56.
105. Afonso, M.L.; Dinis-Almeida, M.; Pereira-de-Oliveira, L.A.; Castro-Gomes, J.; Zoorob, S.E. Development of a semi-flexible heavy duty pavement surfacing incorporating recycled and waste aggregates—Preliminary study. *Constr. Build. Mater.* **2016**, *102*, 155–161. [CrossRef]
106. Zhang, L.; Wang, H.; Li, R. Laboratory study on performance of semi-flexible mixture. In Proceedings of the 4th International Conference on Civil, Architectural and Hydraulic Engineering, ICCAHE 2015, Guangzhou, China, 20–21 June 2015; pp. 989–994.
107. Manikantha Raju, G.; Sita Rami Reddy, D.; Sudhakar Reddy, K. Performance evaluation of cement grouted bituminous mixes. In Proceedings of the 12th International Conference on Asphalt Pavements, ISAP 2014, Raleigh, NC, USA, 1–5 June 2014; pp. 1047–1056.
108. Zhang, W.; Shen, S.; Goodwin, R.D.; Wang, D.; Zhong, J.J.M. Performance Characterization of Semi-Flexible Composite Mixture. *Materials* **2020**, *13*, 342. [CrossRef]
109. Ding, Q.J.; Zhao, M.Y.; Shen, F.; Zhang, X.Q. Mechanical Behavior and Failure Mechanism of Recycled Semi-flexible Pavement Material. *J. Wuhan Univ. Technol.* **2015**, *30*, 981–988. [CrossRef]

110. Hong, J.X.; Wang, K.J.; Xiong, Z.J.; Gong, M.H.; Deng, C.; Peng, G.; Zhu, H.R. Investigation into the freeze-thaw durability of semi-flexible pavement mixtures. *Road Mater. Pavement Des.* **2019**, *21*, 2198–2214. [CrossRef]
111. Hirato, T.; Murayama, M.; Sasaki, H.J. Development of high stability hot mix asphalt concrete with hybrid binder. *J. Traffic Transp. Eng.* **2014**, *1*, 424–431. [CrossRef]
112. Liu, Z.K. Experimental study on the properties of single gradation asphalt crushed stone with cement slurry filling hole. *J. Highw.* **2015**, *60*, 219–221.
113. Li, Y.; Wang, J.W.; Zhang, Y. Study on impact resistance of perfusion semi-flexible pavement materials. *J. Highw. Eng.* **2016**, *41*, 55–59. [CrossRef]
114. Karami, M. Application of the Cementitious Grouts on Stability and Durability of Semi Flexible Bituminous Mixtures. *AIP Conf. Proc.* **2017**, *1903*, 020012. [CrossRef]
115. Oliveira, J.; Thom, N.; Zoorob, S. Fracture and Fatigue Strength of Grouted Macadams. Available online: http://hdl.handle.net/1822/7208 (accessed on 14 July 2021).
116. Da Rios, G.; Agostinacchio, M.; Fiori, F. Performance and durability of grouted open grade asphalt concretes. In Proceedings of the 2021 5th International Conference on Maintenance and Rehabilitation of Pavements and Technological Control, MAIREPAV 2007, Park City, UT, USA, 8–10 August 2007; pp. 505–510.
117. De Oliveira, J.R.M. *Grouted Macadam: Material Characterisation for Pavement Design*; University of Nottingham: Nottingham, UK, 2006.
118. Oliveira, J.; Pais, J.C.; Thom, N.H.; Zoorob, S.E. The effect of using rest periods in 4PB tests on the fatigue life of grouted macadams. In Proceedings of the C-TAC International Conference Communications, Universidade do Minho, Guimarães, Portugal, 24–25 September 2009; pp. 277–283.
119. Ling, T.Q.; Zhao, Z.J.; Wu, D. Experimental study on fatigue resistance of semi-flexible pavement materials. In Proceedings of the Annual Meeting of China Highway Society and Technical Seminar of Highway Construction in Mountainous Areas, Chongqing, China, 10 September 2009.
120. Huang, C. *Study on Volume Stability and Crack Resistance of Semi-Flexible Pavement Materials*; Wuhan University of Technology: Wuhan, China, 2010.
121. Yang, B.; Weng, X. The influence on the durability of semi-flexible airport pavement materials to cyclic wheel load test. *Constr. Build. Mater.* **2015**, *98*, 171–175. [CrossRef]
122. Yang, B.; Weng, X.; Liu, J.; Jiang, L.; Zhang, J.; Liu, P.; Wen, X. The influence of on semi-flexible airport pavement material working performance by porosity. In Proceedings of the Advanced Materials and Structural Engineering: Proceedings of the International Conference on Advanced Materials and Engineering Structural Technology (ICAMEST 2015), Qingdao, China, 25–26 April 2015; p. 157.
123. Oliveira, J.; Sangiorgi, C.; Fattorini, G.; Zoorob, S. Investigating the fatigue performance of grouted macadams. *Proc. Inst. Civil Eng.-Transp.* **2009**, *162*, 115–123. [CrossRef]
124. Wu, D. *Experimental Study on Fatigue Resistance of Semi-Flexible Pavement Materials*; Chongqing Jiaotong University: Chongqing, China, 2009.
125. Corradini, A.; Cerni, G.; D'Alessandro, A.; Ubertini, F.J.C.; Materials, B. Improved understanding of grouted mixture fatigue behavior under indirect tensile test configuration. *Constr. Build Mater.* **2017**, *155*, 910–918. [CrossRef]
126. Oliveira, J.R.; Thom, N.H.; Zoorob, S. Design of pavements incorporating grouted macadams. *J. Traffic Transp. Eng.* **2008**, *134*, 7–14. [CrossRef]
127. Oliveira, J.; Pais, J.C.; Thom, N.H.; Zoorob, S. A study of the fatigue properties of grouted macadams. *Int. J. Pavements* **2007**, *6*, 112–123.
128. Oliveira, J.; Zoorob, S.; Thom, N.; Pereira, P.A. A simple approach to the design of pavements incorporating grouted macadams. Available online: http://hdl.handle.net/1822/8002 (accessed on 14 July 2021).
129. Yu, Z.H. *Study on the Structure and Performance of Semi-Flexible Asphalt Mixture Pavement*; Hunan University: Changsha, China, 2007.
130. Huang, F. *Research on Structural Design Theory and Method of Semi-Flexible Composite Pavement*; Chongqing Jiaotong University: Chongqing, China, 2008.
131. Chen, C. *Study on Semi-Flexible Structure of Highway Tunnel Pavement*; Chongqing Jiaotong University: Chongqing, China, 2011.
132. Xu, G.; Li, G.; Xu, W.; Li, C.; Li, Y. Mechanics Numerical Analysis of Semi-flexible Compound Pavement for Large Aircraft. *J. Highw. Eng.* **2015**, *40*, 122–126. [CrossRef]
133. Wang, W.P. *Research on Semi-Flexible Composite Pavement Material and Structure*; Changsha University of Science and Technology: Changsha, China, 2012.
134. Gao, J.H.; Wang, Z.L. The Analysis of Layer Cohesive Force of Semi-flexible Composite Pavement Based on the Software of BISAR. *J. Highw. Eng.* **2013**, *38*, 17–20.
135. Wang, W.M.; Wu, K.H. Research on crack resistance of semi-flexible pavement material based on J integral. *J. Highw.* **2016**, *61*, 63–67.
136. Setyawan, A. Asessing the Compressive Strength Properties of Semi-Flexible Pavements. *Procedia Eng.* **2013**, *54*, 863–874. [CrossRef]
137. Huang, F.; Wu, G.-X.; Wang, Y.; Zhang, D. Study on Compacted Rebound Modulus of Semi-Flexible Pavement Compound Material. *J. Chongqing Jiaotong Univ.* **2008**, *1*, 65–68.

138. Wu, G.-X.; Wang, Y.; Wang, R.-Y. A study of the strength mechanism of semi-flexible composite pavement material and a modified method for the determination of its compressive resilient modulus. In *Integrated Transportation Systems: Green, Intelligent, Reliable, Proceedings of the 10th International Conference of Chinese Transportation Professionals, Beijing, China, 4–8 August 2010*; American Society of Civil Engineers: Beijing, China, 2010; pp. 3055–3062.
139. Cai, X.; Zhang, H.; Zhang, J.; Chen, X.; Yang, J.; Hong, J. Investigation on reinforcing mechanisms of semi-flexible pavement material through micromechanical model. *Constr. Build. Mater.* **2019**, *198*, 732–741. [CrossRef]
140. De Beer, M.; Maina, J.W.; Netterberg, F. Mechanistic modelling of weak interlayers in flexible and semi-flexible road pavements: Part 2. *J. S. Afr. Inst. Civ. Eng.* **2012**, *54*, 43–54.
141. Highways Agency. *Design Manual for Roads and Bridges*; HM Stationery Office: London, UK, 2006.
142. Pożarycki, A.; Fengier, J.; Górnaś, P.; Wanatowski, D. Backcalculation of pavements incorporating Grouted Macadam technology. *Road Mater. Pavement Des.* **2017**, *19*, 1372–1388. [CrossRef]

applied sciences

MDPI

Article

Evaluation and Correction Method of Asphalt Pavement Rutting Performance Prediction Model Based on RIOHTrack Long-Term Observation Data

Yang Wu, Xingye Zhou *, Xudong Wang and Zhimin Ma

National Observation and Research Station of Corrosion of Road Materials and Engineering Safety in Dadushe Beijing, Research Institute of Highway Ministry of Transport, Beijing 100088, China; yang.w@rioh.cn (Y.W.); xd.wang@rioh.cn (X.W.); mzm1755709496@163.com (Z.M.)

* Correspondence: xy.zhou@rioh.cn

Abstract: In order to improve the accuracy and reliability of an existing rutting performance prediction model, based on the long-term observation data of the RIOHTrack's full-scale pavement structure, the rutting performance prediction model in China's Specifications for Design of Highway Asphalt Pavement was evaluated, and the model correction method was proposed, which improves the model's reliability and makes it more suitable for rutting estimation in the region. The research found that the rutting model in China's Specifications for Design of Highway Asphalt Pavement has significant structural dependence. The model with the highest prediction accuracy and the smallest error is the semi-rigid base asphalt pavement structure with an asphalt concrete layer thickness of 12 cm; the prediction accuracy of other structures is not high. In order to improve the accuracy and reliability of the rutting prediction model, a new model is established by introducing local correction coefficients into the existing model. After local correction, the accuracy of the rutting prediction models for all structures has been greatly improved, and the determination coefficient R^2 is greater than 0.87. Since the basic data has already reflected the characteristics of different pavement structures and materials, as well as the impact of local climate environment and traffic load conditions, the new model is more suitable for rutting prediction of various pavement structures in the region where the RIOHTrack is located.

Keywords: rutting performance prediction model; long-term observation data; full-scale pavement structure; model accuracy evaluation; local correction coefficient; rutting prediction

Citation: Wu, Y.; Zhou, X.; Wang, X.; Ma, Z. Evaluation and Correction Method of Asphalt Pavement Rutting Performance Prediction Model Based on RIOHTrack Long-Term Observation Data. *Appl. Sci.* **2022**, *12*, 6805. https://doi.org/10.3390/app12136805

Academic Editors: Amir Tabakovic, Jan Valentin and Liang He

Received: 10 June 2022
Accepted: 2 July 2022
Published: 5 July 2022

1. Introduction

Rutting is one of the main diseases of asphalt pavement, and it is also a key design index in the structural design of asphalt pavement in many countries such as: the American Mechanistic-Empirical Pavement Design Guide (MEPDG) [1], the Shell Pavement Design Manual [2], China's Specifications for Design of Highway Asphalt Pavement [3], etc.; all control the structural design through the allowable rutting deformation within the design life. In these design methods, the rutting performance prediction model is mainly used to predict the rutting deformation of the asphalt concrete structure at the end of the design life, and through structural combination and material performance optimization, the rutting deformation at the end of the design life is less than the allowable value, so as to ensure the rationality of pavement structure design. It can be seen that how to ensure the accuracy and reliability of the rutting performance prediction model is one of the key issues of these design methods.

The core basis of the rutting performance prediction model is the conversion relationship between the permanent strain accumulation of the laboratory repeated load test and the actual road rutting accumulation [4,5]. At present, two methods are mainly used to establish the rutting performance prediction model. The first is the layered strain method [6–11], which combines the mechanical analysis of pavement structure, material performance tests, and actual rutting measurement results of pavement structures, and predicts the rutting deformation according to the layered strain accumulation method. For example, in the MEPDG research [12], Kaloush [13,14] first established the relationship between the permanent strain of each layered asphalt mixture and the number of loading times. Then 387 sets of observation data from 88 LTPP observation sections in 28 states were used to calibrate and verify the relationship, and finally the prediction model for asphalt layer rutting performance [1] was obtained. As for China's Specifications [3], considering the difference of stress distribution at different depths of asphalt pavement and the rutting resistance performance of different layers, the permanent deformation is calculated layer by layer, and based on 229 rutting test results of various asphalt mixtures, a rutting performance model based on the layered strain accumulation method was established, and the model was revised and verified using the rutting data of more than 10 highways and 5 test sections [5]. The Shell Pavement Design Manual [2] uses simplified linear elastic stress analysis and laboratory static load creep tests without confining pressure to establish a model for predicting the permanent deformation of the asphalt layer. The model considers the thickness of the asphalt layer, the average stress in the layer, and the asphalt concrete stiffness, which could compare the resistance to permanent deformation of different asphalt mixtures. Yang [15] introduced the generalized stiffness modulus of the viscosity part of the asphalt layer on the basis of the Shell model, and Finn and Monismith [10] suggested to directly use the asphalt mixture stiffness from the static load creep test under given load stress and temperature conditions, further optimizing the Shell model's rutting estimation equation. Verstvaeten, Ververka, and Francken et al. [16] established a rutting model considering the intermittent time of load action. Zhang et al. [17] established a high-grade asphalt pavement rutting prediction model by using the elastic layered system theory combined with the rheological model of asphalt mixture. Kenis [18] considered that the permanent deformation depends on parameters such as stress, loading time, and temperature, and assumed that the viscoelastic deformation of the asphalt mixture had sufficient recovery time within the load interval, and established a corresponding rutting model accordingly.

The second is the empirical method [19], which directly uses the observation data accumulated over several years to establish the regression relationship between the rutting amount and the relevant influencing variables through statistical analysis. Considering the actual conditions of test sections or field roads and laboratory tests, this method establishes an empirical equation for predicting the relationship between the permanent strain of the asphalt mixture layer and the properties of load and material. Finally, the rutting of the asphalt pavement under long-term repeated loads is determined. For example, Kim et al. [20] used data collected from 930 test points in 39 test road sections in Michigan from 1991 to 1997, and established a rutting prediction model through structure analysis and nonlinear regression analysis, which took into account the correlation between the amount of rutting and the pavement structure, material parameters, and axle loads. Based on the laboratory asphalt pavement analyzer (APA) test, Shami et al. [21] extended to APA test prediction research under different temperatures and axis loads. Through a large number of experiments, Monismith et al. [22,23] obtained an empirical method for rutting prediction; that is, the relationship between rutting and influencing factors. A. Wijeratane et al. [24] established a double-logarithmic model of the relationship between permanent strain and cumulative axis loads. Huang [25] conducted regression analysis on the data through a laboratory loop test, and established an empirical model including the thickness of the asphalt layer and the cumulative axis loads. Because the second method is limited by observation objects and regional conditions, the model is less applicable. Therefore, the first method is mainly used in the design of asphalt pavement in many countries to establish a rutting performance prediction model.

For the prediction model of rutting performance established by the first method, due to the natural environment and loads that the actual pavement structure bears, there are certain differences from the test conditions imposed in the laboratory. It is necessary to continuously accumulate actual observation data during the service process to make periodic corrections and improvements to the established prediction model [26–33]. Taking the American MEPDG as an example, the first edition of the guide [1] was officially promulgated in 2008, and the rutting prediction model was established based on the observation data and research results at that time. Afterwards, in the process of using this guide, in order to improve the accuracy and reliability of the rutting model, in the second edition of the guide [34] in 2015 and the third edition of the guide [35] in 2020, both of them have made a relatively large revision to the rutting prediction model. Although the model expression remains unchanged, the model parameters are adjusted greatly, and the prediction effect of the rutting performance prediction model after the correction has been greatly improved. The rutting model established by the first method is also used, as well as China's Specifications [3] promulgated in 2017. Due to the late promulgation of the specification, the asphalt pavement designed and opened to traffic in China in accordance with this specification has a longest service life of only 2 years. Due to the lack of long-term observation data of rutting in actual engineering, the effect of using the rutting prediction model is still unknown, and the model cannot be revised and improved periodically.

Therefore, this research uses the long-term observation data of the RIOHTrack's full-scale pavement structures to carry out the accuracy analysis of the rutting performance prediction model in China's Specifications, and proposes a model correction method for the structure with poor model prediction to improve the reliability of the model, and make it more suitable for rutting prediction in the local region. As of April 2022, according to the rutting equivalent conversion principle, the RIOHTrack has completed 51.6 million equivalent single-axle loads (ESALs), which is equivalent to the traffic load level of the expressway heavy traffic level in China's Specification for more than 30 years. The service time of the simulated pavement structure is long enough, and can be used as the basic data to evaluate the accuracy of the rutting prediction model of China's asphalt pavement and propose a correction method.

This research is relatively new work. As mentioned above, after the rutting performance prediction model is established, it is necessary to continuously correct the model according to the long-term observation data accumulated during actual service life, which is very important and practical work. However, there are few asphalt pavements designed and put into operation according to this the newest design specification in China, and long-term observation data cannot be accumulated in the short term, so it is difficult to carry out the work to correct the rutting model. In order to solve this problem, this research analyzes the long-term rutting observation data of the RIOHTrack, and explores a new correction method for the rutting model used by China's Specification. By introducing correction coefficients to locally correct the existing rutting model, the accuracy of the model can be significantly improved. This is of great significance for rutting prediction, and also has important reference value for application in other regions.

2. Objective and Scope

This research aims to evaluate the accuracy of the rutting performance predication model in China's Specifications by using the long-term observation data of full-scale pavement structures and present a model correction method to improve the reliability of the model and make it more suitable for the rutting prediction in the region. The research is mainly carried out on the full-scale test track of the Research Institute of Highway Ministry of Transport (RIOHTrack). A total of 14 kinds of asphalt pavement test sections from 7 categories, including semi-rigid base asphalt pavement, rigid base asphalt pavement, inverted asphalt pavement, and full-depth asphalt pavement with different thicknesses of asphalt concrete materials, are selected as the research object.

3. Methods

This research is mainly carried out on the RIOHTrack, which was completed in October 2015, and has been officially in operation for loading tests since December 2016. To date, the RIOHTrack has been in operation for more than 5 years for accelerated loading tests using real vehicles. As heavy trucks are used for the accelerated loading test, according to China's Specifications, each axle load of the truck should be converted into the cumulative number of equivalent standard axle loads (10 t) to characterize the traffic load level and correspond to the design service life of the pavement structure. During the test, heavy trucks are used for accelerated loading. Given that the pavement structure simulated by the RIOHTrack accelerated loading test has been in service for long enough, the accuracy of the rutting prediction model of asphalt pavement in China can be evaluated on the basis of the evolution data of the pavement structure rutting depth obtained during accelerated loading tests, and the correction method is proposed.

4. Overview of RIOHTrack Full-Scale Pavement Test Track

The RIOHTrack, for which this research based, is the first full-scale pavement test track of road fields in China. It is located in Beijing, where the average air temperature of the coldest month is −4.6 °C and the average air temperature of the hottest month is 25.8 °C. The RIOHTrack, with a total length of 2039 m, is an enclosed curve composed of straight lines and circular curves, with a north–south trend and a symmetrical arrangement. The long-term observation data of each pavement structure of the RIOHTrack is selected to carry out the model accuracy analysis, for several reasons. First, various pavement structures paved by the RIOHTrack are very representative in China, basically covering more than 90% of the commonly used pavement structure types on the asphalt pavement of China's expressways. Moreover, each test section of the RIOHTrack was constructed with the same structure thickness, pavement material, and construction technology as the actual project. The width of each test section is 7.5 m, and two inner and outer lanes are set; each lane is 3.75 m wide, the section length is 50–60 m, and the test size is also completely consistent with the actual engineering project. Therefore, these test structures can better reflect the real situation of actual engineering projects, and the performance observations carried out on this basis are also well representative. Second, the evaluation and accuracy analysis of the rutting model requires the support of long-term service performance observation data. We know that the evolution of pavement service performance is a long-term process. If calculated according to natural life, the service life of pavement usually reaches several decades, which is too long for model verification, and it is difficult and impractical to obtain continuous observation data for decades. The RIOHTrack is a very efficient accelerated loading test platform with the ability to simulate long-term performance in short-term tests. It accelerates the service performance of pavement by increasing the test load of the vehicle. According to the axle load conversion equation, it can be known that each time the test load of the existing RIOHTrack acts once, it is equivalent to dozens or hundreds of times the ESAL of 10 t acts. In this way, the simulation of the long-term service performance of pavement can be realized in the short term. These long-term performance observations can be used to verify the model.

Structure I is a semi-rigid base pavement with an asphalt layer thickness of 12 cm, and the structure numbers are STR1 and STR2. Structure II is rigid base asphalt pavement, the structure numbers are STR4 and STR5. Structure III is a semi-rigid base pavement with an asphalt layer thickness of 18 cm, and the structure numbers are STR7 and STR8. Structure IV is inverted asphalt pavement, and the structure numbers are STR10 and STR12. Structure V is a semi-rigid base pavement with an asphalt layer thickness of 24–28 cm, and the structure numbers are STR11 and STR13. Structure VI is a semi-rigid base pavement with an asphalt layer thickness of 36 cm, and the structure numbers are STR16 and STR17. Structure VII is full-depth asphalt pavement, and the structure numbers are STR18 and STR19 [36–39].

The RIOHTrack layout is shown in Figure 1, and the structure is shown in Figure 2. In the base layer shown in Figure 2, CBG25-I and CBG25-II are cement-bonded graded aggregate material, and the 7-day unconfined compressive strengths are 6 MPa and 4.5 MPa, respectively. CS is cement-stabilized soil, and the 7-day unconfined compressive strength is 2 MPa. LCC and CC are lean cement concrete and cement concrete, respectively. GA is graded aggregate.

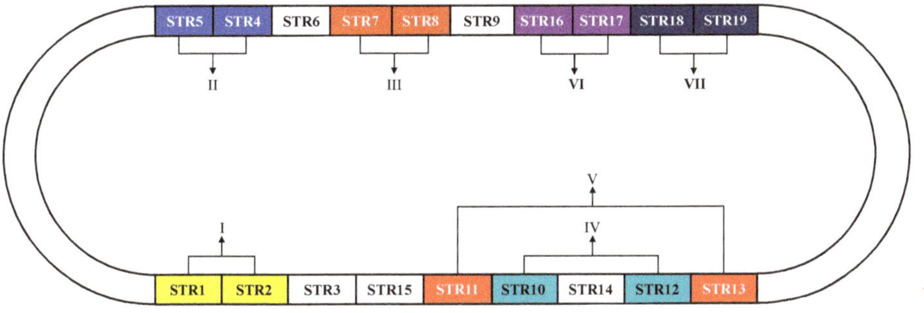

Figure 1. The RIOHTrack layout.

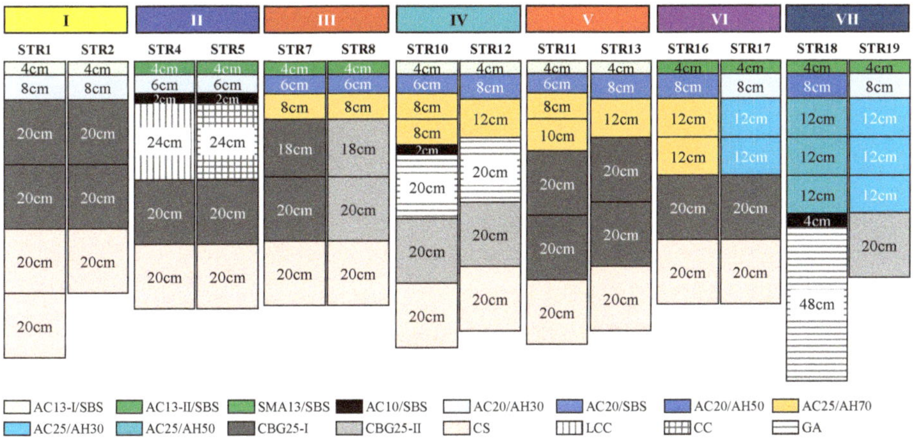

Figure 2. The RIOHTrack structure.

To ensure test efficiency, heavy trucks are used for the accelerated loading on the RIOHTrack. From December 2016 to December 2018, loading mode I is adopted for loading. The loading vehicles include four three-axle trucks (Figure 3). Since January 2019, loading mode II is adopted for loading. The loading vehicles, including six six-axle trucks, are updated, as shown in Figure 4. The test loading efficiency is increased to more than 3 times that of loading mode I. The axle load of the loading vehicles is heavy, more than 10 t, the standard axle load stipulated in China's Specifications. Therefore, it is necessary to convert it into the cumulative number of ESALs taking the rutting as a design indicator based on Equation (1). Based on this, the cumulative number of ESALs in RIOHTrack loading tests has been 51.60 million from December 2016 to April 2022, as shown in Figure 5.

$$N_e = C_1 C_2 \left(\frac{P_i}{10}\right)^4 \tag{1}$$

where C_1 refers to the axle-number coefficient of converted vehicles, converted according to the rutting equivalence principle; when the distance between the front and rear axles is greater than 3 m, take 1. when the distance between the front and rear axles is less than 3 m, take 1.05; C_2 refers to the wheel-set coefficient of converted vehicles. 1.0 for double wheels and 4.5 for single wheels; and P_i refers to the axle load of the converted vehicles (t).

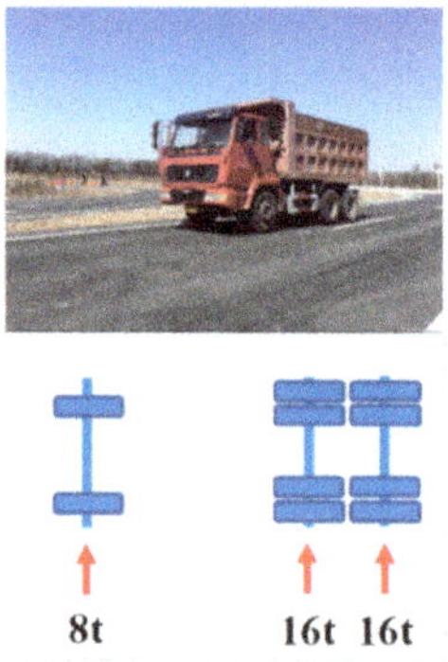

Figure 3. Axle load and axial distribution of trucks in loading mode I.

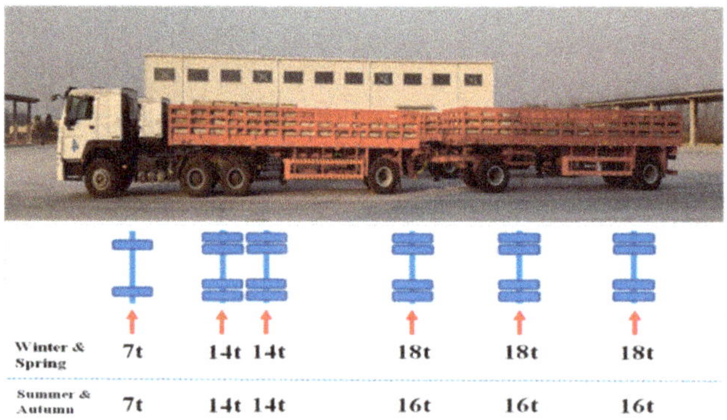

Figure 4. Axle load and axial distribution of trucks in loading mode II.

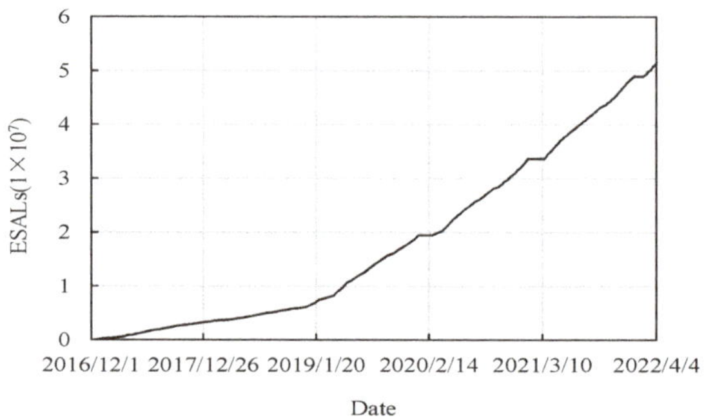

Figure 5. Curves of the cumulative number of equivalent single-axle loads.

5. Rutting Prediction Model and Parameter Values of China's Asphalt Pavement

The rutting performance prediction model in China's Specifications is established by the cumulative permanent deformation. First, each asphalt mixture layer is layered, and then according to the rutting test under standard conditions, the rutting permanent deformation of each layer of asphalt mixture is obtained, and the permanent deformation of each layer and the total permanent deformation of the asphalt mixture layer are calculated, so as to achieve the purpose of controlling the structure design. The rutting performance prediction model expressions are shown in Equations (2) and (3).

$$R_a = \sum_{i=1}^{n} R_{ai} \quad (2)$$

$$R_{ai} = 2.31 \times 10^{-8} k_{Ri} T_{pef}^{2.93} p_i^{1.80} N_e^{0.48} R_{0i} \quad (3)$$

where R_a is the permanent deformation of the asphalt mixture layer (mm); R_{ai} is the permanent deformation of the i-th layer (mm); R_{0i} is the permanent deformation of the i-th layer of the asphalt mixture in the laboratory rutting test at 60 °C (mm); N_e is the cumulative number of ESALs; T_{pef} is the rutting equivalent temperature of the asphalt mixture layer (°C); p_i is the vertical compressive stress on the top surface of the i-th layer (MPa); k_{Ri} is the comprehensive correction coefficient, calculated according to Equations (4)–(6); z_i is the depth (mm) from the midpoint of the i-th layer to the road surface; and h_a is the thickness of the asphalt mixture layer (mm); when it is greater than 200 mm, take 200 mm.

$$k_{Ri} = (d_1 + d_2 \cdot z_i) \cdot 0.9731^{z_i} \quad (4)$$

$$d_1 = -1.35 \times 10^{-4} h_a^2 + 8.18 \times 10^{-2} h_a - 14.50 \quad (5)$$

$$d_2 = 8.78 \times 10^{-7} h_a^2 - 1.50 \times 10^{-3} h_a + 0.90 \quad (6)$$

Since the long-term evolution data of rutting in this research is obtained based on the RIOHTrack full-scale pavement test track, the traffic loads on each pavement structure are exactly the same, so the N_e of each structure should take the same value for rutting prediction, which can be taken according to Equation (1) and Figure 5.

According to the regulations in China's Specifications, T_{pef} is related to the temperature in the region where the pavement structure is located and the thickness of the asphalt mixture layer. It can be seen from the calculation that for the STR1, STR2, STR4, and STR5 in this research, the equivalent temperature of the rutting is 22.02 °C; the equivalent temperature of rutting of STR7 and STR8 is 22.98 °C; and the equivalent temperature of rutting of STR10, STR12, STR11, STR13, STR16, STR17, STR18, and STR19 is 23.30 °C.

There are 10 kinds of asphalt mixtures in the structures of this research. The test results of the permanent deformation of each asphalt mixture in the laboratory rutting test at 60 °C are shown in Table 1. The value of R_{0i} can be selected according to Table 1. In addition, p_i and k_{Ri} can be calculated according to the layered model of each pavement structure, and calculated according to the mechanics of the elastic layered system and Equations (4)–(6), and will not be repeated here.

Table 1. Permanent deformation of asphalt mixture rutting test.

AC Type	AC10	AC13-I	AC13-II	SMA13	AC20	AC20	AC20	AC25	AC25	AC25
Asphalt Type	SBS	SBS	SBS	SBS	AH30	AH50	SBS	AH30	AH50	AH70
Permanent Deformation (mm)	2.40	1.54	1.65	1.82	1.70	2.67	1.30	1.96	2.12	2.82

6. Results and Discussion

6.1. Evaluation on Asphalt Pavement Rutting Prediction Model

Figure 6 shows the comparison between the predicted value of the model and the measured value during the long-term evolution of rutting of seven categories of structures. To evaluate the accuracy and prediction effect of the rutting prediction model quantitatively, the mean square error (MSE), root mean square error (RMSE), mean absolute error (MAE), mean relative error (MAPE), coefficient of determination (R^2), and other indicators between the predicted value of the rutting model and the measured value are calculated and summarized in Table 2.

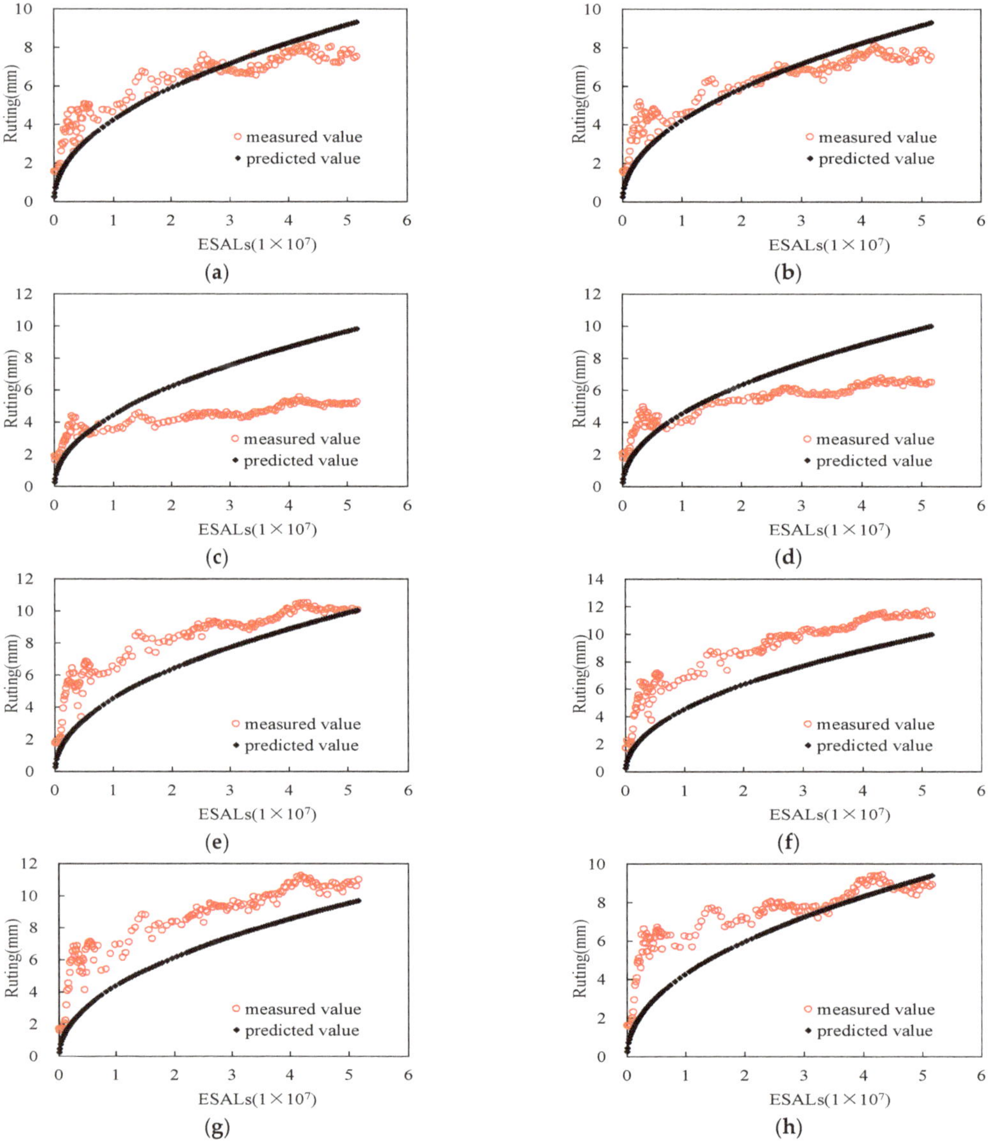

Figure 6. *Cont.*

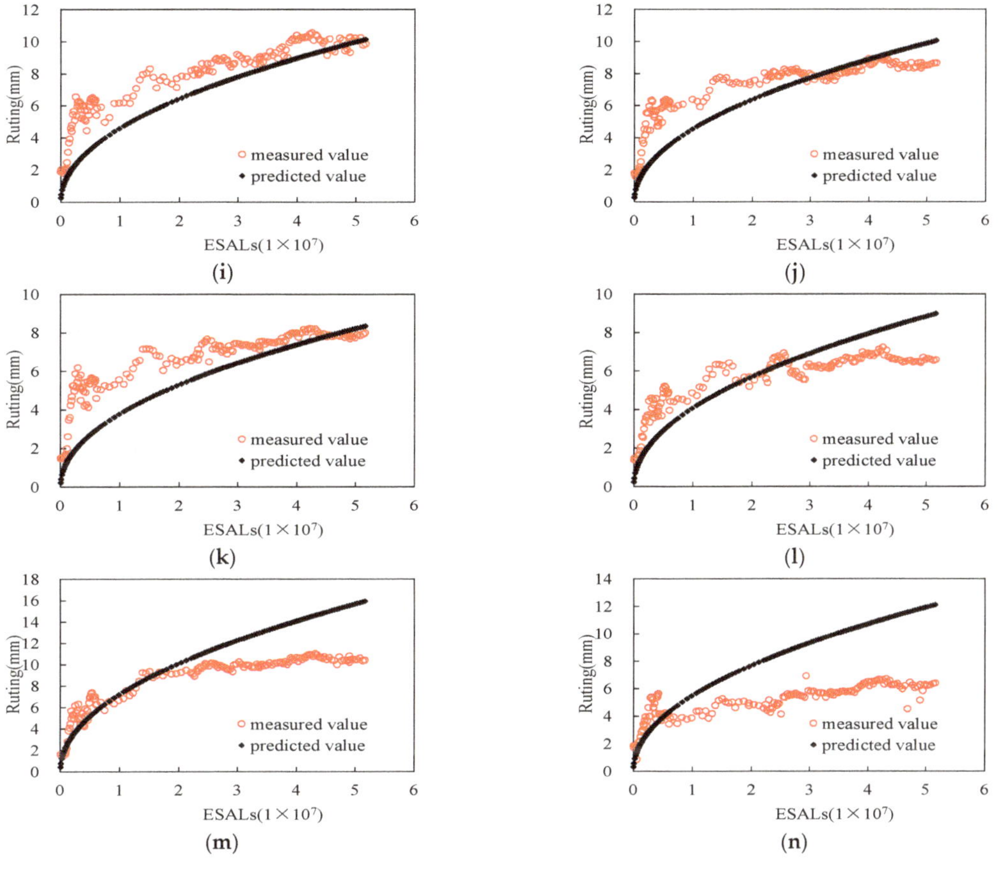

Figure 6. Comparison between the predicted value of the model and the measured value. (**a**) Structure I-STR1. (**b**) Structure I-STR2. (**c**) Structure II-STR4. (**d**) Structure II-STR5. (**e**) Structure III-STR7. (**f**) Structure III-STR8. (**g**) Structure IV-STR10. (**h**) Structure IV-STR12. (**i**) Structure V-STR11. (**j**) Structure V-STR13. (**k**) Structure VI-STR16. (**l**) Structure VI-STR17. (**m**) Structure VII-STR18. (**n**) Structure VII-STR19.

Table 2. Error between the predicted value of the rutting model and the measured value.

Structure Type		MSE	RMSE	MAE	MAPE	R^2
Unit		mm^2	mm	mm	-	-
I	STR1	1.16	1.08	0.91	18.3%	0.6346
	STR2	1.12	1.06	0.85	17.6%	0.6223
II	STR4	7.70	2.78	2.40	53.7%	−8.2491
	STR5	3.67	1.92	1.69	32.1%	−1.2133
III	STR7	3.52	1.88	1.64	25.2%	0.3337
	STR8	5.68	2.38	2.29	30.9%	0.2222
IV	STR10	5.63	2.37	2.20	30.5%	0.0469
	STR12	3.09	1.76	1.35	23.0%	0.0860
V	STR11	2.89	1.70	1.43	23.1%	0.4220
	STR13	2.45	1.56	1.26	21.8%	0.2500
VI	STR16	2.67	1.63	1.31	24.4%	0.0444
	STR17	1.72	1.31	1.15	23.1%	0.2084
VII	STR18	7.23	2.69	2.20	25.3%	−0.0792
	STR19	5.51	2.35	2.01	31.8%	−1.0787

It can be seen from Figure 6 and Table 2 that:

(1) For Structure I, the predicted value of the rutting model is relatively close to the measured value. The measured result fluctuates up and down near the prediction curve, and the error of the model prediction is small. For STR2 with the smallest error, the MSE is 1.12 mm^2; the RMSE is 1.06 mm; the MAE is 0.85 mm; and the MAPE is 17.6%. The R^2 of STR1 and STR2 rutting models are 0.6346 and 0.6223, respectively. According to [12], the R^2 of the predicted value of the model and the measured value in MEPDG is 0.6425. For Structure I, the accuracy of the rutting model in China's Specifications is basically equivalent to that in American MEPDG, and the model has a similar prediction effect.

(2) For Structures II and VII, the correlation between the predicted value of the rutting model and the measured value is similar. A great difference is observed between the predicted value and the measured value of the rutting model with the two categories of structures. The predicted value is significantly larger than the measured value. The predicted curve is above the measured value. In addition, with the increase in the cumulative number of ESALs, the error in the model prediction rises. For STR5 with the smallest error, the MSE is 3.67 mm^2; the RMSE is 1.92 mm; the MAE is 1.69 mm; and the MAPE is 32.1%. The R^2 of STR4, STR5, STR18, and STR19 structural rutting models, namely, −8.2491, −1.2133, −0.0792, and −1.0787, respectively, are all negative. According to the meaning of the determination coefficient R^2, when R^2 is negative, the effect of the simulated prediction is poor and even cannot reach the prediction accuracy of the average value curve of the measured value. Therefore, for Structures II and VII, the accuracy of the rutting model in China's Specifications is relatively poor. This model is not suitable for the rutting prediction. The reason for this phenomenon may be that the characteristics of the two structures are not fully considered when Equation (3) is established, thereby resulting in the poor prediction effect of the model. Figure 2 shows that Structure II belongs to the rigid base asphalt pavement. The lean concrete or cement concrete is adopted for the base, and the modulus and strength of the base are relatively high in value. Structure VII belongs to the full-depth asphalt pavement structure with the thickness of asphalt materials reaching 48–52 cm. According to [3], the rutting model in China's Specifications is established on the basis of the rutting data of 10 highways and 5 test sections. Semi-rigid materials are used as the base or subbase of more than 95% of asphalt pavement in China, and the thickness of the asphalt concrete structural layer generally does not exceed 30 cm. Therefore, to make the rutting model in the design specification more widely applicable, semi-rigid materials are used as the base in the construction of 10 highways and 5 test sections used in the model, and the thickness of the asphalt concrete is in the range of 16–28 cm. When establishing Equation (3), Structure II, which has lean concrete or cement concrete as the base, and Structure VII, which has 48–52 cm-thick asphalt material layer thickness, are not considered, which is also the main reason for the poor prediction effect of Equation (3) on the two kinds of structures.

(3) For Structure III and STR10 in Structure IV, the predicted value is significantly larger than the measured value. The predicted curve is below the measured value. After the initial loading stage (N_e is about 5 million), the predicted curve is basically parallel to the measured curve, and the difference between the predicted value and the measured value is basically constant with the increase in cumulative ESALs. For STR7 with the smallest error, the MSE is 3.52 mm^2; the RMSE is 1.88 mm; the MAE is 1.64 mm; and the MAPE is 25.2%. The R^2 of STR7, STR8, and STR10 structural rutting models are 0.3337, 0.2222, and 0.0469, respectively. R^2 is in the range of 0–0.4. The prediction effect is not ideal.

(4) For the remaining structures, namely, STR12 in Structures IV, V, and VI, an intersection between the model prediction curve and the measured curve of the rutting model exists with the increase in ESALs. Therefore, it can be obviously divided into two stages.

The first stage is the first section of the whole rutting evolution curve. When the ESALs reach a certain critical value ($N_e{}'$) from 0 at the beginning, the predicted value of the rutting model is obviously less than the measured value. However, with the increase in ESALs, the error of the model prediction gradually becomes small. When it reaches the critical value ($N_e{}'$) of ESALs, the predicted value is almost the same as the measured value. The second stage is the latter section of the whole rutting evolution curve. When the number of ESALs is greater than the critical value ($N_e{}'$), the predicted value of the rutting model is higher than the measured value. With the increase in ESALs, the model prediction error gradually rises. For different structures, the critical values ($N_e{}'$) of ESALs vary greatly. For STR17 with the smallest error, the MSE is 1.72 mm^2; the RMSE is 1.31 mm; the MAE is 1.15 mm; and the MAPE is 23.1%. The critical values ($N_e{}'$) of STR12, STR11, STR13, STR16, and STR17 are 35 million, 50 million, 30 million, 45 million, and 25 million, respectively. The R^2 of five structural rutting models are 0.0860, 0.4220, 0.2500, 0.0444, and 0.2084. R^2 is in the range of 0–0.5. The prediction effect is not ideal. From the meaning of the R^2, although the rutting model can be used to predict these structures with R^2 between 0 and 0.5, the accuracy is not high. The main reason is that the rutting observation data in this research are based on the RIOHTrack accelerated loading test. Compared with 10 highways and 5 test sections used for establishing the model in China's Specification, some differences are observed in the geographical location, climate environment, and traffic load, etc. A local correction coefficient should be introduced into the rutting model by referring to [26–33], and the model must be modified to obtain a rutting prediction model suitable for the location of the RIOHTrack, thereby improving the accuracy of model prediction.

According to the equivalence relationship between the design life and the cumulative ESALs in China's Specification, under the heavy traffic load level, the expressway with a 30-year design life is equivalent to more than 50 million cumulative ESALs of pavement. Table 3 summarizes the measured rutting values and model-predicted values of each pavement structure of the RIOHTrack under the cumulative 50 million ESALs in Figure 6. From the data in Table 3, it can be seen that, at the end of design life, the structures with the largest difference between the measured rutting values and the model predicted values are Structure II and VII, which are also predicted poorly in Table 2. Furthermore, STR7 and STR11 have the smallest difference between the measured rutting values and the model predicted values. However, from the rutting evolution curve in Figure 6, the rutting model has the highest prediction effect only when it reaches the end of design life. The prediction effect of the whole life cycle is not ideal, which is basically consistent with the results of the previous analysis.

Table 3. Rutting values at the end of design life.

Structure Type	I		II		III		IV		V		VI		VII	
	STR1	**STR2**	**STR4**	**STR5**	**STR7**	**STR8**	**STR10**	**STR12**	**STR11**	**STR13**	**STR16**	**STR17**	**STR18**	**STR19**
RIOHTrack Measured values (mm)	7.70	7.89	5.15	6.40	9.99	11.60	10.69	8.90	10.05	8.50	7.89	6.63	10.45	7.36
Model Predicted Values (mm)	9.20	9.19	9.69	9.87	9.91	9.85	9.57	9.28	10.02	9.91	8.25	8.86	15.72	11.96
MAPE (%)	19.5	16.5	88.1	54.2	0.8	15.1	10.5	4.3	0.4	16.6	4.5	33.8	50.5	62.5

The aforementioned analysis results show that the rutting model in China's Specifications has significant structural dependence. Some problems are encountered in using a unified rutting model for the prediction of different structures, especially for rigid base and full-depth asphalt pavements. The rutting evolution process of the two structures is not considered when the model is established, so the prediction effect of the model on both structures is poor. To ensure the prediction accuracy of the rutting model, we should correct, improve, and optimize the established rutting model combined with the long-term evolution data of rutting deformation of pavement structure with load and environment according to the characteristics of the pavement structures and materials in different areas, as well as the geographical location, climate environment, and traffic load, to improve the prediction accuracy. Finally, a high-precision rutting prediction model suitable for different pavement structures and areas is obtained.

6.2. Correction of Asphalt Pavement Rutting Prediction Model

To ensure the wide applicability of the rutting prediction model in China's Specifications and the model in MEPDG, the measured rutting data from actual projects and test tracks in different regions of the country were used in establishing the model. As a result, the applicability is ensured, but the prediction accuracy of the model is limited. As shown in [12], when R^2 is higher than 0.60, the prediction effect of the model is acceptable and can be used for design. After verifying the rutting prediction models of different pavement structures based on the RIOHTrack measured results, the predicted value of only the rutting with Structure I is close to the measured value. The R^2 of the rutting model can reach 0.60 and above, and the accuracy of other structures cannot reach the value. The reason for this phenomenon is that, on the one hand, the structure and material characteristics of some sections, such as rigid base structure and full-depth asphalt concrete layer structure, are not considered when the model is established, which leads to the poor prediction effect of the model on these structures. On the other hand, the RIOHTrack, which provides the measured data in this research, has certain differences from the model in climate and traffic environments. The rutting model can only be used with local correction.

At present, in terms of the rutting model, the commonly used correction method aims to introduce a local correction coefficient based on the existing model expression to improve the model, thereby improving the prediction accuracy. In [26–33], the local correction is carried out for the rutting model in MEPDG. The main method is that three local correction coefficients, namely, β_1, β_2, and β_3, are introduced to correct the constant term of the model, the exponential term of temperature, and the exponential term of cumulative ESALs, respectively. Subsequently, the model parameters are calibrated using the rutting observation results of local actual projects or test tracks. The rutting model in China's Specifications is similar to that in the MEPDG in its principles and expression. Therefore, this method can be used to make the local correction of the rutting model to establish a high-precision rutting model suitable for the region where the RIOHTrack is located. According to this idea, we propose a local correction method for the prediction model of the rutting performance of China's asphalt pavement, which is briefly described as follows:

(1) On the basis of the existing rutting performance prediction model Equation (3) in China, the local correction coefficients β_1, β_2, and β_3 are introduced to improve a new rutting prediction model; see Equation (7). Among them, β_1 is used to correct the constant term of the existing rutting model, β_2 is used to correct the exponential term of the temperature, and β_3 is used to correct the exponential term of the cumulative ESALs.
(2) Relying on local projects and test sections, long-term observation of rutting deformation of asphalt pavement is carried out, and the measured rutting data with the cumulative ESALs are obtained. These data can reflect the long-term evolution of pavement rutting under the combined effect of multiple factors such as local climate environment, traffic load, pavement structure, and material type.
(3) Based on the long-term observation results of rutting deformation, a new rutting prediction model after correction is used to fit the rutting data according to the least squares method, so the local correction coefficients β_1, β_2, and β_3 can be obtained, and the model prediction errors are obtained to judge the model accuracy. Finally, taking these coefficients into Equation (7), the new rutting prediction model after local correction can be obtained.

$$R_a = \sum_{i=1}^{n} 2.31 \times 10^{-8} \cdot \beta_1 k_{Ri} T_{pef}^{2.93 \cdot \beta_2} p_i^{1.80} N_e^{0.48 \cdot \beta_3} R_{0i} \tag{7}$$

Equation (7) is adopted to fit the measured results shown in Figure 6. The local correction coefficients of different structures and all kinds of errors in the model prediction can be obtained (Table 4). Table 4 shows that the accuracy of rutting prediction models of all structures is greatly improved after local correction. The determination coefficients R^2 are higher than 0.87, which indicates that the new rutting prediction model after local correction is suitable for the rutting prediction of all kinds of pavement structures where the RIOHTrack is located.

Table 4. Local correction coefficients and errors of rutting model.

Structure Type		β_1	β_2	β_3	MSE	RMSE	MAE	MAPE	R^2
I	STR1	6.1319	1.2123	0.5440	0.21	0.46	0.38	8.6%	0.9331
	STR2	6.1201	1.2120	0.5425	0.18	0.42	0.31	7.4%	0.9400
II	STR4	9.4764	1.2749	0.3677	0.11	0.33	0.24	6.6%	0.8722
	STR5	7.5312	1.2573	0.4409	0.13	0.36	0.26	6.4%	0.9237
III	STR7	7.0638	1.2254	0.5330	0.31	0.56	0.40	8.2%	0.9415
	STR8	5.4336	1.1970	0.6079	0.30	0.55	0.39	7.5%	0.9591
IV	STR10	7.2853	1.2287	0.5340	0.47	0.69	0.49	10.1%	0.9203
	STR12	10.0245	1.2674	0.4357	0.41	0.64	0.48	10.6%	0.8787
V	STR11	7.0344	1.2237	0.5301	0.34	0.59	0.46	8.7%	0.9311
	STR13	9.6484	1.2624	0.4367	0.31	0.56	0.42	9.2%	0.9045
VI	STR16	9.9075	1.2660	0.4412	0.31	0.55	0.39	10.0%	0.8901
	STR17	8.2984	1.2436	0.4598	0.24	0.49	0.38	9.6%	0.8913
VII	STR18	4.8346	1.1941	0.5643	0.46	0.68	0.52	10.8%	0.9312
	STR19	6.9198	1.2416	0.4625	0.25	0.50	0.37	7.9%	0.9073

Figure 7 shows the comparison of results before and after the local correction of the rutting models of several representative pavement structures. After local correction, the predicted values of the rutting models with these structures are relatively close to the measured values. The measured result fluctuates up and down near the prediction curve, the error of the model prediction decreases, and the accuracy is improved significantly. For Structures II and VII, the prediction curve is no longer above the measured curve but fluctuates up and down near the measured curve. R^2 becomes higher than 0.87 from a negative value. For STR5 with the smallest error, the MSE is 0.13 mm^2; the RMSE is 0.36 mm; the MAE is 0.26 mm; and the MAPE is 6.4%. The rutting model can fit the measured value of rutting better, and the prediction effect of the model is better. For Structures III, IV, V, and VI, the prediction curve is no longer parallel to the measured curve and no longer crosses in the middle but fluctuates up and down near the measured curve. R^2 increases to 0.87 and above from 0–0.50. For STR8 with the smallest error, the MSE is 0.30 mm^2; the RMSE is 0.55 mm; the MAE is 0.39 mm; and the MAPE is 7.5%. The accuracy of the model has been improved significantly.

Comparing the results in Figures 6 and 7, it can be seen that the prediction effect of the new rutting model after correction is significantly better than that of the model before, which is mainly related to the introduction of local correction coefficients β_1, β_2, and β_3. The evolution curves of the measured rutting values of 14 kinds of pavement structures in Figure 6 with the cumulative ESALs basically show a similar phenomenon; that is, the rutting deformation in the initial loading stage (N_e is about 0 to 5 million ESALs) increases rapidly, and the rutting evolution curve is relatively steep. When the initial stage is over, the rutting deformation begins to increase slowly, and the rutting evolution curve also becomes relatively flat. The reason for this phenomenon may be that in the initial loading stage, the voids inside the mixture of the newly paved asphalt pavement are compressed under the load, resulting in obvious compaction deformation of the asphalt concrete structure layer, as can be seen from Figure 6. At the end of the initial loading stage (generally N_e is about 5 million ESALs), the compaction deformation of the asphalt pavement can reach about 4–6 mm. After that, due to the limited compressible voids inside the asphalt mixture, when the voids can no longer be compressed, the asphalt mixture begins to produce plastic permanent deformation. Compared with the compaction deformation, the increase of the plastic deformation is relatively slow and not as rapid as the compaction deformation in the initial stage.

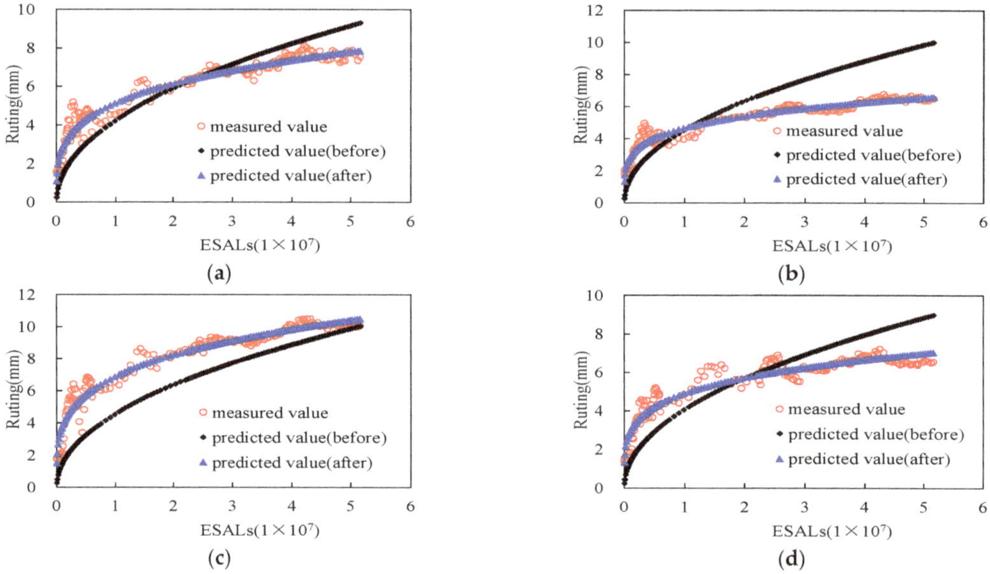

Figure 7. Comparison of predicted and measured values before and after rutting model correction. (**a**) STR2. (**b**) STR5. (**c**) STR7. (**d**) STR17.

As can be seen from Figure 7, although the overall change in trend of the new and old models is in the form of a monotonically increasing power function, the new model has a steeper curve shape in the initial loading stage, which can better simulate the initial compaction deformation of asphalt mixture when N_e is from 0 to 5 million ESALs. As seen in Equation (7), correction coefficients β_1 and β_2 respectively correct the constant term of the model and the exponential term of the equivalent temperature, and the equivalent temperature is fixed for the same structure, so the coefficient β_2 can also be regarded as a correction to the constant term. According to the results in Table 4, it can be seen that when correction coefficients β_1 and β_2 are introduced, the constant term of the new model is numerically at least approximately 6 times higher than that of the old model. In addition, when the initial loading stage is over, the evolution curve of the second stage of the new model is smoother than that of the old model, and it can better simulate the slowly increasing plastic permanent deformation, and the model prediction effect is also better. It can be seen from Equation (7) that β_3 mainly corrects the exponential term of N_e, and N_e is also the only independent variable in Equation (7), and the evolution curve in the second stage of the model mainly depends on the value of the exponential term. The smaller the value, the smoother the curve. From the values in Table 4, it can be seen that the average value of β_3 is about 0.5, the exponential term of the new model is half of that of the old model, and its curve is gentler, which can better simulate the slowly increasing plastic permanent deformation after the initial loading stage.

It should be noted that the measured rutting data of 14 kinds of structures in Figure 6 are obtained based on the accelerated loading test of the RIOHTrack's full-scale pavement structure. Since the test section is built in the wild, it has always been bearing the long-term effect of the local climate environment, and the pavement structure and material type are quite different. In other words, the measured rutting data in Figure 6 is the test result under the combined effect of local climate environment, traffic load, pavement structure, and material type, which also comprehensively reflects the influence of these factors. On the basis of this test data, the new rutting prediction model obtained after local correction will have higher regional applicability, and is more suitable for rutting prediction of various pavement structures in the area where the RIOHTrack is located, and the accuracy is also higher.

7. Conclusions

Based on the measured value of the rutting model with the pavement structure represented by the RIOHTrack full-scale test track, the rutting prediction model in China's Specifications is analyzed. Moreover, the following conclusions are drawn:

(1) The rutting model in China's Specifications for Design of Highway Asphalt Pavement has significant structural dependence. Different structures are predicted using a unified rutting model with different prediction accuracies.

(2) The existing rutting model is used for prediction. Structure I has the highest accuracy and the smallest error, and the R^2 is greater than 0.6. Rigid base and full-depth asphalt pavement structures have the worst accuracy, and the R^2 is negative. The prediction accuracy of other structures is not high, and the R^2 is in the range of 0–0.5.

(3) To improve prediction accuracy, local correction coefficients are introduced in the existing rutting prediction model to establish a new one. The RIOHTrack rutting measured value is adopted to fit the new rutting prediction model, and the local correction coefficients of different structures are obtained.

(4) After local correction, the accuracy of the rutting prediction model of all structures is greatly improved. The R^2 values are greater than 0.87. The predicted values of the rutting models with these structures are relatively close to the measured values. The measured result fluctuates up and down near the prediction curve, the error of the model prediction declines, and the accuracy is improved significantly.

(5) The measured rutting data in this research are the test results under the combined effect of local climate environment, traffic load, pavement structure, and material type. On the basis of this test data, the new rutting prediction model obtained after local correction will also have higher regional applicability, and is more suitable for rutting prediction of various pavement structures in the area where the RIOHTrack is located, and the accuracy is also higher.

In this research, a local correction method of the rutting prediction model is proposed. Although this method is discussed for the rutting model in China's Specification, the basic principle of this method is to introduce correction coefficients. The constant terms, exponential terms of loads, and environment parameters in the model are modified respectively to improve the prediction accuracy of the model. The ideas and methods of this model correction also have a certain reference value for the correction of rutting models in other regions. When the accuracy of the estimation of the existing rutting models in other regions is poor, you can refer to this research's method to make local correction on key parameters to improve accuracy and enhance regional applicability. In addition, the RIOHTrack full-scale pavement test track, which this research is based on, is still carrying out accelerated loading tests. It is expected that by the end of 2022, cumulative ESALs will reach 65 million ESALs. The related work in this research will also carry out further tracking research and analysis with the update of long-term observation data of rutting.

Author Contributions: The authors confirm contribution to the paper as follows: conceptualization and methodology: X.Z.; data curation: Y.W. and Z.M.; formal analysis and validation: Y.W., X.Z. and X.W.; writing and visualization: Y.W., X.Z. and Z.M. All authors have read and agreed to the published version of the manuscript.

Funding: This work was funded by the National Key of Research and Development Plan under Grant number 2020YFA0714300.

Institutional Review Board Statement: Not applicable.

Informed Consent Statement: Not applicable.

Data Availability Statement: Not applicable.

Conflicts of Interest: The authors declare no conflict of interest.

References

1. AASHTO. *Mechanistic-Empirical Pavement Design Guide: A Manual of Practice, Interim Edition*; American Association of State Highway and Transportation Officials: Washington, DC, USA, 2008.
2. Shell. *Pavement Design Manual*; Shell International Petroleum Company, Ltd.: London, UK, 1978.
3. Ministry of Transport of the People's Republic of China. *Specifications for Design of Highway Asphalt Pavement*; China Communications Press: Beijing, China, 2017.
4. Tseng, K.H.; Lytton, R.L. *Prediction of Permanent Deformation in Flexible Pavement Materials*; American Society for Testing and Materials: West Conshohocken, PA, USA, 1989.
5. Yao, Z.K. *Structural Design of Asphalt Pavements*; China Communications Press: Beijing, China, 2011.
6. Van de Loo, P.J. Practical approach to the prediction of rutting in asphalt pavements: The Shell method. *Transp. Res. Rec.* **1976**, *616*, 15–21.
7. Van de Loo, P.J. Creep test: A key tool in asphalt mix evaluation and in prediction of rutting. *J. Assoc. Asph. Paving Technol.* **1978**, *47*, 522–557.
8. Zhang, J.; Cooley, L.A., Jr.; Kandhal, P.S. Comparison of fundamental and simulative test methods for evaluating permanent deformation of hot mix asphalt. *Transp. Res. Rec.* **2002**, *1789*, 91–100. [CrossRef]
9. Monismith, C.L.; Hicks, R.G.; Finn, F.N.; Sousa, J.; Harvey, J.; Weissman, S.; Deacon, J.; Coplantz, J.; Paulsen, G. *Permanent Deformation Response of Asphalt Aggregate Mixes*; Report SHRP-A-415; Strategic Highway Research Program, National Research Council: Washington, DC, USA, 1994.
10. Monismith, C.L.; Epps, J.A.; Finn, F.N. Improved asphalt mix design. *J. Assoc. Asph. Paving Technol.* **1985**, *54*, 340–406.
11. El-Basyouny, M.M.; Witczak, M.; Zapata, C.E. Development of asphalt concrete rutting model using permanent strain database. In Proceedings of the Transportation Research Board 86th Annual Meeting, Washington, DC, USA, 21–25 January 2007.
12. National Cooperative Highway Research Program. *Guide for Mechanistic-Empirical Design of New and Rehabilitated Pavement Structures, Final Document, Appendix GG-1: Calibration of Permanent Deformation Models for Flexible Pavements*; National Cooperative Highway Research Program: Champaign, IL, USA, 2004.
13. Kaloush, K.E.; Witczak, M.W. Tertiary flow characteristics of asphalt mixtures. *J. Assoc. Asph. Paving Technol.* **2002**, *71*, 248–280.
14. Kandhal, P.S.; Mallick, R.B. Effect of mix gradation on rutting potential of dense-graded asphalt mixtures. *Transp. Res. Rec.* **2001**, *1767*, 146–151. [CrossRef]
15. Yang, Z.; He, D. Summarizing and Improving of Rut Pre Evaluating Method in Asphalt Pavement. *J. East North Highw.* **2001**, *24*, 12–15.
16. Verstraeten, J.; Veverka, V.; Francken, L. Rational and Practical Designs of Asphalt Pavements to Avoid Cracking and Rutting. In Proceedings of the 5th International Conference on the Structural Design of Asphalt Pavements, Delft, The Netherlands, 23–26 August 1982.
17. Zhang, D.; Li, J. Prediction of Rutting in Asphalt Pavement. *China J. Highw. Transp.* **1995**, *8*, 23–29.
18. Kenis, W.; Sherwood, J.; McMahon, T. Verification and Application of the Vesys Structural Subsystem. In Proceedings of the 5th International Conference on the Structural Design of Asphalt Pavements, Delft, The Netherlands, 23–26 August 1982.
19. Kim, H.B.; Buch, N.; Park, D. Mechanistic-empirical rut prediction model for in-service pavements. *Transp. Res. Rec.* **2000**, *1730*, 99–109. [CrossRef]
20. Kim, M.; Mohammad, L.; Elseifi, M. Effects of Various Extrapolation Techniques for Abbreviated Dynamic Modulus Test Data on the MEPDG Rutting Predictions. *J. Mar. Sci. Technol.* **2015**, *23*, 353–363.
21. Shami, H.I.; Lai, J.S.; D'angelo, J.A.; Harman, T.P. Development of Temperature-Effect Model for Predicting Rutting of Asphalt Mixtures Using Georgia Loaded Wheel Tester. *Transp. Res. Rec.* **1997**, *1590*, 17–22. [CrossRef]
22. Monismith, C.L.; McLean, D.B.; Ogawa, N. *Design Considerations for Asphalt Pavements*; FHWA-CA-77-D-6-1; Federal Highway Administration: Berkley, CA, USA, 1978.
23. Monismith, C.L.; Tayebali, A.A. Permanent deformation (rutting) consideration in asphalt concrete pavement sections (with discussion and closure). *J. Assoc. Asph. Paving Technol.* **1988**, *57*, 414–463.
24. Wijeratne, A.; Sargious, M. Prediction of Rutting in Virgin and Recycled Asphalt Mixtures for Pavement Using Triaxial Tests. *J. Assoc. Asph. Paving Technol.* **1987**, *56*, 111–129.
25. Huang, X.; Zhang, X.; Deng, X. Asphalt Pavement Rutting Prediction of High-Grade Highway. *J. Southeast Univ. (Nat. Sci. Ed.)* **2000**, *30*, 96–101.
26. Wang, H.; Zhang, C.; You, Z. Calibration of rutting prediction model in MEPDG based on mathematical statistics method. *J. Chang. Univ. (Nat. Sci. Ed.)* **2013**, *33*, 1–7.

27. Liu, W.; Lin, Z. Study on area correction factor of rutting prediction model for flexible base asphalt pavement based on MEPDG. *J. Highw. Transp. Res. Dev.* **2017**, *34*, 67–70.
28. Retherford, J.; Mcdonald, M. Reliability Methods Applicable to Mechanistic-Empirical Pavement Design Method. *Transp. Res. Rec. J. Transp. Res. Board* **2010**, *2154*, 130–137. [CrossRef]
29. Hoegh, K.; Khazanovich, L.; Jensen, M. Local Calibration of MEPDG Rutting Model for MnROAD Test Sections. *Transp. Res. Rec. J. Transp. Res. Board* **2010**, *2180*, 130–141. [CrossRef]
30. Muthadi, N.; Kim, Y. Local Calibration of Mechanistic-Empirical Pavement Design Guide for Flexible Pavement Design. *Transp. Res. Rec. J. Transp. Res. Board* **2008**, *2087*, 131–141. [CrossRef]
31. Seitllari, A.; Lanotte, M.; Kutay, M. Recommended Calibration Procedure of MEPDG Asphalt Rutting Model Using Repeated Load Permanent Deformation and Confined and Unconfined Dynamic Modulus Data. *J. Transp. Eng. Part B Pavements* **2020**, *147*, 04020079. [CrossRef]
32. Jannat, G.; Yuan, X.; Shehata, M. Development of regression equations for local calibration of rutting and IRI as predicted by the MEPDG models for flexible pavements using Ontario's long-term PMS data. *Int. J. Pavement Eng.* **2016**, *17*, 166–175. [CrossRef]
33. Hassan, H.; Baaj, H.; Kadhim, H. Calibration of MEPDG permanent deformation models using Hamburg Wheel Rut Tester and field data. *Int. J. Pavement Eng.* **2021**. [CrossRef]
34. AASHTO. *Mechanistic-Empirical Pavement Design Guide: A Manual of Practice*, 2nd ed.; American Association of State Highway and Transportation Officials: Washington, DC, USA, 2015.
35. AASHTO. *Mechanistic-Empirical Pavement Design Guide: A Manual of Practice*, 3rd ed.; American Association of State Highway and Transportation Officials: Washington, DC, USA, 2020.
36. Wang, X. Design of Pavement Structure and Material for Full-scale Test Track. *J. Highw. Transp. Res. Dev.* **2017**, *34*, 30–37.
37. Wang, X.D.; Zhang, L.; Zhou, X.Y.; Xiao, Q.; Wei, G. Review of Researches of RIOHTrack in 2017. *J. Highw. Transp. Res. Dev.* **2017**, *35*, 5–13.
38. Zhang, L.; Zhou, X.; Wang, X. Research progress of long-life asphalt pavement behavior based on the RIOHTrack full-scale accelerated loading test. *Chin. Sci. Bull.* **2020**, *65*, 3247–3258. [CrossRef]
39. Wang, X.D.; Zhang, L.; Zhou, X.Y.; Xiao, Q.; Guan, W.; Shan, L.Y. Research progress of RIOHTrack in China. In *Accelerated Pavement Testing to Transport Infrastructure Innovation*; Springer: Cham, Switzerland, 2020.

applied sciences

MDPI

Article

An Application of Android Sensors and Google Earth in Pavement Maintenance Management Systems for Developing Countries

Abdullah I. Al-Mansour * and Abdulraaof H. Al-Qaili

Civil Engineering Department, College of Engineering, King Saud University, Riyadh 11421, Saudi Arabia; aalqaili@ksu.edu.sa
* Correspondence: amansour@ksu.edu.sa

Abstract: Governments and road agencies face the challenging task of maintaining roads. One of the reasons this is challenging is that the maintenance process requires utilizing a substantial amount of road network condition data. There are many approaches for measuring road surface conditions which are very costly and time-consuming, as well as requiring skillful operators. Developing countries have limited budgets for planning and monitoring road maintenance. This research aims to establish a low-cost pavement maintenance management system for intermediate and small cities in developing countries. The system utilizes low-cost sensors embedded in smartphones that can be used to measure road surface conditions. Google Earth is then used to present maintenance data, select a maintenance strategy, and view the maintenance output information. Road Lab Pro, an android application, is used to collect the data and estimate the surface condition of roads by using accelerometers, gyroscopes, and a GPS. The road network is divided into segments and the road surface conditions are estimated for each segment using the smartphone application and a suspension vehicle. The required maintenance activities for these segments are then established. A priority index is determined for each segment to decide which segments should be maintained with the available budget. This effort allows us to investigate the feasibility of assessing road surface roughness using a smartphone to determine the presence of road distresses and the overall road condition, which is taken into account when making maintenance decisions. The application of this system reveals that these new technologies can provide cost-effective, easy-handling, and efficient ways for a road agency to perform good maintenance planning.

Keywords: flexible pavement; pavement maintenance; management system; smartphone sensors; Google Earth

Citation: Al-Mansour, A.I.; Al-Qaili, A.H. An Application of Android Sensors and Google Earth in Pavement Maintenance Management Systems for Developing Countries. *Appl. Sci.* **2022**, *12*, 5636. https://doi.org/10.3390/app12115636

Academic Editors: Amir Tabakovic, Jan Valentin and Liang He

Received: 5 April 2022
Accepted: 30 May 2022
Published: 1 June 2022

1. Introduction

Developing countries lack suitable road maintenance, which leads to the deterioration of roads. The current maintenance program reveals that the system used is insufficient to reflect changing conditions and is lacking in its ability to assist in decision making during pavement design [1]. A good pavement management system is therefore necessary to reduce the rate of deterioration and maintain the pavement in a serviceable state [2]. To plan for maintenance properly, sums of data are needed—particularly data on road conditions. Road condition data change over time, especially since collecting them regularly requires a significant investment of resources and time.

A pavement maintenance management system (PMMS) consists of a series of decision-based components used to determine where and how repairs should be carried out. Often, maintenance decision making is achieved by feedback from assessments conducted by the local authority. To attain the best utilization of available funds, it is essential to have convenient methods for assessing the road conditions and a database of these conditions so that analyses can be performed, which provides useful information for decision making.

Unfortunately, this is not the case in many developing countries due to their limited budgets and less advanced technology. Therefore, most road maintenance decisions are made based on engineering judgments [3].

Road pavement roughness is an acceptable indicator for evaluating the overall condition of a pavement's surface. Road roughness can give agencies useful information for the decision-making processes of maintenance planning and programming [4].

Road roughness is usually measured by an index such as the international roughness index (IRI). Many approaches are available for measuring the IRI, although most of them require sophisticated profilers and tools that are highly accurate but costly to obtain, operate, and maintain, as well as skillful operators. On the other hand, a popular practice in many developing countries is visual inspection. Visual inspection is a relatively cheap option but is very labor-intensive and time-consuming. Therefore, it is necessary to investigate low-cost, practical methods for evaluating roughness.

Nowadays, smartphones usually come with many useful and inexpensive embedded sensors, such as three-axis accelerometers, a gyroscope, and a GPS [5]. These sensors can measure the acceleration in m/s^2 along each of the x, y, and z axes, and can be used to recognize motion. Accelerometers in smartphones are primarily used for detecting the orientation of the screen. The embedded sensors in smartphones are capable of recording useful signals for road surface condition estimation, similar to those recorded by high-tech equipment [6].

The embedded sensors within smartphones have been studied with regard to their ability to measure road roughness. The "SmartRoadSense" system presented by Alessandroni et al. (2014) aimed to measure road surfaces using smartphones. They established a model for determining a road roughness index from data obtained by the system [7]. Douangphachanh and Oneyama (2014) estimated road conditions by using the intelligent vehicle monitoring system (IVMS) as a reference for establishing a road roughness index. They collected data via the AndroSensor smartphone application to determine road profiles and compute the IRI [6]. Islam et al. (2014) processed integrated acceleration data using computer software. The study collected road profile and acceleration data with both an inertial profiler and a smartphone mounted on a vehicle. The results showed that smartphones measured road roughness with an acceptable accuracy when compared with an inertial profiler [8]. Zeng et al. (2015) determined that road roughness depends on a normalized acceleration index. The data were collected by mounting two tablets on a vehicle dashboard. Data were captured on the acceleration in three dimensions, the GPS coordinates, and the speed. This revealed that the recommended index could define road surface deficiencies with a high precision (between 80 and 93 percent) [9]. Hanson et al. (2014) made an effort to correlate road roughness between a smartphone and a conventional profiler. They obtained measurements from 11 road segments on a secondary highway in New Brunswick, Canada, and thereafter concluded that there was a strong correlation between the results of the profiler and the smartphone [10]. To evaluate the feasibility of such a technique, Zeng et al. (2018) created a prototype smartphone app. In Virginia, the prototype app was used for collecting acceleration data from two routes by running the app on tablets. The suggested smartphone application technique enables the gathering of roughness data, according to the evaluation in Virginia. [11]. Yeganeh et al. (2019) used the embedded sensors of a smartphone to calculate pavement roughness and investigate the feasibility of assessing pavement surface conditions with smartphones [12]. Chatterjee and Tsai (2020) used low-cost smartphone sensor data to estimate pavement conditions, which were then entered into a known geographic information system (GIS) [13]. Dong and Li (2021) utilized an android-based application for collecting accelerometer data on a local road in Ireland. Their results revealed the ability of the application to collect data for the identification of road surface defects. This method attempts to ensure road serviceability at lower costs for inspection and maintenance [14]. Pavement roughness is a key indicator of pavement surface performance, since it directly reflects the pavement's serviceability to road users. In recent years, several road agencies have employed the IRI as an objective indicator of their pavement network conditions in their business plans. A significant relationship exists between the IRI and features of pavement surface distress, such as cracking, patching, depression, and raveling [15].

This paper presents an application method for android sensors in addition to the use of the Google Earth platform to aid the decision-making process in pavement maintenance management systems. The main objective of this paper is to provide a simple and cost-effective road maintenance management system that can be used by small and intermediate cities in developing countries. Efforts have been made to obtain a data inventory and prioritize road maintenance using sensors embedded in smartphones, as well as using the Google Earth platform to handle data and visualize the output of reporting.

2. Research Methodology

The suggested low-cost pavement maintenance management system (PMMS) includes the basic elements of a common PMMS: defining the roadway network and breaking it into manageable segments; creating an inventory for each segment; gathering pavement conditions; determining the maintenance activity required; developing a method for prioritizing segments; and documenting and reporting results. This system was implemented on a road network segment in the city of Hajjah, Yemen. Yemen is a developing country that lacks a proper and strategic maintenance plan and has a limited budget for road evaluation and maintenance. Hajjah is a small city in a mountainous area of northern Yemen without any systematic road maintenance strategies. Figure 1 illustrates some of the road segment statuses that were investigated.

Figure 1. The road segment statuses.

In this study, the implementation of maintenance to the Hajjah pavement network was developed with a systematic procedure as described above. The systematic procedure framework is presented in Figure 2.

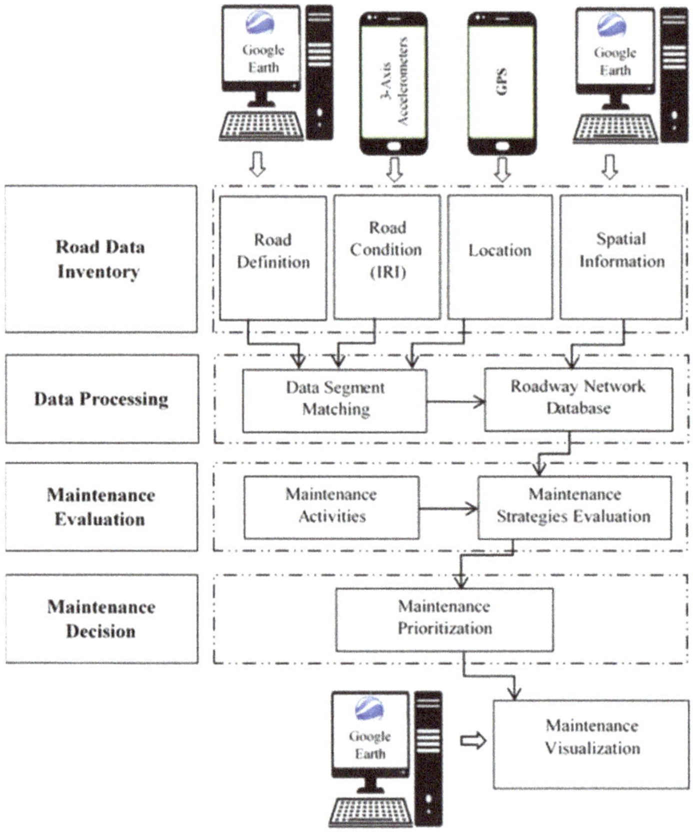

Figure 2. Flowchart of the proposed PMMS.

3. Network Coding

The Hajjah road network was classified functionally into three categories:

Major Roads: high-traffic roads that pass through the whole city.

Main Roads: medium-traffic collector roads.

Local Access Roads: low-traffic local roads.

Each road was divided into smaller manageable sections. A road does not always have consistent characteristics and therefore does not require the same maintenance treatment throughout its entire length. To identify these segments, the boundary between two segments was defined according to one of the following factors:

A change in the number of traffic lanes.

A change in pavement type.

A change in traffic patterns or volume.

Roadway intersections.

4. Road Inventory Data

It is important to collect and keep an accurate and reliable database for a particular road network. Data were collected to characterize each segment accurately. For this study, the data were collected and extracted from the city's General Corporation of Roads and Bridges as well as an internet and field survey. The inventory data for each section included the road name, the coding of the segment, the coordinates for the beginning and end of the segment, the number of lanes, the pavement type, and the segment length and width.

The Google Earth platform can be used to integrate, analyze, and represent a road network's inventory information. The main component that distinguishes Google Earth from other data systems is how data are stored and accessed. The standard Google Earth functions include thematic mapping, charting, manipulation, and simultaneous access to several databases. Visualization helps users deal with problems related to the dimensions of space and time through thematic maps rather than dimensionally restricted data graphs and tables.

A map of the Hajjah road network is presented in Google Earth, with color-coding based on the functional classification of the roads. Each road was divided into segments depending on the aforementioned factors. Road segment coding was then established. Subsequently, the start and end coordinates were defined. Figure 3 shows the Hajjah road network on Google Earth.

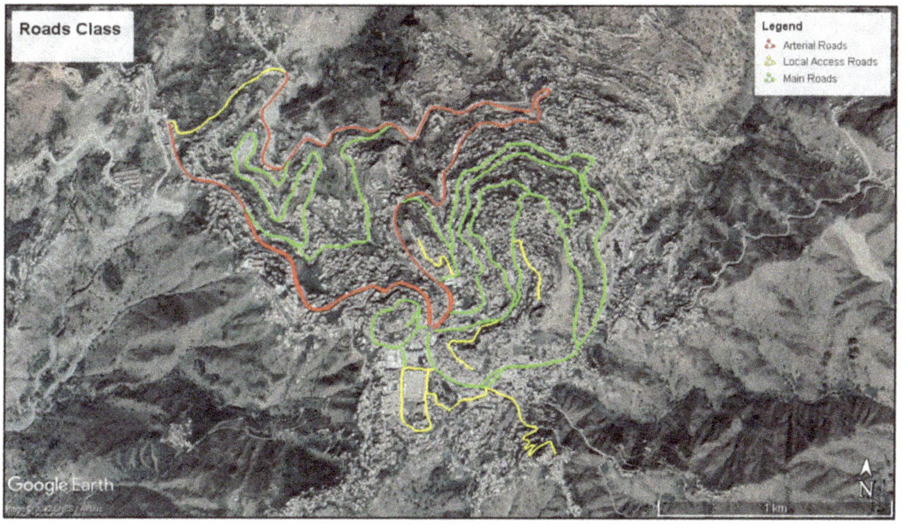

Figure 3. Hajjah road network with road classes shown on the Google Earth platform.

5. Road Condition Data

Since many existing pavement survey methods can be costly and labor-intensive, there is a need for simple and low-cost tools to facilitate the evaluation of pavement conditions with limited resources. To address this need, a smartphone application was utilized to assess pavement conditions. Smartphones contain accelerometers made of piezoelectric sensors, which can measure acceleration on the x, y, and z axes. Capable of measuring a large number of data points per second, accelerometers are very small and require very little power. A typical accelerometer is shown in Figure 4.

Road Lab Pro is a smartphone application created to measure the road roughness and monitor the road surface quality. This application was designed as a data collection tool for engineers by the World Bank in collaboration with the Beldor Center and Softteco [16]. It is used to record data from accelerometers, gyroscopes, and GPS devices to measure road roughness and create maps of road networks. To present the results graphically, a color-coding convention was adopted, as shown in Figure 5.

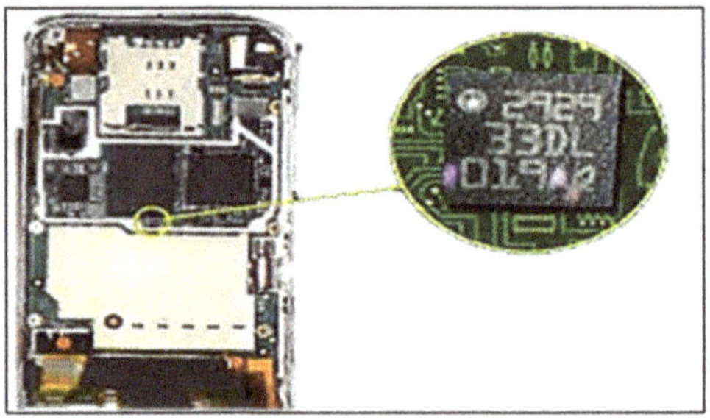

Figure 4. A typical accelerometer used in a smartphone.

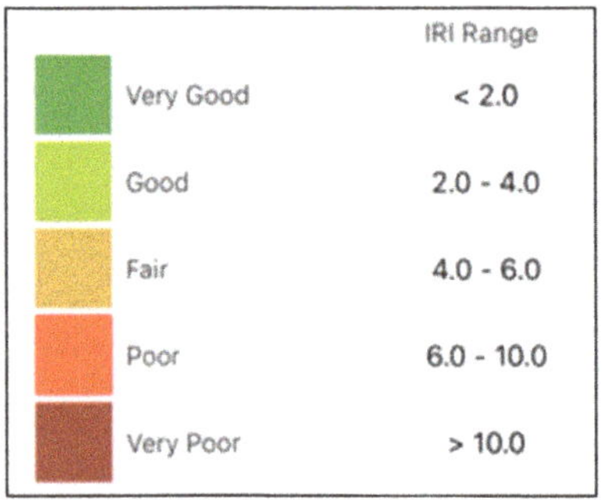

Figure 5. Color coding for road roughness [16].

For this research, a smartphone (Samsung Galaxy S6 (SM-G920x)) was used. The Road Lab Pro application was installed on the smartphone. The type of vehicle and the speed range were selected from the main menu of the application. The selected vehicle was a Toyota Rav4 and the speed was set between 20 and 60 km/h. After the setting process, the smartphone was mounted on the dashboard of the vehicle, as seen in Figure 6.

Figure 6. Mounting of the smartphone on the vehicle dashboard.

The smartphone application was then run, and a driver drove the vehicle at a predetermined speed of 20–60 km/h. A predetermined segment of roads was recognized using Google Earth, which was also used to determine the spatial information.

The application used the three-axis accelerometer and the GPS in the smartphone to measure the IRI automatically. The collected data were stored in comma-separated value (CSV) and keyhole markup language (KML) file formats. The KML files could be used by Google Earth to display the data and the CSV files could be opened by Excel for further analysis. Collected data were then zipped into the smartphone and could be exported from the smartphone by email, by uploading to Dropbox, or via a cable connected to a computer.

Based on the spatial road database that was extracted from Google Earth, there were many options for enhancing the data and adding attributes such as road class, road width, segment length, and others.

The measured road map was uploaded to Google Earth to visualize it and attach the measurement values for analysis. The road segment lengths were adjusted to correspond with the previous classification of road segments and the start/end coordinates were redefined. The IRI values were also aggregated and matched to road segments. Based on this arrangement, the data for each road segment included the coding, road class, number of lanes, length, width, start and end coordinates, and IRI value. All these data were exported to an Excel file, as illustrated in Table 1.

Table 1. Sample of road segment condition data.

Street Name	Section Code	Traffic Level	Road Class	Surface Type	No. of Lanes	Traffic Direction	Length (m)	Width (m)	Area (m^2)	Start Coordinate		End Coordinate		IRI Value	Roughness Road Condition	Importance
Sana'a Road	1R/100	H	Arterial	AC	2	1	92	7.1	656	15°41′32.89″ N	43°36′9.06″ E	15°41′39.20″ N	43°36′9.37″ E	5	Fair	Important
Sana'a Road	2R/100	H	Arterial	AC	2	1	172	7.1	1221	15°41′34.82″ N	43°36′11.22″ E	15°41′39.57″ N	43°36′9.51″ E	6	Fair	Important
Al-Hodaidah Road	1R/200	H	Arterial	AC	2	1	211	5	1055	15°41′33.16″ N	43°36′8.58″ E	15°41′38.67″ N	43°36′5.60″ E	6	Fair	Important
Al-Hodaidah Road	1L/00200	H	Arterial	AC	2	1	396	5	1980	15°41′33.20″ N	43°36′8.39″ E	15°41′38.30″ N	43°35′59.79″ E	4	Fair	Important
Al-Souq Street	1/110	H	Main	AC	2	2	163	7	1141	15°41′31.98″ N	43°36′8.41″ E	15°41′27.06″ N	43°36′9.68″ E	7	Poor	Important
Al-Souq Street	2/110	H	Main	AC	2	2	238	7	1666	15°41′27.06″ N	43°36′9.68″ E	15°41′24.69″ N	43°36′17.11″ E	9	Poor	Important
Al-Dahreen Street	3/120	L	Main	AC	2	2	319	7.2	2297	15°41′36.13″ N	43°36′20.35″ E	15°41′45.84″ N	43°36′20.10″ E	6	Poor	Important
Qirwee Street	1/115	L	Local Access	AC	2	1	431	4	1724	15°41′34.34″ N	43°36′18.52″ E	15°41′27.22″ N	43°36′15.20″ E	11	Bad	Normal

Figure 7 presents the IRI values of the segments that were captured from the field using the smartphone application.

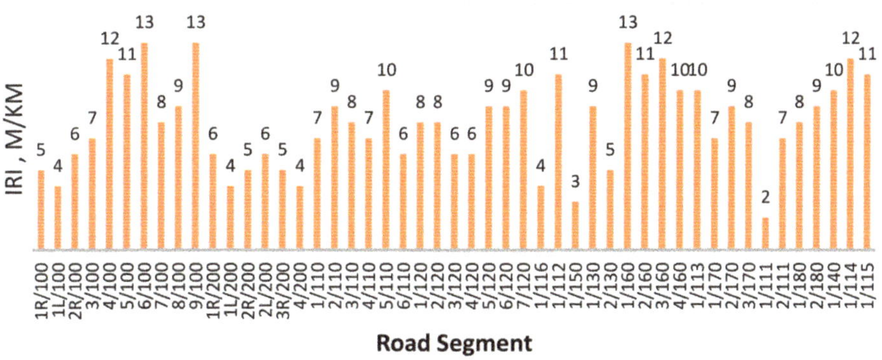

Figure 7. The IRI values of each road segment.

The IRI values for road segments can also be represented by a color-coded map on the Google Earth platform, as shown in Figure 8.

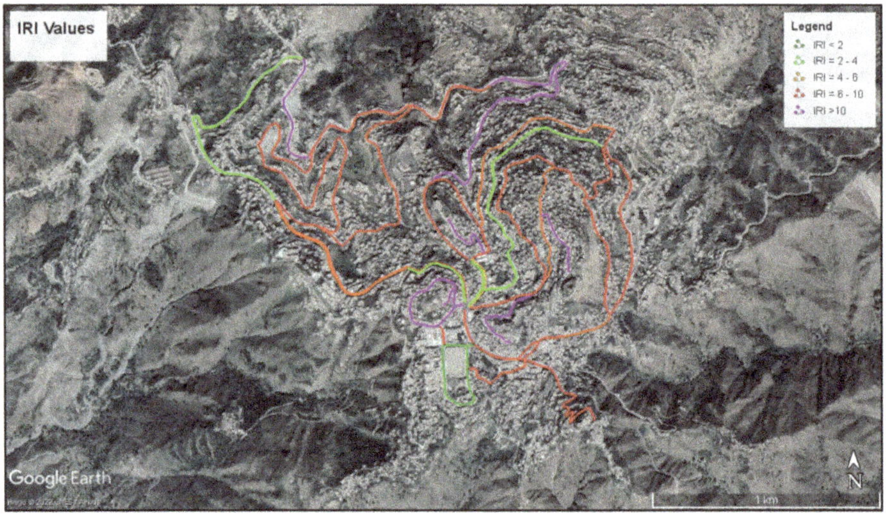

Figure 8. The IRI values in a color-coded Google Earth map.

6. Maintenance Activities

The maintenance decision criteria established for the developed maintenance management system were based on the collected IRI values for each segment. Generally, the maintenance strategies selected ranged from crack seals, slurry seals, and thin asphalt overlays to thick asphalt overlays. In other words, the selected type of treatment depended on the IRI values. The decision criteria for the maintenance activities were established as follows:

- IRI < 2: the deterioration is very slight and there is no need for any treatment.
- 2 < IRI < 4: the pavement deterioration is mainly due to the presence of localized distresses such as cracks and potholes. This requires routine maintenance such as crack sealing and patching.
- 4 < IRI < 6: the pavement deterioration is major and the pavement surface needs treatment that reduces the rate of deterioration; therefore, preventive maintenance such as a slurry seal is selected.
- 6 < IRI < 10: in this case, enhancing ride quality is needed, so thin asphalt overlaying is required.
- IRI > 10: this indicates that the pavement surface has become old; thus, a thick asphalt overlay might be the right choice.

The suggested pavement maintenance activity based on the IRI values is illustrated in Table 2.

Table 2. Suggested pavement maintenance activity vs. IRI ranges.

IRI Range (m/km)	Pavement Condition	Color Indicator	Suggest Maintenance	Code
<2	Very good	Dark green	Do nothing	M-00
2–4	Good	Green	Patching and/or crack sealing	M-01
4–6	Fair	Orange	Slurry sealing	M-02
6–10	Poor	Red	Thin asphalt overlaying (30 mm)	M-03
>10	Bad	Purple	Thick asphalt overlaying (50 mm)	M-04

Google Earth was used to reflect the type of maintenance selected for each segment, as presented in Figure 9.

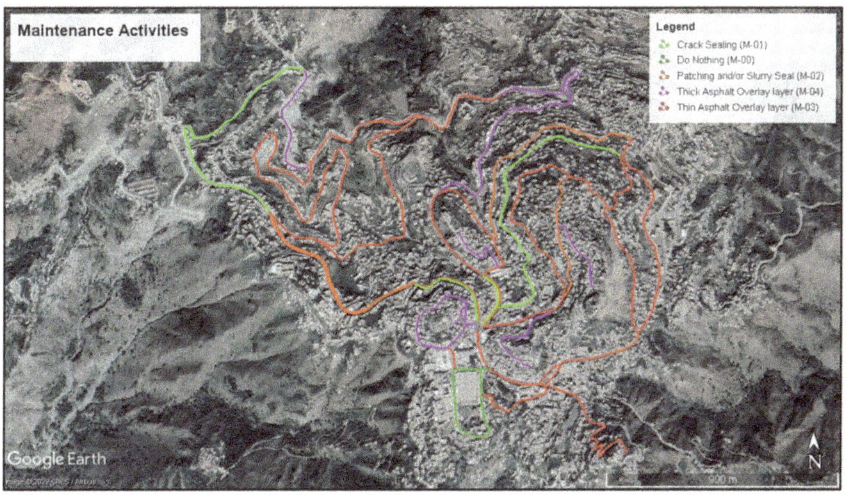

Figure 9. Road maintenance activity map.

The unit costs of maintenance activities include labor, equipment, and materials. These were obtained from the city's General Corporation of Roads and Bridges, as presented in Table 3.

Table 3. Estimated unit cost for maintenance activities (prices were set in 2018).

Maintenance Activity	Cost/Square Meter (USD)
Crack sealing (M-01)	9
Patching and/or slurry sealing (M-02)	35
Thin asphalt overlay layer (M-03)	54
Thick asphalt overlay layer (M-04)	90

The total required maintenance cost calculated for each segment of the road network is presented in Figure 10.

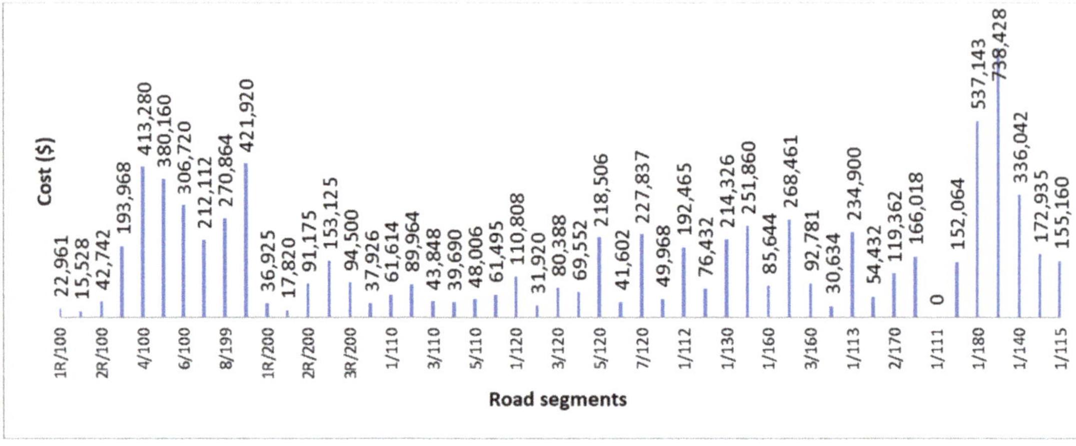

Figure 10. The maintenance cost for selected segments.

7. Priority Setting Procedure

Different types of priority methods are commonly used. These models range from simple subjective rankings to true optimization. It was decided, however, to use a parameter-based priority ranking method. This method is relatively simple and aims to answer the basic questions of where to apply maintenance and in which order. For this purpose, the dependable factors in the decision analysis should be specified. In this study, the selected factors were roughness, traffic, road class, special factors (to emphasize the importance of the selected roads), and the maintenance budget. The simple priority method evaluates the following decision form which has m alternatives (road segments) associated with n attributes (dependable factors):

$$\begin{array}{c} \\ A_1 \\ \vdots \\ A_i \\ \vdots \\ A_m \end{array} \begin{array}{c} \begin{array}{ccccc} B_1 & \dots & B_j & \dots & B_n \end{array} \\ \begin{bmatrix} x_{11} & \dots & x_{1j} & \dots & x_{1n} \\ \vdots & \dots & \vdots & \dots & \vdots \\ x_{11} & \dots & x_{ij} & \dots & x_{in} \\ \vdots & \dots & \vdots & \dots & \vdots \\ x_{m1} & \dots & x_{11} & \dots & x_{mn} \end{bmatrix} \end{array}$$

where:

B_j = the jth alternative (road segment).

A_i = the ith attribute (dependable factor).

For decision-making issues, the priority method usually uses subjective weights defined by the decision-makers. In this situation, a set of weights $W = (w_1, w_2, \ldots, w_n)$, $\Sigma w = 1$, was determined by the decision-makers. One of the techniques for determining objective weights is the Shannon entropy method [17].

The calculation process for the Shannon entropy method is below:

1) Construct the normalized decision r_i,

$$r_i = \frac{x_i}{\sum_{i=1}^{m} x_i}$$

2) Construct the vector of the Shannon entropy $e = (e_1, e_2, \ldots, e_j, \ldots, e_n)$,

$$e_j = -\frac{1}{\ln m}\sum_{i=1}^{m} r_i \ln r_i$$

3) Calculate the vector of diversification degrees $d = (d_1, d_2, \ldots, d_j, \ldots, d_n)$,

$$d_j = 1 - e_j$$

4) Calculate the vector of criteria weights $W = (w_1, w_2, \ldots, w_j, \ldots, w_n)$,

$$w_j = \frac{d_j}{\sum_{j=1}^{n} d_j} \quad (1)$$

For this study, the weight calculation process was performed using the Shannon entropy method and MS Excel software. Table 4 shows the weight factors for the prioritization.

Table 4. Weight factors for prioritization.

Factor	Weights	Mark
Traffic factor weight (w_{TF})	0.127	+
Road class factor weight (w_{FC})	0.065	+
Special factor weight (w_{SR})	0.111	+
Roughness weight (w_{IRI})	0.199	+
Maintenance budget weight (w_{MF})	0.498	−

The following model determined the maintenance priority for each segment:

$$PI = \frac{(w_{IRI} * IRI + w_{TF} * TF + w_{FC} * FC + w_{SR} * SR - w_{MF} * MF)}{5}$$

where:

PI = priority index.

TF = traffic factor.

FC = road classification factor.

SR = special factor to emphasize the importance of selected roads.

MF = maintenance budget factor.

To decide on the specific segments that should be implemented first, the road segments were sorted according to their priority index in descending order. Figure 11 presents the results of the calculated priority indexes of segments in descending order. As can be seen, the (6/100) road section had the highest priority index score, with 0.98, 0.99, 0.98, 0.97, and 0.32 for the traffic factor, road classification factor, special factor, roughness, and budget, respectively. The (1/140) road segment, on the other hand, had the lowest priority index score, with 0.33, 0.66, 0.33, 0.19, and 0.41 for the traffic factor, road classification factor, special factor, roughness, and budget, respectively.

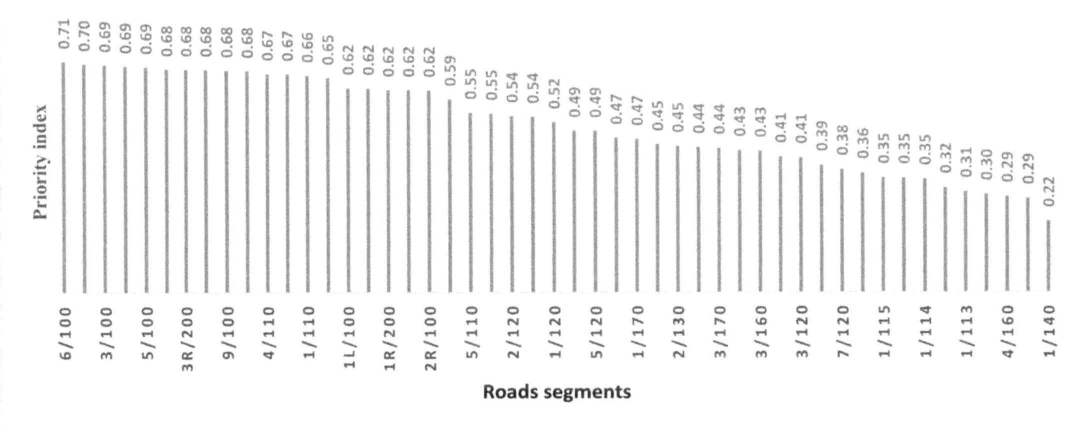

Figure 11. Arrangement of priority indexes for selected segments.

8. Conclusions

This research proposes a simple low-cost PMMS utilizing a smartphone app and Google Earth for small to intermediate cities in developing countries. The system uses an Android application installed on a smartphone to measure the road surface roughness based on GPS and accelerometer sensors. Google Earth is utilized for the visual illustration of the road conditions and other analyses.

The proposed system was implemented in Hajjah, a city in Yemen. The results reveal that IRI measurements performed by the app and Google Earth can provide a low-cost solution for cities such as Hajjah to manage their road network at an acceptable level. Google Earth is an excellent tool that can display road network conditions and all other attributes related to road maintenance management.

The following conclusions can be drawn from these results:

- Smartphones with built-in accelerometers, gyroscopes, and a GPS may be used to gather road data and assess road conditions. Smartphones are being used as probe devices due to their low cost and ease of use, as well as their potential for a large population coverage.
- Google Earth, which is user-friendly, is an excellent application for presenting maintenance data and subsequently making maintenance strategy decisions.
- A priority index can be established for each road segment in order to identify which segments should be implemented initially.
- Thematic maps of the output maintenance information can be visually shown using Google Earth.
- Any country, especially developing countries with limited budgets for producing maintenance studies, can employ this approach, as it allows for the easy and cost-effective production of road maintenance studies.

Author Contributions: Conceptualization, formal analysis, funding acquisition, investigation, methodology, project administration, and resources, A.I.A.-M.; visualization, A.I.A.-M. and A.H.A.-Q.; writing—original draft, A.I.A.-M. and A.H.A.-Q.; writing—review and editing, A.H.A.-Q. and A.I.A.-M. All authors have read and agreed to the published version of the manuscript.

Funding: The authors would like to acknowledge the support provided by the Researchers Supporting Project Number (RSP2022R424), King Saud University, Riyadh, Saudi Arabia.

Institutional Review Board Statement: Not applicable.

Informed Consent Statement: Not applicable.

Data Availability Statement: Available upon request and with regulations.

Acknowledgments: The authors acknowledge King Saud University for providing the funding for this research paper.

Conflicts of Interest: The authors declare no conflict of interest.

References

1. Salih, J.; Edum-Fotwe, F.; Price, A. Investigating the Road Maintenance Performance in Developing Countries. *World Acad. Sci. Eng. Technol. Int. J. Civ. Environ. Struct. Constr. Archit. Eng.* **2016**, *10*, 472–476.
2. Hokam, V.S.; Landge, V.S. Establishment of pavement maintenance management system in industrial area. *Int. J. Res. Publ.* **2015**, *4*, 1–9.
3. Plessis-Fraissard, M. Planning roads for rural communities. *Transp. Res. Rec.* **2007**, *1989*, 1–8. [CrossRef]
4. Gamage, D.; Pasindu, H.R.; Bandara, S. Pavement Roughness Evaluation Method for Low Volume Roads. In Proceedings of the Eighth International Conference on Maintenance and Rehabilitation of Pavements, Singapore, 27–29 July 2016; pp. 1–10.
5. Douangphachanh, V.; Oneyama, H. A study on the use of smartphones for road roughness condition estimation. *J. East. Asia Soc. Transp. Stud.* **2013**, *10*, 1551–1564.
6. Douangphachanh, V.; Oneyama, H. Exploring the use of smartphone accelerometer and gyroscope to study on the estimation of road surface roughness conditions. In Proceedings of the 2014 11th International Conference on Informatics in Control, Automation and Robotics (ICINCO), Vienna, Austria, 1–3 September 2014; Volume 1, pp. 783–787.
7. Alessandroni, G.; Klopfenstein, L.C.; Delpriori, S.; Dromedari, M.; Luchetti, G.; Paolini, B.; Bogliolo, A. Smartroadsense: Collaborative road surface condition monitoring. In Proceedings of the 2nd IEEE SPS Italy Chapter Summer School on Signal Processing, Frascati, Italy, 7–11 July 2014; pp. 210–215.
8. Islam, S.; Buttlar, W.G.; Aldunate, R.G.; Vavrik, W.R. Measurement of pavement roughness using android-based smartphone application. *Transp. Res. Rec.* **2014**, *2457*, 30–38. [CrossRef]
9. Zeng, H.; Park, H.; Fontaine, M.D.; Smith, B.L.; McGhee, K.K. Identifying deficient pavement sections by means of an improved acceleration-based metric. *Transp. Res. Rec.* **2015**, *2523*, 133–142. [CrossRef]
10. Hanson, T.; Cameron, C.; Hildebrand, E. Evaluation of low-cost consumer-level mobile phone technology for measuring international roughness index (IRI) values. *Can. J. Civ. Eng.* **2014**, *41*, 819–827. [CrossRef]
11. Zeng, H.; Park, H.; Smith, B.L.; Parkany, E. Feasibility assessment of a smartphone-based application to estimate road roughness. *KSCE J. Civ. Eng.* **2018**, *22*, 3120–3129. [CrossRef]
12. Yeganeh, S.F.; Mahmoudzadeh, A.; Azizpour, M.A.; Golroo, A. Validation of smartphone-based pavement roughness measures. *arXiv* **2019**, arXiv:1902.10699.
13. Chatterjee, A.; Tsai, Y.C. Training and testing of smartphone-based pavement condition estimation models using 3d pavement data. *J. Comput. Civ. Eng.* **2020**, *34*, 04020043. [CrossRef]
14. Dong, D.; Li, Z. Smartphone sensing of road surface condition and defect detection. *Sensors* **2021**, *21*, 5433. [CrossRef] [PubMed]
15. Al-Mansour, A.; Shukri, A. Correlation of Pavement Distress and Roughness Measurement. *Appl. Sci.* **2022**, *12*, 3748. [CrossRef]
16. Available online: https://apps.apple.com/us/app/roadlabpro/id1125333158 (accessed on 25 May 2022).
17. Ding, S.; Shi, Z. Studies on incidence pattern recognition based on information entropy. *J. Inf. Sci.* **2005**, *31*, 497–502.

applied sciences

MDPI

Article

Bituminous Mixtures Experimental Data Modeling Using a Hyperparameters-Optimized Machine Learning Approach

Matteo Miani [1], Matteo Dunnhofer [2], Fabio Rondinella [1], Evangelos Manthos [3], Jan Valentin [4], Christian Micheloni [2] and Nicola Baldo [1,*]

[1] Polytechnic Department of Engineering and Architecture (DPIA), University of Udine, Via del Cotonificio 114, 33100 Udine, Italy; matteo.miani@phd.units.it (M.M.); fabio.rondinella@phd.units.it (F.R.)
[2] Department of Mathematics, Computer Science and Physics (DMIF), University of Udine, Via delle Scienze 206, 33100 Udine, Italy; matteo.dunnhofer@uniud.it (M.D.); christian.micheloni@uniud.it (C.M.)
[3] Department of Civil Engineering, University Campus, Aristotle University of Thessaloniki, 54124 Thessaloniki, Greece; emanthos@civil.auth.gr
[4] Faculty of Civil Engineering, Czech Technical University, Thákurova 7, 166 29 Prague, Czech Republic; jan.valentin@fsv.cvut.cz
* Correspondence: nicola.baldo@uniud.it; Tel.: +39-0432-558-745

Abstract: This study introduces a machine learning approach based on Artificial Neural Networks (ANNs) for the prediction of Marshall test results, stiffness modulus and air voids data of different bituminous mixtures for road pavements. A novel approach for an objective and semi-automatic identification of the optimal ANN's structure, defined by the so-called hyperparameters, has been introduced and discussed. Mechanical and volumetric data were obtained by conducting laboratory tests on 320 Marshall specimens, and the results were used to train the neural network. The k-fold Cross Validation method has been used for partitioning the available data set, to obtain an unbiased evaluation of the model predictive error. The ANN's hyperparameters have been optimized using the Bayesian optimization, that overcame efficiently the more costly trial-and-error procedure and automated the hyperparameters tuning. The proposed ANN model is characterized by a Pearson coefficient value of 0.868.

Keywords: bituminous mixtures; stiffness modulus; neural network; Bayesian optimization

Citation: Miani, M.; Dunnhofer, M.; Rondinella, F.; Manthos, E.; Valentin, J.; Micheloni, C.; Baldo, N. Bituminous Mixtures Experimental Data Modeling Using a Hyperparameters-Optimized Machine Learning Approach. *Appl. Sci.* **2021**, *11*, 11710. https://doi.org/10.3390/app112411710

Academic Editor: Cesare Oliviero Rossi

Received: 10 November 2021
Accepted: 7 December 2021
Published: 9 December 2021

1. Introduction

The different types of bituminous mixtures (BMs) used for road pavements have to be properly designed as mixtures made of aggregates and bitumen, to withstand traffic loads and climate conditions. Unsuitable mechanical characteristics and volumetric properties of bituminous mixtures may lead to various types of distress in road pavements, generally comprising cracks due to fatigue or low temperature, permanent deformations, stripping, etc. Such failure modes decrease the service life of the pavement and represent serious safety issues for road users. As a result, it is important to properly characterize the mechanical behavior of mixes with respect to their composition to allow a performance-based optimization during the mix design phase [1–3]. Experimental methods, which require expensive laboratory tests and skilled technicians, are currently used to evaluate the bituminous mixtures' performance [4–9]. Consequently, any modification of the mixtures' composition, in terms of bitumen type or content, rather than of aggregate gradation, requires new laboratory tests with an increase in time and costs of the design process.

In recent years, many researchers have devoted their efforts to the problem of defining a mathematical or numerical model of BMs' mechanical behavior, which could quickly elaborate a reliable prediction of the bituminous mixture's response. To develop predictive equations, two main types of procedures can be used, namely, advanced constitutive modeling methods rather than statistical or data science approaches. Although the mechanistic constitutive methods allow a rational and in-depth analysis of the material response to be performed [10–17], statistical or machine learning approaches are gaining considerable success in the academic community due to the good reliability of their predictions [18–21], even if they are not physically based. Nevertheless, it has also been reported that statistical regressions of experimental data can produce less accurate predictions than machine learning methods, specifically with regard to Artificial Neural Networks (ANNs) [22–27]. Recently, Lam et al. [28] have found that a multiple regression analysis (coefficient of determination, R^2, equal to 0.790) is less reliable than an ANN approach (R^2 = 0.925) searching the analytical model to infer compressive strength of roller-compacted concrete pavement from steel slags aggregate and fly ash levels replacing cement.

The ANNs simulate simplified models of human brain: their computing power derive from the number of connections established between artificial neurons, fundamental computing units, and their main function is to map patterns between input and output from a representative experimental sample, mimicking the biological learning process. Such neural models are basically based on a nonlinear fitting approach, neither physical nor mechanistic, to correlate experimental data; the mathematical framework has already been widely discussed, e.g., by Baldo et al. [29]. In recent years, an increased number of researchers have used ANNs in many civil engineering applications, producing impressive results even with regards to the evaluation of road pavements' characteristics and performance. Tarefder et al. [30] developed a four-layer feed-forward neural network to correlate mix design parameters and BM samples performance in terms of permeability. Ozsahin and Oruc [31] constructed ANN-based models to determine the relationship between the resilient modulus of emulsified BMs and its affecting factors such as curing time, cement addition level, and residual bitumen content. Tapkın et al. [32] presented an application of ANN for the prediction of repeated creep test results for polypropylene-modified BMs. Accurate predictions (R^2 between 0.840 and 0.970) of the fatigue life of BMs under various loading and environmental conditions were also produced [33,34]. Ceylan et al. [35] discussed the accuracy and robustness of ANN-based models for estimating the dynamic modulus of hot mixes: such models exhibit significantly higher prediction accuracy (also at the input domain boundaries), less prediction bias and better understanding of the influences of temperature and mixture composition than their regression-based counterparts. Recently, Le et al. [36] developed an advanced hybrid model, as it is based both on ANNs and optimization technique, to accurately predict the dynamic modulus of Stone Mastic Asphalt (R^2 = 0.985); also, they use the proposed model to evaluate and discuss the effects of temperature and frequency on the mechanical parameter. Similarly, Ghorbani et al. [37] used a simple ANN approach for modeling experimental test results and examining the impact of different features on the properties of construction and demolition waste, such as the reclaimed asphalt pavement.

Although the documented research has attempted to introduce new approaches to an empirical–mechanical mix design, the Marshall approach is still widely adopted in many laboratories [38–44]. Tapkın et al. [45] have verified the possibility of applying ANNs for the prediction of Marshall test results of BMs. The proposed NN model uses the physical properties of standard Marshall specimens to predict the Marshall stability (MS), flow (MF) and Marshall Quotient values obtained at the end of mechanical tests. Ozgan [46] has studied the effects of varying temperatures and exposure times on the stability of BMs and modeled the test data by using a multilayer ANN. Conversely, Mirzahosseini et al. [23] have validated the efficiency of the multilayer perceptron ANNs for the assessment of the rutting potential of dense BMs: the flow number of Marshall specimens has been correlated to the aggregate and bitumen contents, percentage of voids in mineral aggregate, MS and MF. The mechanical characteristics of the bituminous mixtures depend on the volumetric properties as well as the bitumen content. Such parameters have to meet the limits, set by the current local specifications, for the pavement layer interested from the intervention. Nevertheless, voids in mineral aggregate, voids filled with bitumen and air voids (AV) are determined with a specific test (EN 12697-8), which requires additional time and costs. Khuntia et al. [47] have proposed a neural network model that uses the quantities of bitumen and aggregate in Marshall specimens to predict the MS, MF value and AV obtained from the tests. Likewise, Zavrtanik et al. [48] have used ANNs to estimate air void content in several types of BMs produced according to EN 13108-1. Anyway, the literature presented has in common the need to provide the road engineer with an algorithm that can provide accurate predictions of empirical parameters related to BMs, without the need for sophisticated, time-consuming and expensive laboratory testing.

Despite the fact that ANNs have successfully provided predictive equations to quicken the empirical Marshall mix design, such computational models were usually based on a neural network structure set a priori by the research engineer and trained on a random subset of the available data sample. In case of a relatively small data set, such a practice may involve the risk of leaving out some relevant trends from the training set and leading to a variable prediction error, measured on the validation set, due to data sample variability and selected ANN architecture [49]. These issues can be avoided if an efficient model selection and appropriate data partition are performed. In particular, the search for the optimal network architecture, one of the most difficult tasks in ANN studies, consists of tuning the model settings, called hyperparameters, that yield the best performance score on the validation set. Applications of the trial-and-error procedure, as random or grid search, to find the optimal hyperparameters of a machine learning algorithm for a given predictive modeling problem can be found in the relevant literature [23,24]. Nevertheless, Baldo et al. [50,51] have highlighted the limits of such a time-consuming approach and applied a statistical technique of data partitioning, called k-fold Cross Validation, that allows a more accurate estimation of a model's performance.

An efficient hyperparameters tuning approach, in contrast to random or grid search, is the Bayesian optimization, which has become popular in recent years [52]. Given that evaluating the performance function score for different hyperparameters is extremely expensive, the Bayesian approach builds a probabilistic model, called "surrogate", mapping hyperparameters of past evaluations to a probability of a score on the performance function and uses such a model to find the next set of promising hyperparameters (i.e., that optimize the surrogate function) to evaluate the actual performance function [53,54].

This paper aimed to develop an autonomous and impartial procedure of neural model selection for predictive modeling problems of bituminous mixtures' mechanical behavior, using the Bayesian optimization method, that would replace the more costly trial-and-error procedure. In particular, the ANN approach was used to analyze stiffness modulus (ITSM), MS, MF and AV content of 320 Marshall specimens tested at the Highway Engineering Laboratory of the Thessaloniki Aristotle University. The experimental database includes different types of bitumen and aggregate and covers a wide range of bitumen contents and aggregate gradations. In addition, both laboratory- and plant-prepared mixtures were used and their production site was considered among the feature's variables of the proposed NN model; it correlates mechanical and volumetric properties, collected by means of laboratory tests, to fundamental characteristics of bituminous mixtures, such as bitumen content (% by weight of mix), filler-to-bitumen ratio (%), type of bitumen and aggregate as well as maximum nominal grain size.

The innovative aspect of the presented research is the application of state-of-the-art procedures in the machine learning domain (namely, k-fold Cross Validation and Bayesian optimization) that allow researchers and engineers to solve the problems of classical neural network applications in bituminous mixtures' behavior modeling. However, the procedure is not intended to replace the experimental method for mixture characterization, but to integrate it with a predictive algorithm that allows the road engineer to improve the mix design process, reducing time and operational costs. The major drawback of the proposed approach is that its proper implementation requires human resources with specific skills, such as machine learning expertise, and large training data sets covering the diversity of BMs materials.

2. Materials and Experimental Design

Note that 320 Marshall specimens, having a diameter of 100 mm and an average thickness of 63.7 mm, were produced, both in laboratory and in plant, according to the impact compactor method of test EN 12697-30. These mixtures, designed as part of a research project carried out at the Aristotle University of Thessaloniki, were characterized by different contents of bitumen and aggregate gradations. Aggregates represent the lithic skeleton of a bituminous mixture, while the bitumen is the component binding the aggregate grains together.

2.1. Materials

The aggregates employed were limestone- or diabase-type crushed stones with maximum nominal size of 20 mm or 12.5 mm: the calcareous sedimentary aggregate came from the same Greek quarry, while the mafic igneous one from three different local quarries. To control the physical properties of the aggregates, several tests were conducted. The obtained results are presented in Table 1.

Table 1. Aggregate properties.

Property	Aggregate Type	
	Limestone	Diabase
Los Angeles coeff. (%), EN 1097-2	29	25
Polished stone value (%), EN 1097-8	-	55 to 60
Flakiness index (%), EN 933-3	23	18
Sand equivalent (%), EN 933-8	79	59
Methylene blue v. (mg/g), EN 933-9	3.3	8.3

The bituminous mixtures composed of aggregates with maximum nominal size of 12.5 mm (BM12.5) belong to the category "binder course", while the ones characterized by maximum nominal size of the aggregates equal to 20 mm (BM20) to the "base course" category. In this research, 27 aggregate gradations were considered to meet the gradation limits for binder and base course, set by the current Greek specifications. In each mix category, there are various types of compositions related to the aggregate's maximum nominal size: for lab-prepared mixtures, 4 types of gradations were used to fit the limits for BM12.5 and 4 types for the BM20. The remaining ones concern BM20 mixtures prepared in plant and correspond to different production days. Figures 1–3 show the grading curves involved.

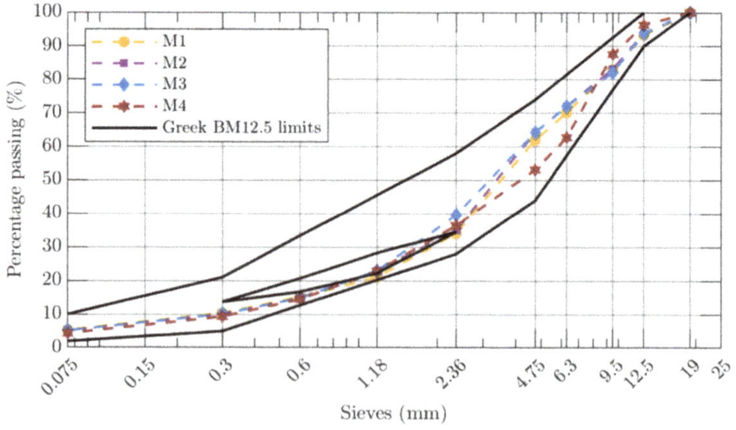

Figure 1. Gradation curves of lab-prepared BM12.5.

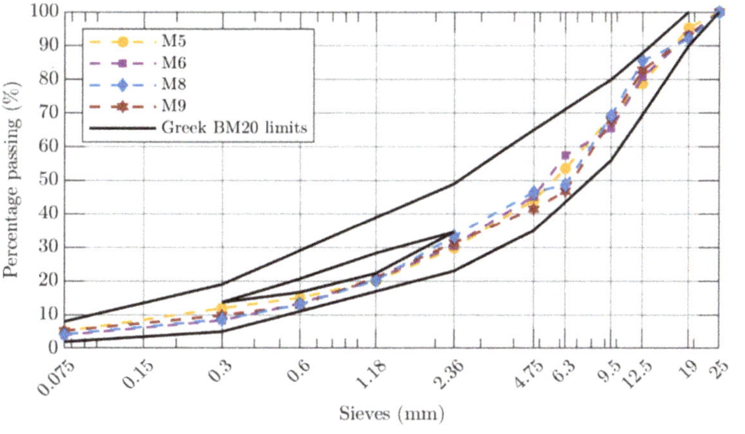

Figure 2. Gradation curves of lab-prepared BM20.

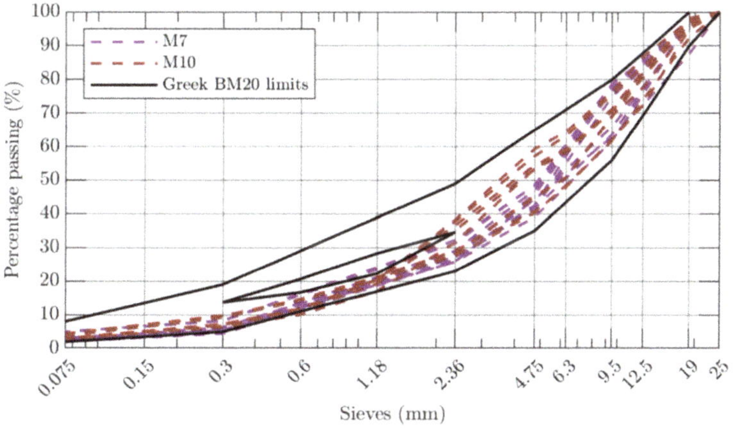

Figure 3. Gradation curves of plant-prepared BM20 (gradations of different production dates).

The standard 50/70 penetration bitumen was used in the preparation of 129 Marshall specimens, while the remaining 191 were produced utilizing a bitumen modified in the laboratory with styrene–butadiene–styrene polymers (SBS). The two types of bitumen were tested to ensure that their physical properties were compliant with specific acceptance requisites. The characteristics of bituminous binders are reported in Table 2. No aging process was performed on bituminous mixtures.

Table 2. Bitumen properties.

Property	Bitumen Type	
	50/70	Modified
Penetration (0.1 × mm), EN1426	64	45
Softening point (°C), EN1427	45.6	78.8
Elastic recovery (%), EN 13398	-	97.5
Fraas breaking point (°C), EN 12593	−7.0	−15.0

2.2. Experimental Design

The Marshall samples were produced with a bitumen percentage between 3.8% and 6.0% (by weight of mix), and in number equal to three for each bitumen content adopted. Table 3 summarizes the number of specimens produced for each combination of bitumen and aggregate; abbreviations coding the mixtures are also reported.

Table 3. Number and codes of Marshall specimens.

Maximum Size (mm)	Aggregate Type	Bitumen Type	Production Site	Mixture ID	Specimens
12.5	Limestone	50/70	Laboratory	M1	30
12.5	Limestone	Modified	Laboratory	M2	30
12.5	Diabase	50/70	Laboratory	M3	30
12.5	Diabase	Modified	Laboratory	M4	30
20	Limestone	50/70	Laboratory	M5	39
20	Limestone	Modified	Laboratory	M6	30
20	Limestone	Modified	Plant	M7	41
20	Diabase	50/70	Laboratory	M8	30
20	Diabase	Modified	Laboratory	M9	30
20	Diabase	Modified	Plant	M10	30

Among the mechanical parameters of bituminous mixtures, the ITSM allows a rational performance-based characterization of the mixes to be performed [4,55]. Therefore, the ITSM test (Figure 4) was executed on all BM samples using the standard testing conditions, defined by EN 12697-26 (temperature of 20 °C, target deformation fixed at 5 μm, and rise time equal to 124 ms). Subsequently, considering that the Marshall parameters are still widely used in road pavement design [38–45], MS and MF were evaluated for the bituminous mixtures produced, according to EN 12697-34. Finally, the specimens' volumetric characteristics have been determined applying EN 12697-8. The test results are reported in Table S1; such experimental data have been already discussed in previous papers [29,50]. Table 4 shows some statistical information (minimum and maximum values along with the mean value and its standard deviation) about mechanical characteristics and volumetric property of the BMs.

Table 4. Statistical information about Marshall specimens.

Mixture ID	Parameter	Minimum Value	Maximum Value	Mean Value	Standard Deviation
M1	ITSM (MPa)	3756	5554	4556.43	567.93
	MS (kN)	7.71	12.17	9.93	1.09
	MF (mm)	1.99	4.70	3.18	0.84
	AV (%)	1.77	6.37	3.99	1.37
M2	ITSM (MPa)	3628	5142	4345.50	486.50
	MS (kN)	8.74	14.00	11.35	1.73
	MF (mm)	2.00	4.20	3.20	0.57
	AV (%)	2.20	6.29	4.18	1.23
M3	ITSM (MPa)	3812	5942	4804.10	725.03
	MS (kN)	10.30	15.20	12.88	1.53
	MF (mm)	2.00	5.00	3.35	0.95
	AV (%)	1.49	8.91	5.22	2.38
M4	ITSM (MPa)	4035	6293	5076.17	759.06
	MS (kN)	11.60	16.43	13.62	1.42
	MF (mm)	2.20	5.00	3.40	0.92
	AV (%)	1.33	8.36	5.05	2.18

Table 4. *Cont.*

Mixture ID	Parameter	Minimum Value	Maximum Value	Mean Value	Standard Deviation
M5	ITSM (MPa)	3215	4919	4252.26	502.24
	MS (kN)	8.91	14.86	11.37	1.51
	MF (mm)	2.18	4.60	3.15	0.50
	AV (%)	2.17	6.75	4.28	1.16
M6	ITSM (MPa)	3907	6043	5243.10	538.97
	MS (kN)	10.40	13.99	11.81	1.21
	MF (mm)	2.24	4.16	3.24	0.40
	AV (%)	1.68	5.21	3.49	1.08
M7	ITSM (MPa)	3103	6399	5065.34	906.93
	MS (kN)	6.60	14.75	9.86	2.20
	MF (mm)	2.10	9.86	3.22	0.62
	AV (%)	3.03	2.20	5.22	1.19
M8	ITSM (MPa)	2304	4900	3829.63	783.23
	MS (kN)	10.45	15.48	13.05	1.36
	MF (mm)	2.20	5.00	3.37	0.83
	AV (%)	0.35	8.44	4.37	2.43
M9	ITSM (MPa)	2930	5994	4911.30	851.15
	MS (kN)	8.92	15.48	12.22	2.03
	MF (mm)	1.85	5.00	3.10	0.72
	AV (%)	1.26	8.44	4.97	1.86
M10	ITSM (MPa)	4049	5968	5309.27	565.88
	MS (kN)	7.80	16.55	11.98	2.15
	MF (mm)	2.70	5.40	3.95	0.76
	AV (%)	4.60	9.70	7.12	1.59

Figure 4. ITSM test setup.

3. Methodology

3.1. Artificial Neural Networks

ANNs are nonlinear mathematical models that aim to simulate the human brain processing schemes [56–59]. In a first approximation, an ANN is the result of the weighted and biased connections of logistic regression units, i.e., the artificial neurons. In a multilayer feedforward network, such neurons are organized into hidden layers between an input layer, consisting of source nodes (the components of the input vector), and an output layer composed of one or more computational neurons that compute the response of the network. Each neuron is characterized by an activation function that limits the range of its response to a continuous value between 0 and 1.

The input, i.e., the feature vector or the hidden layer output, is weighted, meaning that a certain weight is associated to each input signal (the components of the input feature vector or the hidden neurons' individual responses). These network parameters are learned through a supervised training process so that the network itself (at this point interpretable as a nonlinear parametric function) can perform a specific task. In regression problems, the network's task is to predict the experimental target associated with a given feature vector, i.e., to represent the implicit relationship between the input and the ground-truth output.

In formal terms, given a data set of input–output vector pairs:

$$D = \left\{ \left(\boldsymbol{x}^{(d)}, \boldsymbol{y}^{(d)} \right) \right\}^{|D|}{}_{d=1} \tag{1}$$

$$\text{where } \boldsymbol{x}^{(d)} = \left[x^{(d)}{}_0, \ldots, x^{(d)}{}_{F-1} \right] \in R^F \text{ and } y^{(d)} = \left[\boldsymbol{y}^{(d)}{}_0, \ldots, y^{(d)}{}_T \right] \in R^T \tag{2}$$

and the input layer of an ANN is referred as being the input feature vector $\boldsymbol{x}^{(d)}$. The first hidden layer's output is the vector:

$$\boldsymbol{h}^{(1)} = \left[h_0^{(1)}, \ldots, h_{N-1}^{(1)} \right] \tag{3}$$

where each item $h_j^{(1)}$ is obtained as:

$$h_j^{(1)} = \phi\left(\sum_{i=0}^{F-1} w_{ij}^{(1)} \cdot x_i + b_j^{(1)} \right). \tag{4}$$

In this expression, $w_{ij}^{(1)}$ are the weights of the connections between the neurons $x^{(d)}{}_i$ and the neurons $h_j^{(1)}$, $b_j^{(1)}$ are the biases and ϕ is the activation function. The subsequent layer, $\boldsymbol{h}^{(2)}$, is computed in a similar way, by considering the items of the vector $\boldsymbol{h}^{(1)}$ as input layer, with corresponding weights $w_{ij}^{(2)}$ and biases $b_j^{(2)}$:

$$h_j^{(2)} = \phi\left(\sum_{i=0}^{N-1} w_{ij}^{(2)} \cdot h_i^{(1)} + b_j^{(2)} \right). \tag{5}$$

This process is repeated to compute the activations of each layer $\boldsymbol{h}^{(l)}, l \in \{1, 2, \ldots, L\}$ and the output $\boldsymbol{y}^{\hat{(d)}} = \phi^{(out)}\left(\boldsymbol{h}^{(L)}\right)$. All the learnable parameters are considered to form the set:

$$W = \left\{ w_{ij}^{(l)}, b_j^{(l)} \,\middle|\, l \in \{1, 2, \ldots, L\} \right\}. \tag{6}$$

In the proposed ANN: the number of source nodes is equal to the number of input features (F = 3), namely, the bitumen content (BC), the filler-to-bitumen ratio (FtB), along with a categorical variable that distinguishes the bituminous mixture in terms of bitumen type, maximum nominal grain size, aggregate type, and production site (values: 1 for M1,

2 for M2, 3 for M3 and so on). These features are fundamental parameters of the pavement engineering, related to the composition of the bituminous mixtures.

Each hidden layer is equipped with N neurons and passed to $\phi(\cdot)$ being either an ELU, TanH or ReLU function (Figure 5). The output layer consists of $T = 4$ neurons (corresponding to MS, MF, ITSM, and AV) and the identity activation function is considered $\phi^{(out)}(.)$. In previous studies, Marshall parameters, as well as ITSM, rather than volumetric properties, have been always determined separately, by means of different neural models [29,45–48,55], while in the current paper they can be predicted simultaneously by a single multi-output ANN model.

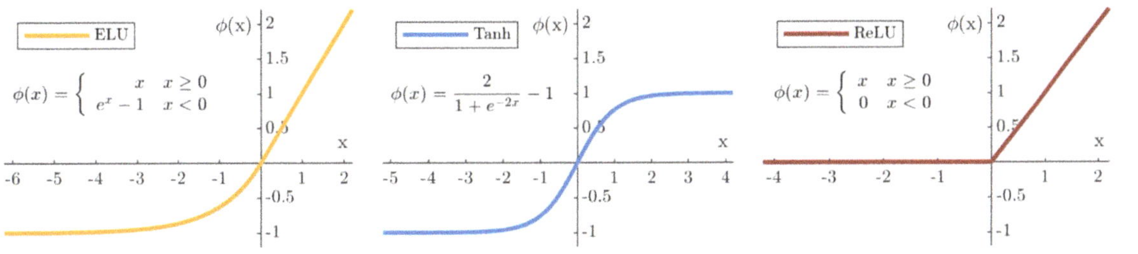

Figure 5. Activation functions: exponential linear (ELU), hyperbolic tangent (TanH), rectified linear (ReLU).

3.2. ANN Optimization

The network parameters W are identified by means of a supervised training process, which is divided into two successive steps, i.e., a forward and a backward phase. In the latter, a backpropagation algorithm [58] is exploited to update the ANN's weight and biases:

$$W^{(e)} = W^{(e-1)} - \alpha \nabla E\left[W^{(e-1)}\right], e \in \{0, \ldots, E-1\} \tag{7}$$

where E is the number of training iterations, α is the learning rate and $W^{(e-1)}$ are the parameters values at iteration $e - 1$. A detailed and comprehensive description of the Equation (7) has already been widely discussed, e.g., by Baldo et al. [29]. At the end of the training process (i.e., after the E iterations have been completed), the parameters W of the e-th iteration that produced the minimum of the loss function L are selected. Such loss function is usually defined as the 2-norm of the difference between the network output $\hat{y^{(d)}}$ and the ground-truth vector $y^{(d)}$, thus being called Mean Squared Error (MSE):

$$L\left(\hat{y^{(d)}}, y^{(d)}\right) = \| \hat{y^{(d)}} - y^{(d)} \|_2^2 \tag{8}$$

The training process just outlined is known as Stochastic Gradient Descent (SGD). Such approach is commonly used in the well-known [60,61] and widely used MATLAB® ANN Toolbox [22,23,29,45,46,50].

With the aim of increasing the convergence speed of the learning algorithm, several improvements have been proposed in recent years. The first remarkable one, called RAdam optimizer [62], faces the issue of implementing warmup heuristics to avoid falling in bad local minima. This problem can occur using adaptive optimizers such as Adam [63]. According to the authors, the variance of the adaptive learning rate in the initial weight updates can be intractable and direct the solution search towards local minima with poor performance. Adding a rectifier operation to the adaptive learning rate was shown to reduce the variance and lead to better solutions. The second remarkable improvement, Lookahead [64], is a suitable method in reducing the need for extensive hyperparameters search, and the variance of the parameter updates. The Lookahead algorithm maintains two copies of the parameters, respectively, the "fast weights" and "slow weights". The first is initially updated for k times using a standard SGD-based optimizer. Then, the second set of weights is updated in the direction (i.e., the gradient) of the last computed fast weights. Intuitively, as the name states, the first kind of update looks ahead to acquire information of the loss function's surface. Once obtained, the gradients used for the actual weight update result in a more accurate direction towards the loss function's minimum. The combination of the strategies is known as Ranger optimizer. In this work, such optimization algorithm was used to calculate the optimal weights of the implemented ANN, in contrast with the common procedures of neural modeling used in previous literature studies [22,23,29,31,45,46,50].

3.3. ANN Regularization

To perform proper neural modeling, the overfitting phenomenon must be avoided: a model becomes overfitted when it starts to excessively adapt to the training data and stops smoothly regressing the selected validation data. In the current study setup, the weight decay technique [65] (also known as L2 Regularization) has been implemented to overcome such problem. It consists in adding an additional term to the optimization objective, which is calculated as the 2-norm of the parameters of the network. In this way, the global optimization objective becomes:

$$L_{opt}\left(\hat{\boldsymbol{y}^{(d)}}, \boldsymbol{y}^{(d)}, W^{(e-1)}\right) = L\left(\hat{\boldsymbol{y}^{(d)}}, \boldsymbol{y}^{(d)}\right) + \beta \parallel W^{(e-1)} \parallel_2^2 \quad (9)$$

where β is a hyperparameter to control the magnitude of the term.

3.4. k-Fold Cross Validation

A fixed training–validation split of data is usually performed [60,61]: this technique, also known as a "hold-out" method, can result in biased model performance as a consequence of the different descriptive statistics characterizing training and validation data sets [49].

Conversely, the k-fold Cross Validation (k-fold CV) represents an effective alternative to evaluate the actual model generalization capabilities. This resampling method suggests splitting the data set of interest into k folds, equally sized. For example, in the current study setup, a 5-fold CV was implemented. It means that 5 alternative partitions of the data set were obtained: in turn, 4 folds were used to train the neural model and the remaining one to validate it. This involved running 5 experiments, at the end of which the obtained validation scores were averaged to evaluate the general performance capabilities of the proposed model [49,51].

3.5. Bayesian Hyperparameters Optimization

Modeling by machine learning (ML) approaches involves the accurate setting of several hyperparameters: these parameters are used to define the topology and size of a neural network (e.g., the number of hidden layers and neurons), and to control the learning process (e.g., the learning rate). Standard procedures involve a grid or random (i.e., based on a sampling method) search for the best combination of hyperparameters within variation intervals accurately defined by the experimenter on the basis of his/her experience [23,29,31,45,46,48,50]. However, to obviate the significant time demands and computational resources of the abovementioned methods, a semi-automatic strategy has been implemented in the current study, namely, the Bayesian Optimization (BO). This method, based on Bayesian statistics, aims to find the maximum (for maximization problems) or the minimum (for minimization problems) of a function: $f(\boldsymbol{x})$, $\boldsymbol{x} = [x_p]$, $p \in \{0, \ldots, P\}$, $P \in N$, where x_p is a parameter in a bounded set $X_p \subset R$. This mathematical problem is solved by optimization algorithms that define a probabilistic model over $f(x)$, to decide at each iteration which is the most likely point in X to maximize (or minimize) $f(\cdot)$. In this context, Snoek et al. [66] were the first to use the BO for the search of ML model hyperparameters: since the trend of the objective function is unknown, the authors treated $f(\cdot)$ as a random function and placed a Gaussian process (GP) prior [67] over it, to capture its behavior. During the optimization process, the prior is updated based on ML experiments results, produced by different hyperparameters combinations, to form the posterior distribution over the function $f(\cdot)$. The latter, in turn, is exploited by an acquisition function to determine the next evaluation point.

In practice (Figure 6), a set of O observations of the form $\left\{x^{(o)}, y^{(o)}\right\}_{o=1}^{O}$, with $y^{(o)} = N\left(f\left(x^{(o)}\right), \nu\right)$ and where ν is the variance of noise in the observation of $f(\cdot)$, is exploited to determine a multivariate Gaussian distribution over R^O through a mean $m: X \to R$ and covariance $K: X \times X \to R$ functions. The next set of hyperparameters $x_{next} \in X$ that should be evaluated during the optimization process is determined by solving the equation $x_{next} = argmax_x a(x)$. The function $a: X \to R^+$ is called acquisition function and generally depends on both the previously observed samples $\left\{x^{(o)}, y^{(o)}\right\}_{o=1}^{O}$ and the GP parameters θ. It is formulated as follows: $a\left(x; \left\{x^{(o)}, y^{(o)}\right\}, \theta\right)$. In the GP prior setting, $a(\cdot)$ relies on the predictive mean $\mu\left(x; \left\{x^{(o)}, y^{(o)}\right\}, \theta\right)$ and variance $\sigma^2\left(x; \left\{x^{(o)}, y^{(o)}\right\}, \theta\right)$ functions of the GP model. There are several existing definitions for $a(\cdot)$ [68,69], but the GP Upper Confidence Bound (UCB) [70] has been proven to efficiently reduce the number of function evaluations needed to determine the global optimum of several black-box multimodal functions. UCB implements the acquisition function as $a_{UCB}\left(x; \left\{x^{(o)}, y^{(o)}\right\}, \theta\right) = \mu\left(x; \left\{x^{(o)}, y^{(o)}\right\}, \theta\right) - \kappa\sigma\left(x; \left\{x^{(o)}, y^{(o)}\right\}, \theta\right)$, where κ is a hyperparameter to control the balance between exploitation (i.e., favoring parts that the model predicts as promising) and exploration.

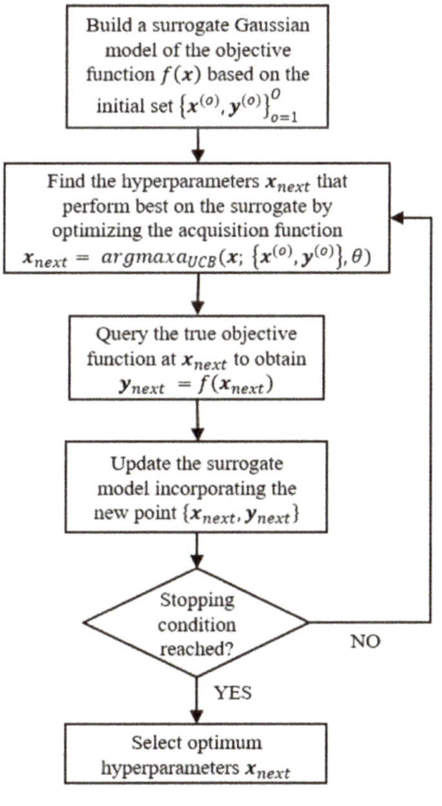

Figure 6. Flow chart of the optimization process.

In the actual problem, namely the prediction of mechanical and volumetric properties of bituminous mixtures for road pavement, $f(\cdot)$ is defined as $f : X_L \times X_N \times X_{act} \times X_\alpha \times X_\beta \times X_E \rightarrow [-1, 1]$ and has to be maximized. Therefore, given the $P = 6$ hyperparameters $L, N, act, \alpha, \beta, E$, $f(\cdot)$ constructs an ANN with L layers, N neurons per layer and performs a 5-fold CV experiment in which the ANN is trained for E iterations with α and β as the learning rate and weight decay parameter, respectively. $f(\cdot)$ returns a scalar value that expresses the average Pearson coefficient (R) obtained by the ANN on the 5 validation folds. The BO algorithm performs 500 iterations: at each iteration, $f(\cdot)$ runs an experiment based on the hyperparameter combination sampled by the UCB algorithm on the posterior distribution; the result is used to update such probabilistic model and improve sampling of next points. The step-by-step procedure is described in Figure 7: the modeling procedure begins with the mixtures design and the execution of laboratory tests for the definition of features and targets variables; these latter are assumed to be representative of the physical problem treated; the input-target fitting is performed by a neural network, whose structure and algorithmic functioning is not known a priori; the search for the topology and the values of the training process parameters that minimize the prediction error of the ANN is handled by the Bayesian optimizer, by comparing network outputs and experimental targets; once the optimal hyperparameters combination has been identified, the model can be put into service for the designed application, by training it on the entire data set. In the current study setup, the hyperparameters to be optimized by the Bayesian approach can range in the following integer or logarithmic intervals:

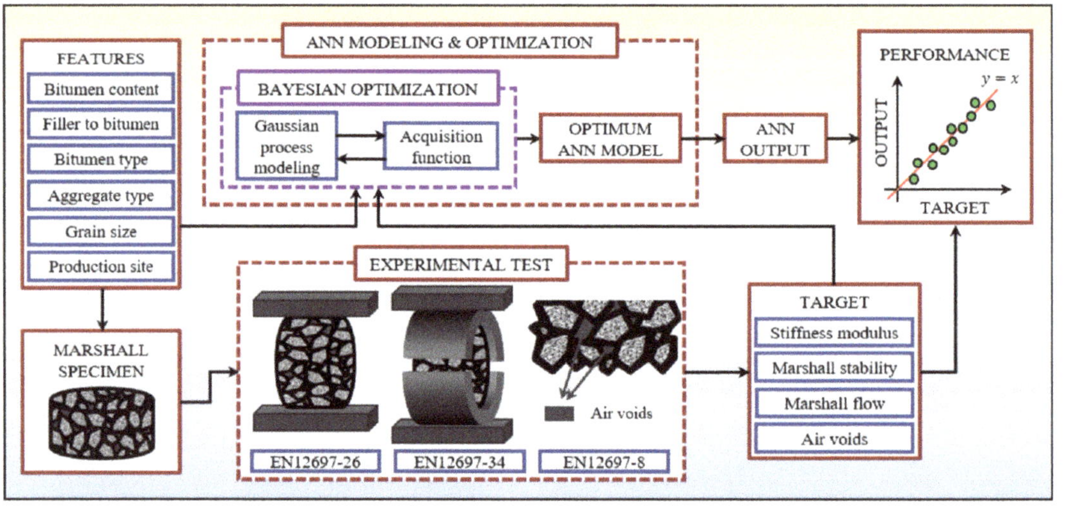

Figure 7. Step-by-step procedure followed in this study: it starts with the mix design (left side) and testing processes (bottom side); these tests define the set of target variables (lower right side); the input-target fitting is performed by a neural network, whose structure and algorithmic functioning are searched by the Bayesian optimizer (upper side) comparing network outputs and experimental targets (right side).

- $X_L = \{1, \ldots, 5\}$, for the number of hidden network layers L;
- $X_N = \{4, \ldots, 64\}$, for the number of neurons N for each hidden layer;
- $X_{act} = \{tanh, ReLU, ELU\}$ for the set of activation functions to be applied after each hidden layer;
- $X_\alpha = [10^{-6}, 10^{-2}]$ for the learning rate α;
- $X_\beta = [10^{-6}, 10^{-2}]$ for the weight decay parameter β;
- $X_E = \{500, \ldots, 5000\}$ for the number of learning process iterations.

These ranges were defined in the optimizer implementation [67] and have been left unchanged.

3.6. Implementation Details

Before being inputted to the ANN, each feature contained in the feature vectors $x^{(d)}$ was standardized, i.e., the respective mean was subtracted and division by the respective standard deviation was applied. The statistics were calculated on D. The same procedure was performed for the target feature vectors $y^{(d)}$, where each target variable was subtracted by its mean and divided by its standard deviation computed on D. Also, the BO observations $x^{(o)}$ were transformed in a similar fashion. In this case, each hyperparameter was standardized by the mean and standard deviation of the respective range.

The source code was written with the Python language. ANNs were implemented using the PyTorch machine learning framework, while the hyperparameters BO procedure was realized with the Bayesian optimization package [71]. The code was run on a machine provided with an Intel(R) Xeon(R) W-2125 4GHz CPU and 32GB of RAM running Ubuntu 18.04. Each experiment lasted circa 24 h.

4. Results and Discussion

Figure 8 shows the k-fold CV score of neural models that were designed in the 500 iterations of the BO algorithm. The Pearson's R coefficient averaged over the 5 validation folds tended to become uniform in value among the different combinations tested by the BO optimizer at least during the first 350 iterations (Figure 8) but showed more marked variations afterwards. The algorithm has detected a region of the search space where the validation error was not significantly changing and, having assessed an over-exploration of a specific zone, it has decided to focus its search on an unexplored area that might have high-performing solutions (however, identifying worse combinations). This result highlights the regulation between exploitation and exploration performed by the UCB algorithm.

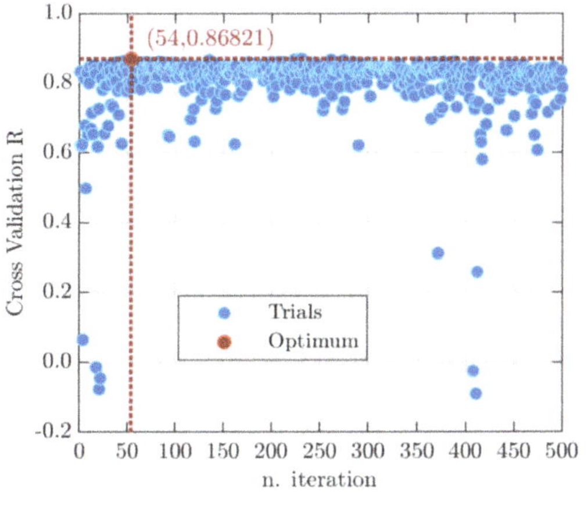

Figure 8. Average R-score on the 5 validation folds for the 500 algorithm iterations.

Despite the large size of the search space, the optimal model (i.e., showing the minimum validation loss MSE_{CV} = 0.249 and then the maximum value of the Pearson coefficient R_{CV} = 0.868) was found at iteration 54 (Figure 8). The hyperparameters discovered by the BO algorithm defined an ANN with L = 5 layers, N = 37 neurons, and hyperbolic tangent activation function (*tanh*), that was trained for E = 3552 iterations, with a learning rate of $\alpha = 0.01$ and weight decay $\beta = 1 \cdot 10^{-6}$. Table 5 shows the validation MSE (second column) and the R-score of the optimal model for each of the 5 folds (last column). In addition, the final average results for each mechanical characteristic and volumetric property are reported in the last row of Table 5. Figure 9 shows the relation between network output and experimental target for each fold.

Table 5. Number and codes of Marshall specimens.

Fold	MSE_{mean}	R—Pearson Correlation Coefficient				R_{mean}
		ITSM	MS	MF	AV	
0	0.219	0.837	0.866	0.842	0.964	0.877
1	0.203	0.963	0.835	0.725	0.973	0.874
2	0.254	0.872	0.836	0.799	0.917	0.856
3	0.223	0.918	0.826	0.917	0.956	0.905
4	0.346	0.841	0.731	0.834	0.912	0.829
CV_{result}	0.249	0.886	0.819	0.824	0.944	0.868

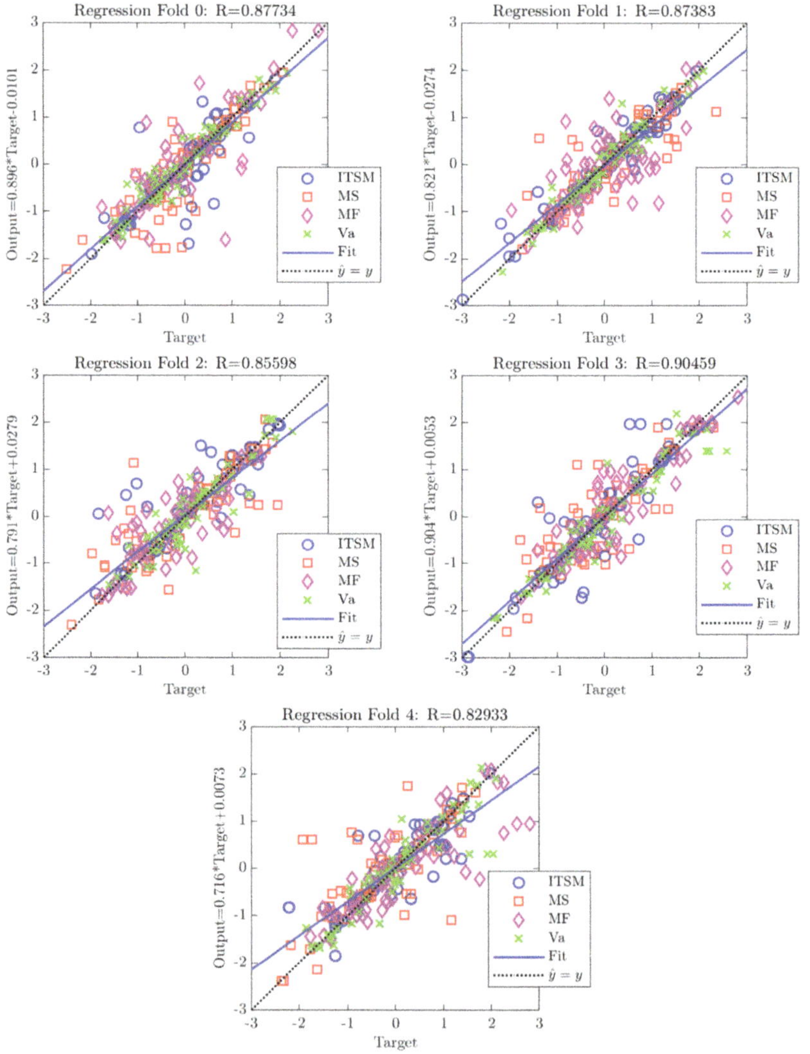

Figure 9. Regression analysis on the 5 folds.

Although there is a high variability in the data set, which can be explained by considering the different properties of the mixtures analysed, the optimal BO model returns very successful results (Figure 9). The 5-folds averaged Pearson coefficient for each mechanical and physical parameter is always greater than 0.819 (last row of Table 5). Fluctuations in values of the MSE and R parameters (second and seventh column in Table 5) can be attributed to the different distribution of training and test data that characterize each fold. This result shows how a fixed split of the data set can lead to an incorrect assessment of the prediction error, which can be worse (R_4 = 0.829) or better (R_3 = 0.906) than the most likely situation represented by the k-fold CV (R_{CV} = 0.868). Random and grid searches based on the prediction error by a fixed split training test may find solutions that are not optimal [49], due to fluctuations resulting from considering one partition rather than another. Figure 10 shows the comparison between experimental targets and predicted outputs, for the four parameters analyzed, as regards fold 4. Values calculated by the ANN model characterized by the highest prediction error (MSE_{mean} = 0.346, Table 5) are very close in value to the experimental data, whatever variable is considered. This result is relevant from an engineering point of view, because it proves that ANNs can be an accurate method to model (even simultaneously) the mechanical response and physical properties of bituminous mixtures, also very different in terms of composition.

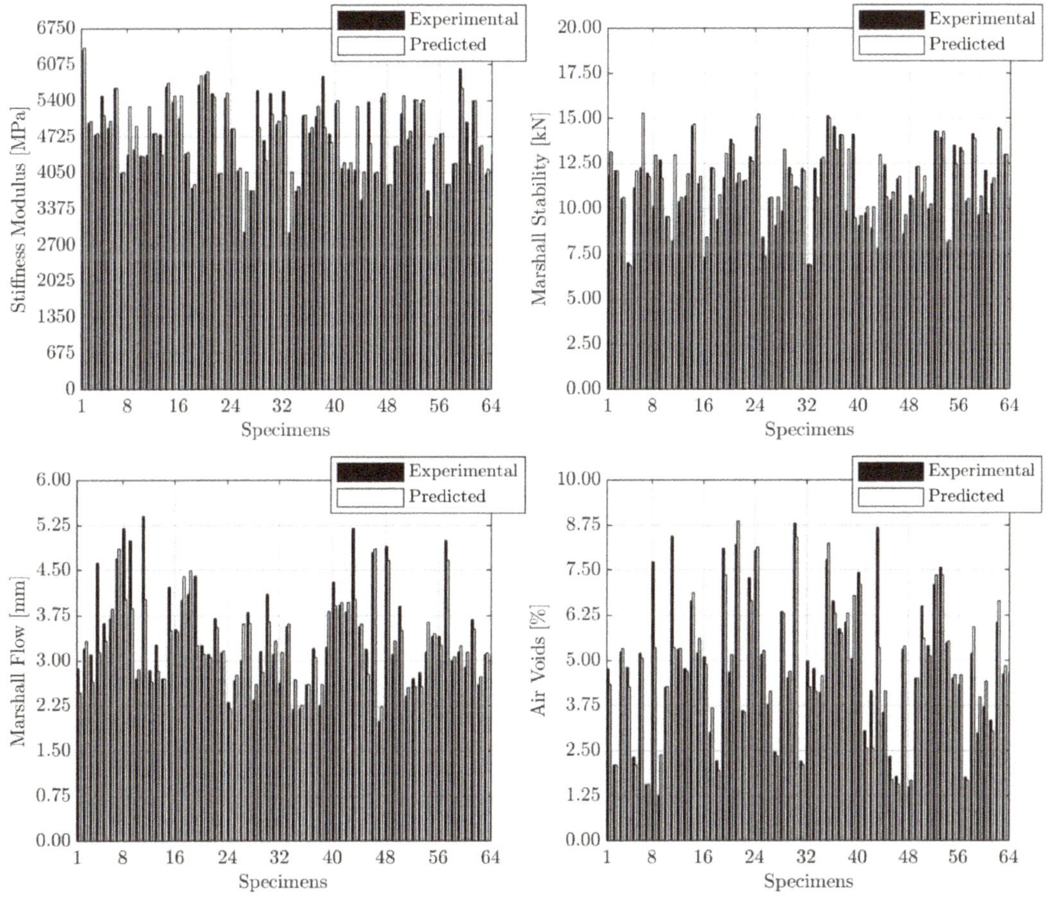

Figure 10. Experimental vs. predicted data, related to fold 4.

Although the presented modeling procedure is conceptually and computationally more complex than a simple grid search, it overcomes some of the problems (such as biased performance evaluations and sub-optimal hyperparameters selection) inherent in traditional methods (i.e., grid and random search) or in the most frequently used toolboxes (such as the MATLAB® ANN Toolbox), as the above reported results suggest.

5. Conclusions

The main goal of this study was to present and discuss a semi-automatic and unbiased procedure to select the most reliable ANN model for a given predictive modeling problem, which can overcome the limitations of conventional approaches. In particular, the focus was on predicting at once volumetric properties and mechanical characteristics of bituminous mixtures prepared using different types of bitumen and aggregate, binder content and maximum nominal grain size, to support the mix design phase, providing numerical estimations of the investigated parameters without any other costly laboratory test. The study results can be summarized as follows:

- To perform proper neural modeling, the evaluation of the several network structures resulting from the selection of different model hyperparameters values is required. The procedure developed in this article allowed the limitations of the most widely used ANN toolbox to be overcome.
- The proposed approach with the k-fold CV produces more reliable results in terms of model validation error, with respect to the standard grid search based on a random data set partition: in fact, if the procedure were based on a fixed random split of the available data set, different results are possible, worse (R_4 = 0.829) or better (R_3 = 0.906) than the most likely situation represented by the k-fold CV (R_{CV} = 0.868), due to the different distribution of the training and validation data.
- The BO algorithm has shown to be successful in solving the challenging problem of properly setting the model hyperparameters: it has identified the optimal solution, in terms of algorithmic and structural configuration of the ANN, in only 54 iterations. The hallmark of such a technique lies in the ability to take past evaluations into account so as to limit the loss function recalls. Nonetheless, the reader should be aware that the BO procedure results may be linked to the constraints set by the research engineer in terms of hyperparameters' variability.
- In the current paper, Marshall parameters, ITSM, as well as AV content have been determined simultaneously by a single multi-output ANN, unlike previous studies; therefore, such approach represents an integrated predictive model of the selected mechanical and volumetric properties.
- The neural network structure best suited (MSE_{CV} = 0.249, R_{CV} = 0.868) to model experimental mixtures data is defined by 5 layers, 37 neurons in each hidden layer and *tanh* transfer function. A learning step size α equal to 0.01 and weight decay β equal to 1×10^{-6} are implemented in the Ranger training algorithm.
- The algorithms applied and the analytical steps taken by the artificial networks have been illustrated in detail to make the procedure followed replicable to the reader. If it is desired to put the proposed model into service for the designed application (e.g., use in a laboratory or plant for estimates of mechanical parameters and volumetric properties of bituminous mixtures), then the optimized ANN must be trained with all available data.

It is worth recalling that the proposed model is applicable only to the specific types of aggregate, bitumen and mixture structure considered. This study did not account for the influence of different aggregate gradation on the results for a specific bitumen content. An interesting further development would be to investigate such an effect, integrating new input variables related to aggregate gradation or mix proportioning. Alternatively, the modeling variables relating to the mechanical characteristics of the mixture could be replaced by those relating to the road pavement performance, such as fatigue and permanent deformation resistance. This replacement would represent further progress towards a performance-based mix design.

Supplementary Materials: The following are available online at https://www.mdpi.com/article/10.3390/app112411710/s1, Table S1: Laboratory test results and specimens' characteristics.

Author Contributions: Conceptualization, M.M., M.D., E.M., C.M. and N.B.; data curation, M.M., M.D., F.R., E.M., J.V., C.M. and N.B.; formal analysis, M.M., M.D., F.R., E.M., J.V., C.M. and N.B.; investigation, E.M. and N.B.; methodology, M.M., M.D., F.R., E.M., J.V., C.M. and N.B.; software, M.M. and M.D.; supervision, E.M., C.M. and N.B.; validation, M.M., M.D., F.R., E.M., J.V., C.M. and N.B.; visualization, M.M., M.D., F.R., E.M., J.V., C.M. and N.B.; writing—original draft, M.M., M.D., F.R., E.M., J.V., C.M. and N.B.; writing—review and editing, M.M., M.D., F.R., E.M., J.V. and N.B.; resources, E.M. and N.B. All authors have read and agreed to the published version of the manuscript.

Funding: This research received no external funding.

Data Availability Statement: The data that support the findings of this study are available from the Supplementary Materials.

Conflicts of Interest: The authors declare no conflict of interest.

References

1. Zhou, F.; Scullion, T.; Sun, L. Verification and modeling of three-stage permanent deformation behavior of asphalt mixes. *J. Transp. Eng.* **2004**, *130*, 486–494. [CrossRef]
2. Gandomi, A.H.; Alavi, A.H.; Mirzahosseini, M.R.; Nejad, F.M. Nonlinear genetic-based models for prediction of flow number of asphalt mixtures. *J. Mater. Civ. Eng.* **2011**, *23*, 248–263. [CrossRef]
3. Alavi, A.H.; Ameri, M.; Gandomi, A.H.; Mirzahosseini, M.R. Formulation of flow number of asphalt mixes using a hybrid computational method. *Constr. Build. Mater.* **2011**, *25*, 1338–1355. [CrossRef]
4. Dias, J.L.F.; Picado-Santos, L.; Capitão, S. Mechanical performance of dry process fine crumb rubber asphalt mixtures placed on the Portuguese road network. *Constr. Build. Mater.* **2014**, *73*, 247–254. [CrossRef]
5. Liu, Q.T.; Wu, S.P. Effects of steel wool distribution on properties of porous asphalt concrete. Key engineering materials. *Trans. Tech Publ.* **2014**, *599*, 150–154. [CrossRef]
6. Garcia, A.; Norambuena-Contreras, J.; Bueno, M.; Partl, M.N. Influence of steel wool fibers on the mechanical, termal, and healing properties of dense asphalt concrete. *J. Test. Eval.* **2014**, *42*, 1107–1118. [CrossRef]
7. Pasandín, A.; Pérez, I. Overview of bituminous mixtures made with recycled concrete aggregates. *Constr. Build. Mater.* **2015**, *74*, 151–161. [CrossRef]
8. Zaumanis, M.; Mallick, R.B.; Frank, R. 100% hot mix asphalt recycling: Challenges and benefits. *Transp. Res. Procedia* **2016**, *14*, 3493–3502. [CrossRef]
9. Wang, L.; Gong, H.; Hou, Y.; Shu, X.; Huang, B. Advances in pavement materials, design, characterisation, and simulation. *Road Mater. Pavement Des.* **2017**, *18*, 1–11. [CrossRef]
10. Erkens, S.; Liu, X.; Scarpas, A. 3D finite element model for asphalt concrete response simulation. *Int. J. Geomech.* **2002**, *2*, 305–330. [CrossRef]
11. Giunta, M.; Pisano, A.A. One-dimensional visco-elastoplastic constitutive model for asphalt concrete. *Multidiscip. Modeling Mater. Struct.* **2006**, *2*, 247–264. [CrossRef]
12. Underwood, S.B.; Kim, R.Y. Viscoelastoplastic continuum damage model for asphalt concrete in tension. *J. Eng. Mech.* **2011**, *137*, 732–739. [CrossRef]
13. Yun, T.; Kim, Y.R. Viscoelastoplastic modeling of the behavior of hot mix asphalt in compression. *KSCE J. Civ. Eng.* **2013**, *17*, 1323–1332. [CrossRef]
14. Pasetto, M.; Baldo, N. Computational analysis of the creep behaviour of bituminous mixtures. *Constr. Build. Mater.* **2015**, *94*, 784–790. [CrossRef]
15. Di Benedetto, H.; Sauzéat, C.; Clec'h, P. Anisotropy of bituminous mixture in the linear viscoelastic domain. *Mech. Time Depend. Mater.* **2016**, *20*, 281–297. [CrossRef]
16. Pasetto, M.; Baldo, N. Numerical visco-elastoplastic constitutive modelization of creep recovery tests on hot mix asphalt. *J. Traffic Transp. Eng.* **2016**, *3*, 390–397. [CrossRef]
17. Darabi, M.K.; Huang, C.W.; Bazzaz, M.; Masad, E.A.; Little, D.N. Characterization and validation of the nonlinear viscoelastic-viscoplastic with hardening-relaxation constitutive relationship for asphalt mixtures. *Constr. Build. Mater.* **2019**, *216*, 648–660. [CrossRef]
18. Kim, S.H.; Kim, N. Development of performance prediction models in flexible pavement using regression analysis method. *KSCE J. Civ. Eng.* **2006**, *10*, 91–96. [CrossRef]

19. Laurinavičius, A.; Oginskas, R. Experimental research on the development of rutting in asphalt concrete pavements reinforced with geosynthetic materials. *J. Civ. Eng. Manag.* **2006**, *12*, 311–317. [CrossRef]
20. Shukla, P.K.; Das, A. A re-visit to the development of fatigue and rutting equations used for asphalt pavement design. *Int. J. Pavement Eng.* **2008**, *9*, 355–364. [CrossRef]
21. Rahman, A.A.; Mendez Larrain, M.M.; Tarefder, R.A. Development of a nonlinear rutting model for asphalt concrete based on Weibull parameters. *Int. J. Pavement Eng.* **2019**, *20*, 1055–1064. [CrossRef]
22. Specht, L.P.; Khatchatourian, O.; Brito, L.A.T.; Ceratti, J.A.P. Modeling of asphalt-rubber rotational viscosity by statistical analysis and neural networks. *Mater. Res.* **2007**, *10*, 69–74. [CrossRef]
23. Mirzahosseini, M.R.; Aghaeifar, A.; Alavi, A.H.; Gandomi, A.H.; Seyednour, R. Permanent deformation analysis of asphalt mixtures using soft computing techniques. *Expert Syst. Appl.* **2011**, *38*, 6081–6100. [CrossRef]
24. Androjić, I.; Marović, I. Development of artificial neural network and multiple linear regression models in the prediction process of the hot mix asphalt properties. *Can. J. Civ. Eng.* **2017**, *44*, 994–1004. [CrossRef]
25. Alrashydah, E.I.; Abo-Qudais, S.A. Modeling of creep compliance behavior in asphalt mixes using multiple regression and artificial neural networks. *Constr. Build. Mater.* **2018**, *159*, 635–641. [CrossRef]
26. Ziari, H.; Amini, A.; Goli, A.; Mirzaiyan, D. Predicting rutting performance of carbon nano tube (CNT) asphalt binders using regression models and neural networks. *Constr. Build. Mater.* **2018**, *160*, 415–426. [CrossRef]
27. Montoya, M.A.; Haddock, J.E. Estimating asphalt mixture volumetric properties using seemingly unrelated regression equations approaches. *Constr. Build. Mater.* **2019**, *225*, 829–837. [CrossRef]
28. Lam, N.-T.-M.; Nguyen, D.-L.; Le, D.-H. Predicting compressive strength of roller-compacted concrete pavement containing steel slag aggregate and fly ash. *Int. J. Pavement Eng.* **2020**, *2020*, 1–14. [CrossRef]
29. Baldo, N.; Manthos, E.; Pasetto, M. Analysis of the mechanical behaviour of asphalt concretes using artificial neural networks. *Adv. Civ. Eng.* **2018**, *2018*, 1650945. [CrossRef]
30. Tarefder, R.A.; White, L.; Zaman, M. Neural network model for asphalt concrete permeability. *J. Mater. Civ. Eng.* **2005**, *17*, 19–27. [CrossRef]
31. Ozsahin, T.S.; Oruc, S. Neural network model for resilient modulus of emulsified asphalt mixtures. *Constr. Build. Mater.* **2008**, *22*, 1436–1445. [CrossRef]
32. Tapkın, S.; Çevik, A.; Uşar, Ü. Accumulated strain prediction of polypropylene modified marshall specimens in repeated creep test using artificial neural networks. *Expert Syst. Appl.* **2009**, *36*, 11186–11197. [CrossRef]
33. Xiao, F.; Amirkhanian, S.; Juang, C.H. Prediction of fatigue life of rubberized asphalt concrete mixtures containing reclaimed asphalt pavement using artificial neural networks. *J. Mater. Civ. Eng.* **2009**, *21*, 253–261. [CrossRef]
34. Ahmed, T.M.; Green, P.L.; Khalid, H.A. Predicting fatigue performance of hot mix asphalt using artificial neural networks. *Road Mater. Pavement Des.* **2017**, *18*, 141–154. [CrossRef]
35. Ceylan, H.; Schwartz, C.W.; Kim, S.; Gopalakrishnan, K. Accuracy of predictive models for dynamic modulus of hot-mix asphalt. *J. Mater. Civ. Eng.* **2009**, *21*, 286–293. [CrossRef]
36. Le, T.-H.; Nguyen, H.-L.; Pham, B.T.; Nguyen, M.H.; Pham, C.-T.; Nguyen, N.-L.; Le, T.-T.; Ly, H.-B. Artificial intelligence-based model for the prediction of dynamic modulus of stone mastic asphalt. *Appl. Sci.* **2020**, *10*, 5242. [CrossRef]
37. Ghorbani, B.; Arulrajah, A.; Narsilio, G.; Horpibulsuk, S.; Bo, M.-W. Thermal and mechanical properties of demolition wastes in geothermal pavements by experimental and machine learning techniques. *Constr. Build. Mater.* **2021**, *280*, 122499. [CrossRef]
38. Aksoy, A.; Iskender, E.; Kahraman, H.T. Application of the intuitive k-NN Estimator for prediction of the Marshall Test (ASTM D1559) results for asphalt mixtures. *Constr. Build. Mater.* **2012**, *34*, 561–569. [CrossRef]
39. Van Thanh, D.; Feng, C.P. Study on marshall and rutting test of SMA at abnormally high temperature. *Constr. Build. Mater.* **2013**, *47*, 1337–1341. [CrossRef]
40. Abdoli, M.; Fathollahi, A.; Babaei, R. The application of recycled aggregates of construction debris in asphalt concrete mix design. *Int. J. Environ. Res.* **2015**, *9*, 489–494. [CrossRef]
41. Sarkar, D.; Pal, M.; Sarkar, A.K.; Mishra, U. Evaluation of the properties of bituminous concrete prepared from brick-stone mix aggregate. *Adv. Mater. Sci. Eng.* **2016**, *2016*, 2761038. [CrossRef]
42. Xu, B.; Chen, J.; Zhou, C.; Wang, W. Study on Marshall design parameters of porous asphalt mixture using limestone as coarse aggregate. *Constr. Build. Mater.* **2016**, *124*, 846–854. [CrossRef]
43. Zumrawi, M.M.; Khalill, F.O. Experimental study of steel slag used as aggregate in asphalt mixture. *Am. J. Constr. Build. Mater.* **2017**, *2*, 26–32. [CrossRef]
44. Al-Ammari, M.A.S.; Jakarni, F.M.; Muniandy, R.; Hassim, S. The effect of aggregate and compaction method on the physical properties of hot mix asphalt. *IOP Conf. Ser. Mater. Sci. Eng.* **2019**, *512*, 012003. [CrossRef]
45. Tapkın, S.; Çevik, A.; Uşar, Ü. Prediction of Marshall test results for polypropylene modified dense bituminous mixtures using neural networks. *Expert Syst. Appl.* **2010**, *37*, 4660–4670. [CrossRef]
46. Ozgan, E. Artificial neural network based modelling of the Marshall Stability of asphalt concrete. *Expert Syst. Appl.* **2011**, *38*, 6025–6030. [CrossRef]
47. Khuntia, S.; Das, A.K.; Mohanty, M.; Panda, M. Prediction of marshall parameters of modified bituminous mixtures using artificial intelligence techniques. *Int. J. Transp. Sci. Technol.* **2014**, *3*, 211–227. [CrossRef]

48. Zavrtanik, N.; Prosen, J.; Tušar, M.; Turk, G. The use of artificial neural networks for modeling air void content in aggregate mixture. *Autom. Constr.* **2016**, *63*, 155–161. [CrossRef]
49. James, G.; Witten, D.; Hastie, T.; Tibshirani, R. Resampling methods. In *An Introduction to Statistical Learning*; Springer: Berlin/Heidelberg, Germany, 2013; Volume 112, Chapter 5; pp. 176–186.
50. Baldo, N.; Manthos, E.; Miani, M. Stiffness modulus and marshall parameters of hot mix asphalts: Laboratory data modeling by artificial neural networks characterized by cross-validation. *Appl. Sci.* **2019**, *9*, 3502. [CrossRef]
51. Baldo, N.; Miani, M.; Rondinella, F.; Celauro, C. A machine learning approach to determine airport asphalt concrete layer moduli using heavy weight deflectometer data. *Sustainability* **2021**, *13*, 8831. [CrossRef]
52. Shahriari, B.; Swersky, K.; Wang, Z.; Adams, R.P.; de Freitas, N. Taking the human out of the loop: A review of Bayesian optimization. *Proc. IEEE* **2015**, *104*, 148–175. [CrossRef]
53. Bergstra, J.; Bardenet, R.; Bengio, Y.; Kégl, B. Algorithms for hyper-parameter optimization. Proceeding of the 24th Advances in Neural Information Processing Systems (NIPS 2011), Granada, Spain, 12–17 December 2011; Shawe-Taylor, J., Zemel, R., Bartlett, P., Pereira, F., Weinberger, K.Q., Eds.; Curran Associates, Inc.: New York, NY, USA, 2011.
54. Bergstra, J.; Yamins, D.; Cox, D. Making a science of model search: Hyperparameter optimization in hundreds of dimensions for vision architectures. *Int. Conf. Mach. Learn. PMLR* **2013**, *28*, 115–123.
55. Xiao, F.; Amirkhanian, S.N. Artificial neural network approach to estimating stiffness behavior of rubberized asphalt concrete containing reclaimed asphalt pavement. *J. Transp. Eng.* **2009**, *135*, 580–589. [CrossRef]
56. Widrow, B.; Hoff, M.E. *Adaptive Switching Circuits*; Technical Report; Stanford University, Ca Stanford Electronics Labs: Stanford, CA, USA, 1960.
57. Rosenblatt, F. *Principles of Neurodynamics: Perceptrons and the Theory of Brain Mechanisms*; Technical Report; Cornell Aeronautical Lab Inc.: Buffalo, NY, USA, 1961.
58. Rumelhart, D.E.; Hinton, G.E.; Williams, R.J. Learning representations by back-propagating errors. *Nature* **1986**, *323*, 533–536. [CrossRef]
59. McCulloch, W.S.; Pitts, W. A logical calculus of the ideas immanent in nervous activity. *Bull. Math. Biophys.* **1943**, *5*, 115–133. [CrossRef]
60. Demuth, H.B.; Beale, M.H.; de Jess, O.; Hagan, M.T. *Neural Network Design*; Martin Hagan: Stillwater, OK, USA, 2014.
61. Hagan, M.T.; Menhaj, M.B. Training feedforward networks with the Marquardt algorithm. *IEEE Trans. Neural Netw.* **1994**, *5*, 989–993. [CrossRef] [PubMed]
62. Liu, L.; Jiang, H.; He, P.; Chen, W.; Liu, X.; Gao, J.; Han, J. On the variance of the adaptive learning rate and beyond. *arXiv* **2019**, arXiv:1908.03265.
63. Kingma, D.P.; Ba, J. Adam: A method for stochastic optimization. *arXiv* **2014**, arXiv:1412.6980.
64. Zhang, M.R.; Lucas, J.; Hinton, G.; Ba, J. Lookahead optimizer: K steps forward, 1 step back. *arXiv* **2019**, arXiv:1907.08610.
65. Ng, A.Y. Feature selection, L 1 vs. L 2 regularization, and rotational invariance. In Proceedings of the 21th International Conference on Machine learning, Banff, AB, Canada, 4–8 July 2004; pp. 615–622. [CrossRef]
66. Snoek, J.; Larochelle, H.; Adams, R.P. Practical bayesian optimization of machine learning algorithms. *arXiv* **2012**, arXiv:1206.2944.
67. Rasmussen, C.E. *Gaussian Processes in Machine Learning. Summer School on Machine Learning*; Springer: Berlin/Heidelberg, Germany, 2003; pp. 63–71.
68. Kushner, H.J. A New Method of locating the maximum point of an arbitrary multipeak curve in the presence of noise. *J. Basic Eng.* **1964**, *86*, 97–106. [CrossRef]
69. Mockus, J.; Tiesis, V.; Zilinskas, A. The application of Bayesian methods for seeking the extremum. In *Towards Global Optimization*, 2nd ed.; Dixon, L.C.W., Szego, G.P., Eds.; North Holland Publishing Co.: Amsterdam, The Netherlands, 1978; pp. 117–129.
70. Srinivas, N.; Krause, A.; Kakade, S.M.; Seeger, M. Gaussian process optimization in the bandit setting: No regret and experimental design. *arXiv* **2009**, arXiv:0912.3995.
71. Nogueira, F. Bayesian Optimization: Open Source Constrained Global Optimization Tool for Python. Available online: https://github.com/fmfn/BayesianOptimization (accessed on 25 November 2021).

applied sciences

MDPI

Article

Using a Random Forest Model to Predict the Location of Potential Damage on Asphalt Pavement

Xiaogang Guo and Peiwen Hao *

School of Highway, Chang'an University, Xi'an 710064, China; gangxo@outlook.com
* Correspondence: pwhao@chd.edu.cn

Featured Application: This article provides guidance for the future studies of new pavement prediction technology, which may be the concern of many fields in road construction, design, and maintenance. It can really help engineers to make decisions and predictions, in order to save money, time, and lives.

Abstract: Potential damage, eventually demonstrated as moisture damage on inner and in-situ road structures, is the most complex problem to predict, which costs lots of money, time, and natural resources for maintenance and even leads to safety problems. Traditional linear regression analysis cannot fit well with this multi-factor task in such in-field circumstances. Random Forest (RF) is a progressive nonlinear algorithm, which can combine all relative factors to gain accurate prediction and good explanation. In this study, an RF model is constructed for the prediction of potential damage. In addition, relative variable importance is analyzed to obtain the correlations between factors and potential damage separately. The results show that, through the optimization, the model achieved a good average accuracy of 83.33%. Finally, the controlling method for moisture damage is provided by combining the traditional analysis method and the RF model. In a word, RF is a prospective method in predictions and data mining for highway engineering. Trained with effective data, it can be multifunctional and powerful to solve hard problems.

Keywords: moisture damage; random forest; machine learning; factor importance; prediction

Citation: Guo, X.; Hao, P. Using a Random Forest Model to Predict the Location of Potential Damage on Asphalt Pavement. *Appl. Sci.* **2021**, *11*, 10396. https://doi.org/10.3390/app112110396

Academic Editors: Luis Picado Santos, Amir Tabakovic, Jan Valentin and Liang He

Received: 15 September 2021
Accepted: 4 November 2021
Published: 5 November 2021

1. Introduction

Moisture damage is one of the main problems of asphalt pavement in service. Road networks are afflicted by this problem in the world for decades [1]. It is hard to detect the distress as it always happens underneath the surface in the middle and lower layers initially as potential damage [2]. As soon as the response emerges on the surface, the surface course will fail within just a few days, which may lead to serious safety problems. In another aspect, the unexpected distress on the road should need rapid maintenance, which will suspend traffic, profuse raw materials, and emit harmful smoke [3,4]. This is a huge carbon-consuming process, which not only costs massive time and money, but also natural resources [5]. To solve such a problem, the concise prediction for the positions of potential damages is one promising way [6–8].

Machine learning (ML) is a good way at building a high-performance prediction model. In the practice, we can continue to input the maintaining information into the prediction model. Then, if there is a section of the road that has already lost strength in the inner or middle of the structure but nothing or little response occurred on the surface, we can find it by the model. When the positions can be determined, potential distress can be eliminated in time instead of breaking out. Then, a significant problem can be weakened to a minor one, which can save lives and increase road value. Additionally, the performance and interpretability of the model are both important for the evaluation and application of the model. The propose of this study is using the actual detection data to construct a high practical model to solve the daily problem of road maintenance. It can be the complementary method to really help engineers to more precisely judge the state of the pavement. It is also important that the construction process and hyperparameters optimization of the model provide an example to support for development and improvement ML models in highway engineering.

1.1. Moisture Damage and Potential Distress

Moisture damage is one of the main forms of potential distress that results in strength loss, stripping, and deformation of pavements [9]. It is generally caused by segregation in the construction process, which is presented as poorly bonding and compaction [10]. This category of distress is so-called potential damage because it is hard to detect with little change in its extrinsic feature at the initial phase but the strength has been loosened [11]. When it bears certain loading, a sudden breakdown may happen on the road surface [12]. Because of this hidden distress, maintenance work and driving safety burden more pressure than the average routine [1].

In order to solve the problem, traditional methods, such as geological radar and falling weight deflectometer (FWD), are used for in-field projects to detect potential failure inter road structures [13–15]. Lots of effort has been made in making a long-term detective routine [16]. However, accuracy of these methods is not achievable. Because moisture damage and potential distress cause multiple driving problems, they cannot be determined by tests from a one-directional aspect under the wide-range and complex in-field conditions [9]. A simple linear relationship cannot construct and explain the correlations between properties and in-field failure of pavements. Furthermore, predictions and classifications derived from the traditional methods may be unstable because of the personal experiences of judgment for potential damage by most nondestructive detection methods [17]. For example, even though the geological radar technology has developed more automatic and precise for pavement detection and the radar images are processed by high advanced software, the detection results are easily affected by the different detection conditions, which will lead to misjudgment from the engineers with less experience.

1.2. Machine Learning and Random Forest

Machine learning (ML) is a technology using algorithms to let computers analyze data and process other affairs stimulating the way of humans learning, which can continue improving their accuracy and capability by the algorithms themselves [18]. ML acts a significant role in statistical research with the rapid development of computational speed and artificial intelligence (AI) algorithms [19]. A ML model can be trained like a project manager to perform classification, prediction, and mining interrelationships on data [20]. Almost every scientific discipline is driven by AI in this big data era, which is growing hugely day by day [21]. Subsequently, science research aided by computers has become more popular and there is a higher general demand of researchers nowadays.

Highway engineering is a traditional discipline of applied science. Its basement is also built by tests and data. With instrument automation in recent years, the data from pavement detection grows furiously [22]. To understand the inner relationships of the big data in highway engineering, two major methods can be used: simplification and comprehension. As an experiential mechanical science, the first method, which combines simplified data and hypotheses, can make problems easier and solve them with mechanical models [23]. However, it overlooks some parts of the experimental characters and randomness of the data to gain a general result. The light weight factors, which are ignored in experiential mechanical models, also have impact in the results. In fact, some unseen capability loss has already existed before distresses appear, which cannot be measured [24]. That is why the traditional methods can explain the reasons well but cannot predict the results accurately for potential damage [25,26]. Therefore, ML models can be the perfect complementary to traditional methods. Neural networks, gradient-boosted model, random forest, and support vector machine have been used in mining data for the long-term reservation or open assess databases of pavement detection [27–30]. Due to these ML models, the relationships of data can be found and understood more comprehensively than conventional physical models. Furthermore, through the predictions, a better decision can be made by their excellent prediction performance. All in all, ML is an advantageous tool in experimental and theoretical studies for highway projects. In the practice, model adaptability, model structures, and inputted variables are the three key matters we need to consider carefully in a ML construction work.

Random forest (RF) is a promising machine learning algorithm which can help researchers forecast or classify data and information with high performance [31–33]. It is a model that assembles decision trees using a modified bagging method to improve the predictive accuracy [34]. The common strategy of ML to solve a nonlinear problem is to raise data dimension by different weights and biases to discover key features, such as kernel SVM, a neural network. The process of data transition increases the computational complexity. Combining their computational frameworks, it may lead to lower computing productivity as a whole when in a multivariate classification problem. For instance, under the framework of the one vs rest, SVM consumes huge memory with increasing variables (data dimensions), especially using a nonlinear kernel. Comparing to this strategy, RF uses the bagging method to make data into a tree-like 2D structure, which can keep the simplicity of data. Therefore, it has outstanding computing speed and interpretability. In addition, RF can perform as well as kernel SVM and neural network by the bagging method [35]. It has been successfully applied in the predictions of IRI, strength, and cracking on the pavement, and has gained great performance [36–38]. Moreover, the most important advantage is that RF is good at processing multicollinear, imbalanced, missing data with multiple variables [39]. That is the reason the RF model is suitable for the data derived from in-field tests and detections.

In summary, random forest (RF) model can be trained to predict potential distress and moisture damage for flexible pavement. Considered the complex factors of the test environment, it can avoid the deviation in the typical prediction method, which may just be extracted from a linear regression model. In this study, an RF model is trained and constructed based on the data from a full-size track road test for potential damage prediction. The prediction performance and relative factor importance are estimated for the model. Finally, the analysis method of insight relationships and project problems can be developed and promoted with the RF model.

2. Objectives

- To construct a random forest model to fit the principles for the potential deterioration of the typical flexible pavement;
- To predict the process and position of potential damage on asphalt pavement;
- To evaluate the performance and interpretability of the random forest model.

3. Data Collection and Preparation

3.1. Full-Size Track Road Test

A test road was constructed and prepared for the full-size track road test. The track was designed to be a 40 km single-lane of a typical structure (4 cm + 6 cm Superpave surface course and an 8-cm asphalt-treated base) on a 50-cm cement-stabilized base (20 + 30 cm cement-stabilized macadam) (Figure 1). Twenty-five full-load trucks were used to present the accelerated test in an unfavorable season, which was in hot weather and frequent rain days during June that year. The average range of air temperature was 16–28° and there were eight days of rain over 20 testing days. The whole loading process was separated into four stages by every 5 days. After each stage, the quality parameters, such as the international roughness index (IRI), the deflection value, and the rutting depth, were collected by automatic devices. Asphalt in-field samples were cored to test their void rates and splitting strength before and after the process. As 29,820 standardized loading times were achieved in the whole test, some bumps and pits appeared at random surface areas.

These positions of potential and emerged distress obtained by coring and observation are marked as the label for the model.

Figure 1. The structure of the test road.

3.2. Data Collection and Preparation

The data related to deterioration were collected and arranged in Table 1. Thirty-four variables were chosen to build the forest.

Table 1. Description of variables.

Variables	Symbol	Unit
The original rutting depth	R0	mm
The rutting depth on the fifth day	R5	mm
The rutting depth on the tenth day	R10	mm
The rutting depth on the fifteenth day	R15	mm
The rutting depth on the twentieth day	R20	mm
Increase in rutting depth at the first stage of the test	RI1	mm
Increase in rutting depth in the second stage of the test	RI2	mm
Increase in rutting depth in the third stage of the test	RI3	mm
Increase in rutting depth in the fourth stage of the test	RI4	mm
Increase in rutting depth through the whole test	RI20	mm
The original deflection value	D0	0.001 mm
The deflection value on the fifth day	D5	0.001 mm
The deflection value on the tenth day	D10	0.001 mm
The deflection value on the fifteenth day	D15	0.001 mm
The deflection value on the twentieth day	D20	0.001 mm
Increase in the deflection value in the first stage of the test	DI1	0.001 mm
Increase in the deflection value in the second stage of the test	DI2	0.001 mm
Increase in the deflection value in the third stage of the test	DI3	0.001 mm
Increase in the deflection value in the fourth stage of the test	DI4	0.001 mm
Increase in the deflection value through the whole test	DI20	0.001 mm
The original IRI value	IRI0	m/km
The IRI value on the fifth day	IRI5	m/km
The IRI value on the tenth day	IRI10	m/km
The IRI value on the fifteenth day	IRI15	m/km
The IRI value on the twentieth day	IRI20	m/km
Increase in the IRI value in the first stage of the test	IRII1	m/km
Increase in the IRI value in the second stage of the test	IRII2	m/km
Increase in the IRI value in the third stage of the test	IRII3	m/km
Increase in the IRI value in the fourth stage of the test	IRII4	m/km

Table 1. *Cont.*

Variables	Symbol	Unit
Increase in the IRI value through the whole test	IRII20	m/km
The void rate of the surface course	VRS	%
The void rate of the bottom ATB	VTB	%
The splitting strength of the surface course	SSS	Mpa
The splitting strength of the bottom ATB	SSB	Mpa

The data can be separated into two groups, which are initial data before the test running and in-process data during the test. Some variables imply that damage has already occurred. However, when a decision was made to uncover the surface, it was found that the result was not accurate. Some variables are linear with predictions that can be accumulated by different weights, but some variables are non-linear whose margins are hard to decide. That is why an RF model is needed to improve the accuracy of predictions.

The data are collected as much as detected to avoid disregarding any small factor, which may also have influence on the predicting result. Nevertheless, some of the variables are dependent on other inputs. We prefer more information richness than data independence. This is because the RF model is very good at multicollinear problems. Besides, if there are some negative factors introduced in the model, they can be pruned in the procedure of model optimization for computing spend and model strength.

4. Methodology

As stated in the introduction, the RF model is an ensemble learning (parallel learning) model with high accuracy. Based on every decision tree, RF can avoid over-fitting and under-fitting problems by efficiently estimating variables on large databases in most classification problems [40]. Based on a bagging (an abbreviation for the bootstrap aggregation) strategy, the database is split into N groups to build and train multiple decision trees [41]. The great number of de-correlated trees can be scored by the different generated branches to balance and improve predictive performance.

4.1. Decision Trees

In the decision tree algorithm, a set of splitting rules is used to partition data features into smaller spaces with similar responses by asking simple if-else questions about each feature. Every sub-space of data presents a simpler model, which is fitted to obtain predictions. This division-and-conquer technique can produce simple rules that can easily be understood and visualized by tree diagrams. In the classification trees of this study, the gini impurity and information gain criteria are computed to evaluate the possibility and performance of each tree.

4.2. Bagging

Bootstrap aggregation or bagging is a powerful procedure to improve the bagged decision trees learning behavior to achieve low root mean squared error (RMSE) by reducing the high variance from a single-tree structure [42].

The training data can be split into multiple subsets at random, which are fitted and trained by independent decision tree models separately. The aggregation of the predictions across all the trees is averaged to minimize the correlation effects between each couple of trees. The process is shown as follows:

(1) Set B as the number of the generated trees;
(2) Build bth prediction tree model as $\hat{f}^{b*}(x)$ by bootstrapping;
(3) Average all predictions of trees, as shown as Expression (1) [42].

$$\hat{f}_{bag}(x) = \frac{1}{B}\sum_{b=1}^{B} \hat{f}^{b*}(x) \quad (1)$$

According to this procedure, though, every single tree model has high variance. The averaged B trees, which combine hundreds of trees, can reduce the value as a whole.

4.3. Out-of-the-Box Performance

For the classification problem, which has qualitative outcomes, a voting strategy is adapted to record the predicted class and pick the most frequently occurring class. It is a straightforward method to assess the error performance for a bagged prediction model.

Out-of-bag (OOB) observations are used to predict and evaluate the results by trained model. Comparing the results and observations, the classification error or test error can be accumulated.

Set the testing data space as T, which has n trees. The data can be presented as:

$$T = \{(x_1, y_1), (x_2, y_2), \ldots, (x_n, y_n)\}$$

Fed T to the given RF model, we obtain another data set, such as:

$$T_f = \left\{\left(x_1, y_{f1}\right), \left(x_2, y_{f2}\right), \ldots, \left(x_n, y_{fn}\right)\right\}$$

Therefore,

$$OBB\ error = \frac{number\left(y_i \neq y_{fi}\right)}{n} \tag{2}$$

4.4. The State-of-the-Art Method

The state-of-the-art (SOTA) methods are applied to check if the RF model can achieve the best performance in this learning work. Decision tree and support vector machines (SVM) models are selected to compare with the trained RF model. The decision tree model, which has good strength, is the base unit of the RF model. The SVM is also a high-performance classification algorithm. They are both commonly used in data mining.

Decision trees are constructed by the significance measurement of data. In addition, the SVM is built based on the liner kernel. The accuracy of the models is set to be the baseline for the comparison.

4.5. Relative Importance of Variables

Even though the structure of bagged trees grows bigger to gain significant improvement from a single tree, the whole model becomes harder to interpret. For computing the relative importance of each variable in the RF model, the importance value of each predictor in every single tree is recorded and accumulated to realize the comparison process. Thus, the most effective factor will be gained in the given predicted result. The high value of the relative importance means a significant weight in their relationships, which is a more important factor in the road deterioration process. Each variable importance can be summed by the reduction in the loss function, which is attributed to each split in a given tree.

5. Model Construction

5.1. Model Structure Design

Four key steps are organized, as shown in Figure 2. A cyclic process is selected to train the RF model and optimize the model parameters repeatedly to obtain a minimal OOB error. Firstly, the quality of the given database is the most important as the basement in the whole model structure. Next, the main body of the RF model for model training is built by R language and its packages. The model will be several times to return to this step for the procedure of model optimization. This interaction of these two steps determines the final model structure and parameters, which will be applied in testing data. Finally, the prediction error rate will be estimated to assess the model performance. If insufficient performance is found, the cyclic process must run again and again after adding new data and checking the data effectiveness and correlations until achieving the best fitness of the RF model.

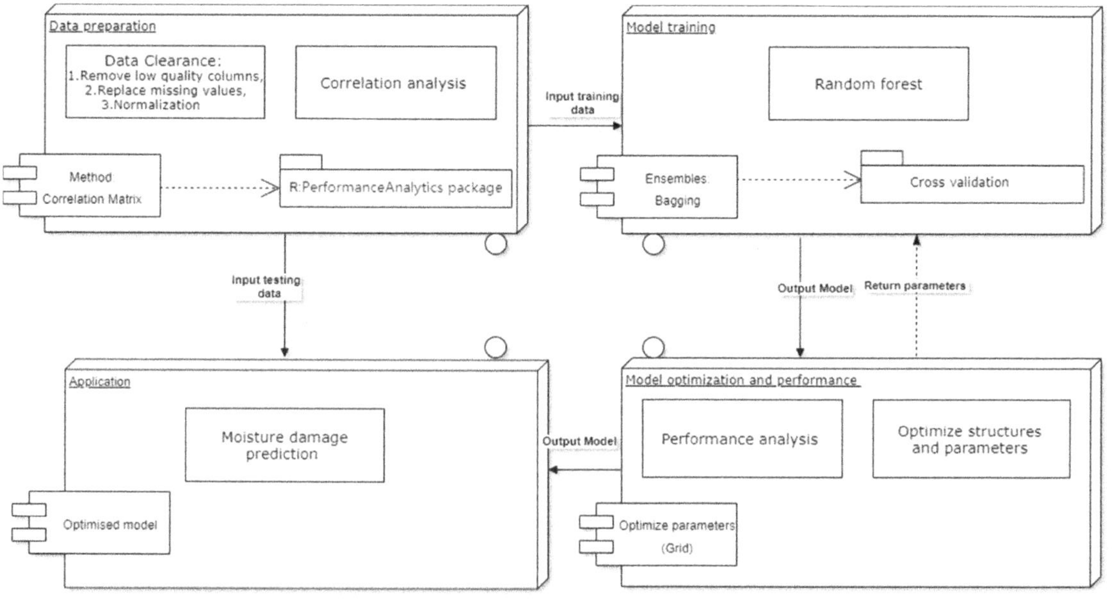

Figure 2. Structure design of the RF model.

5.2. RF Model Construction

When the structure of the RF model for potential damage is decided, the training set data are input to fit and grow every single tree with two key hyperparameters, including *mtry* and *ntree*. The *mtry* is the number of variables tried at each split. The *ntree* is the total number of trees the forest will grow. To obtain classification, every tree is run down in the forest with a number *m* of variables, which is used to split the node. With no pruning, trees are grown as large as possible. Random forest cannot be overfit. Therefore, the number of single trees can be grown as many as the computer capability can do. With the increase in the tree number, the OOB error will keep decreasing. When all the data are run down the trees, the proximities, OOB error, and variable importance are computed. Finally, the most possible result is voted by majority voting to obtain the confident prediction. The process is shown in Figure 3.

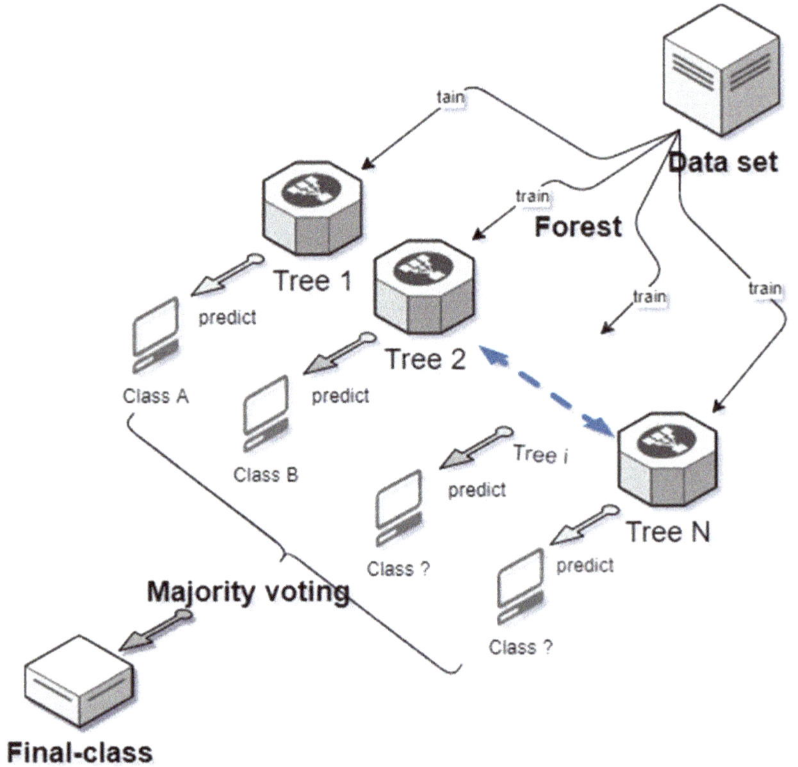

Figure 3. RF model construction process.

6. Results and Discussions

6.1. Data Characteristics and Correlations

Related variables are considered as many as possible in this study for comprehensive understanding. Therefore, thirty-four categories of data about the properties of the in-field road are prepared to train. Data collection is the most important step before a model is constructed. The resources and data features are matters of the prediction results. The details of the data used in the training process cannot be exhibited due to the large data group. A general view of characters and correlations of data sets is plotted in a matrix in Figure 4.

The diagonal line of plots is the distribution status of variables, which shows that all datasets collected from road properties are almost on or can be standardized into normal distribution. Therefore, the training data are effective to work reasonably in the model. A dataset on normal distribution means it fits with the principles of the average detection data. There is no need to delete a low or abnormal variable.

The plots on the intersection between every two properties are their correlation index and fitting curves. It is clear that some of them have obvious linear correlations, which are always desirable and easy to evaluate in a typical numerical analysis. However, the other data with non-linear relationships are hard to obtain rules for. Therefore, there are no consistent principles for these factors that can determine the occurrence of potential damage. This RF model can help to combine and follow all hints of variables, even those that are not important to achieve the best prediction.

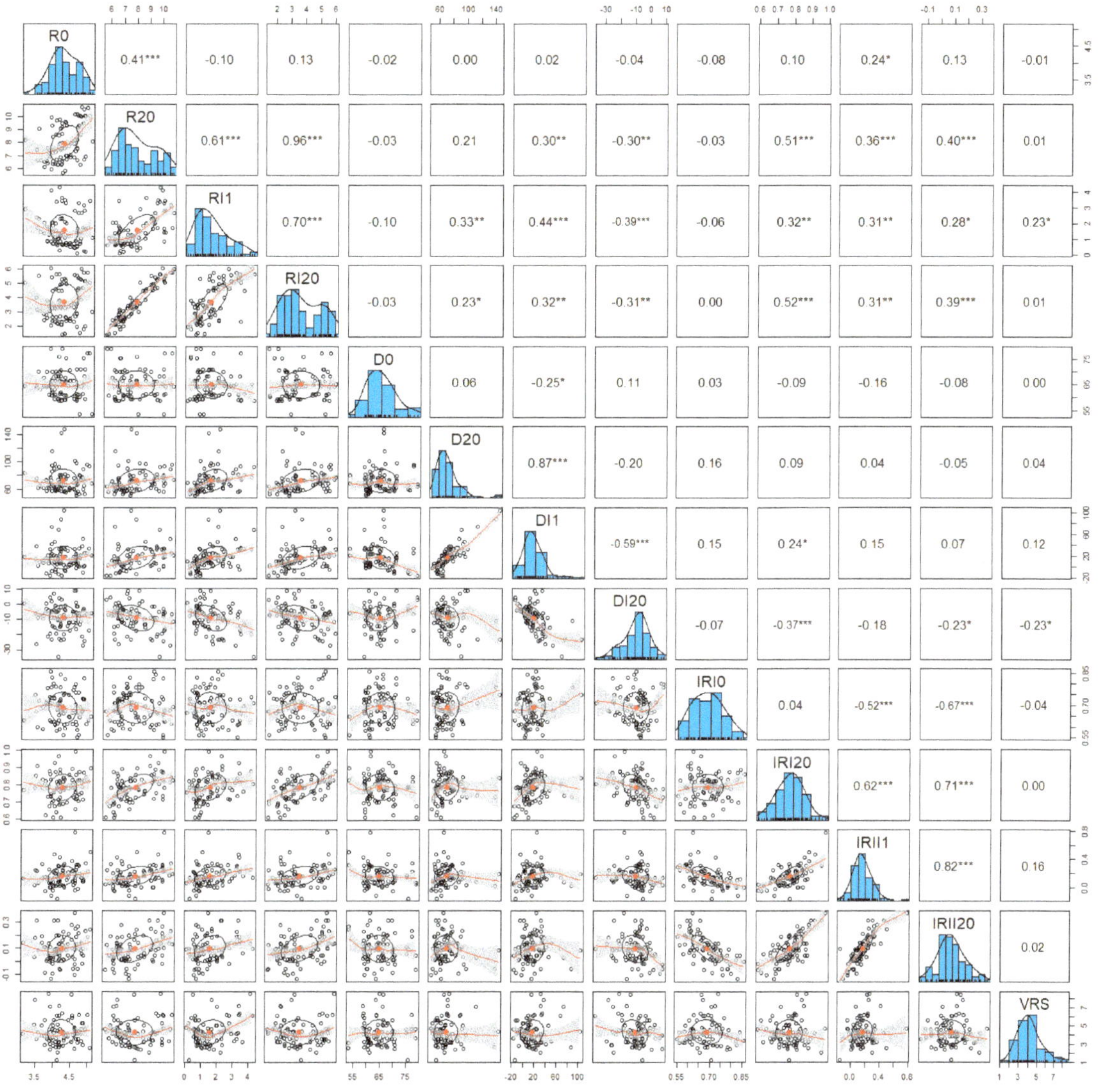

Figure 4. Data characteristics and correlations matrix: the * represents the significance level.

6.2. Number of Trees and Number of Variables Tried at Each Split

The two key hyperparameters, *ntree* and *mtry*, are determined by the exhaustive method. At first, an RF forest was constructed with the following default settings: *ntree* = 500 and *mtry* = 5. The OOB estimate of the error rate of the RF model is 20.24% and the confusion matrix is shown in Table 2.

Through the exhaustive method, *mtry* is assigned for 1 to 35 in the default RF model with the other parameters fixed to gain the minimal error rate. According to the same method, *ntree* is taken to traversal algorithm again by fixing *mtry* value. The results and the processes are presented in Figures 5 and 6.

Table 2. Confusion matrix of the default RF model.

	Ture N	Ture Y	Class. Error
Pred. N	39	7	0.1521739
Pred. Y	10	28	0.2631579

Y presents the points marked as distress; N presents the points marked as in good condition.

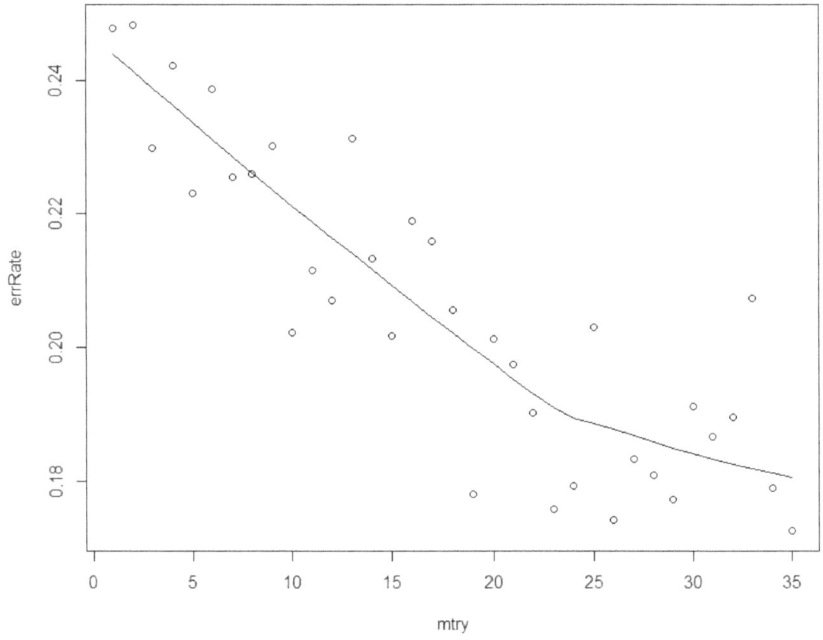

Figure 5. The relationship between *mtry* and the error rate.

With the increase in *mtry* from 1 to 35, the error rate keeps decreasing. Generally, the number of variables tried at each split in an RF model, namely the maximal deep of a tree the model grows, is random in the range between one to the number of variables. It always needs a balance for lower single-tree correlations and a certain prediction strength. Therefore, it is not a general law for an RF model as the error rate can be reduced by introducing more variables unless the variables are all effective for the model with little correlations.

In our given RF model, there are some correlated factors. This is not the main reason which affects the model accuracy until *mtry* equals 23. Before that, the model does not consider enough variables, which strongly helps with the increase in variables. After that or even after 19, the model is improved little when there are less independent residual factors. However, the optimal *mtry* value is 35 with the lowest error rate. That means that all factors have their own weights in the model even though some parts are subject to dependency.

With the increase in *ntree*, i.e., the number of trees generated by the model, the noise can be reduced in the model. When *ntree* arrives at a certain number, the error rate of the model will maintain stability. However, for the calculating speed of a computer, the best *ntree* value is determined. When *ntree* passed over 400, the prediction error rates for Y, N, and the average of the model achieved the lowest value and kept the trend. Therefore, the *ntree* is selected as 400 for the RF model.

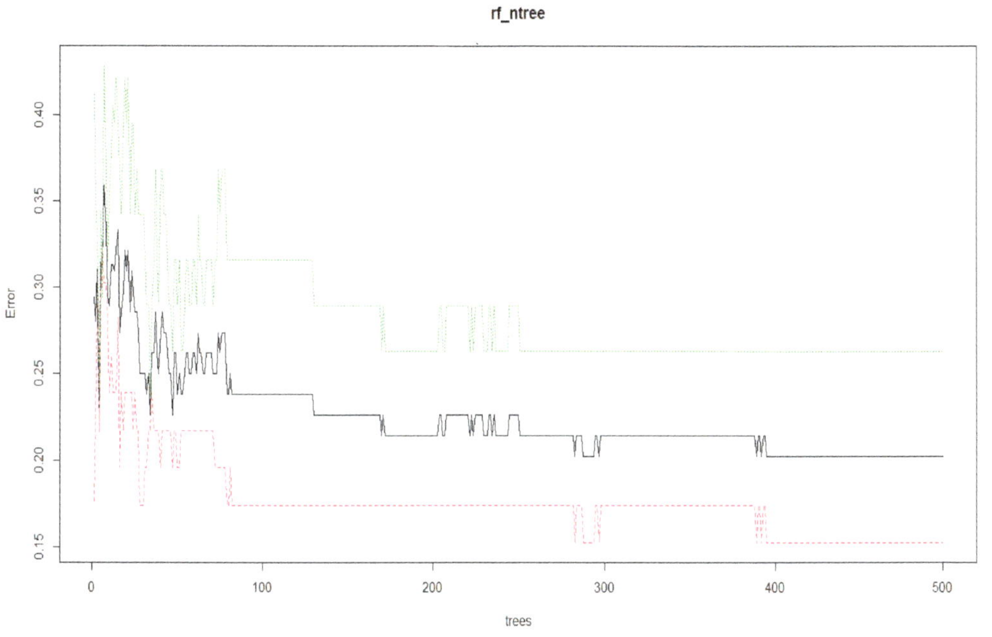

Figure 6. The relationship between *ntree* and the error rate.

6.3. The Optimized RF Model

The final RF model used in training and predicting is gained through the two steps of optimization for the hyperparameters. The main tree sizes, and the node numbers of every tree in the forest, are distributed, as shown in Figure 7. The most frequent occurrence in tree sizes is six, which presents the major samples the trees in the forest look like.

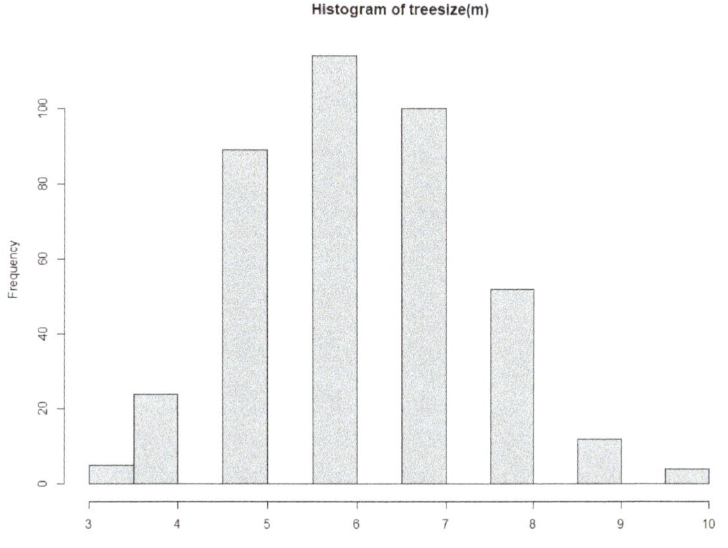

Figure 7. The tree size and its occurrence frequency.

The optimized RF model is estimated by the bagged testing data. The performance of the accuracy for the model is shown in matrix Table 3. The average OOB of error rate is 16.67%, which is improved greater from 20.24%. For an in-field project, prediction accuracy higher than 76% is thought to be a good performance. Compared with some other studies in highway or road topics, there are more variables in this program, which may accelerate the accuracy by considering more comprehensively. In particular, the accuracy of Y prediction, which means that the road has a potential failure at this position, has arrived at 85.13%. It is very important for road maintenance and safety in the application for saving money and lives.

Table 3. Confusion matrix of the optimized RF model.

	Ture N	Ture Y	Class. Error
Pred. N	26	6	0.1875000
Pred. Y	4	24	0.148671

Y presents the points marked as distress; N presents the points marked as in good condition.

A decision tree model was constructed to compare with the RF model. The result is shown in Figure 8 and Table 4. Moreover, a support vector machine (SVM) model was built, and the relative confusion matrix is shown in Table 5.

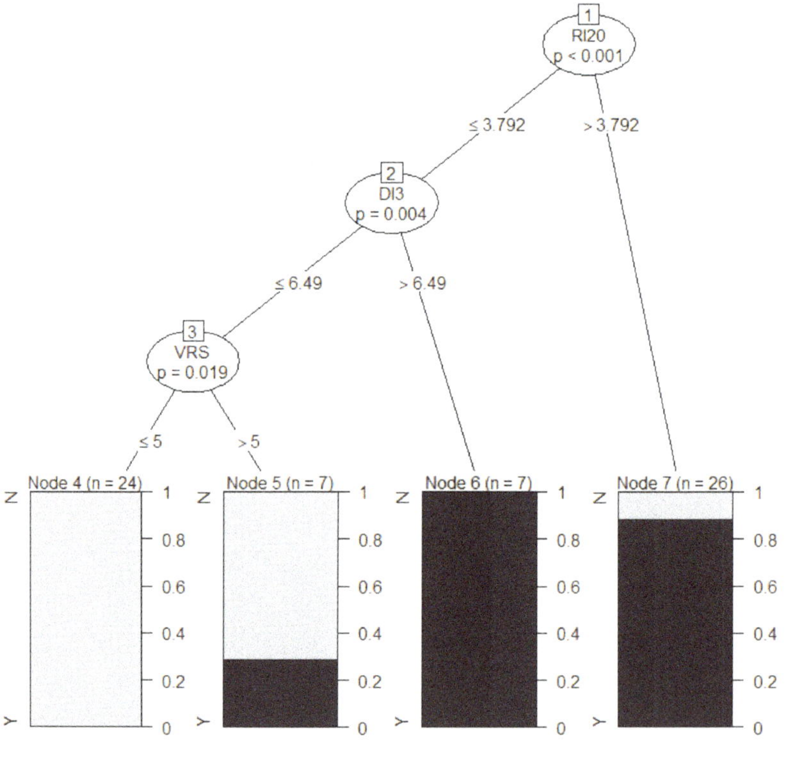

Figure 8. The decision tree model.

In general, the accuracy of the decision tree model and the SVM model is 65% and 65.52%, respectively, for the separate predictions of the potential damage. Nevertheless, the decision tree model is more logical and easier to interpret. In the tree, the RI20, DI3, and VRS are the three most important factors to classify the data, and the prediction probability is given. All in all, the performance of the RF model is outstanding among the three models.

Table 4. Confusion matrix of the decision tree model.

	Ture N	Ture Y	Class. Error
Pred. N	10	3	0.230769
Pred. Y	4	3	0.571428

Y presents the points marked as distress; N presents the points marked as in good condition.

Table 5. Confusion matrix of the SVM model.

	Ture N	Ture Y	Class. Error
Pred. N	10	7	0.411765
Pred. Y	3	9	0.250000

Y presents the points marked as distress; N presents the points marked as in good condition.

6.4. Model Application and Prediction Evaluation

The RF model is applied and the classification performance is examined by margins and multidimensional scaling (MDS) analysis. If a margin value of a test point is higher than zero, it is identified as the right prediction. As shown in Figure 9, the prediction data meets the normal distribution. In addition, the major body of the predictions is in the upper area, which shows a good model performance, despite some abnormal points.

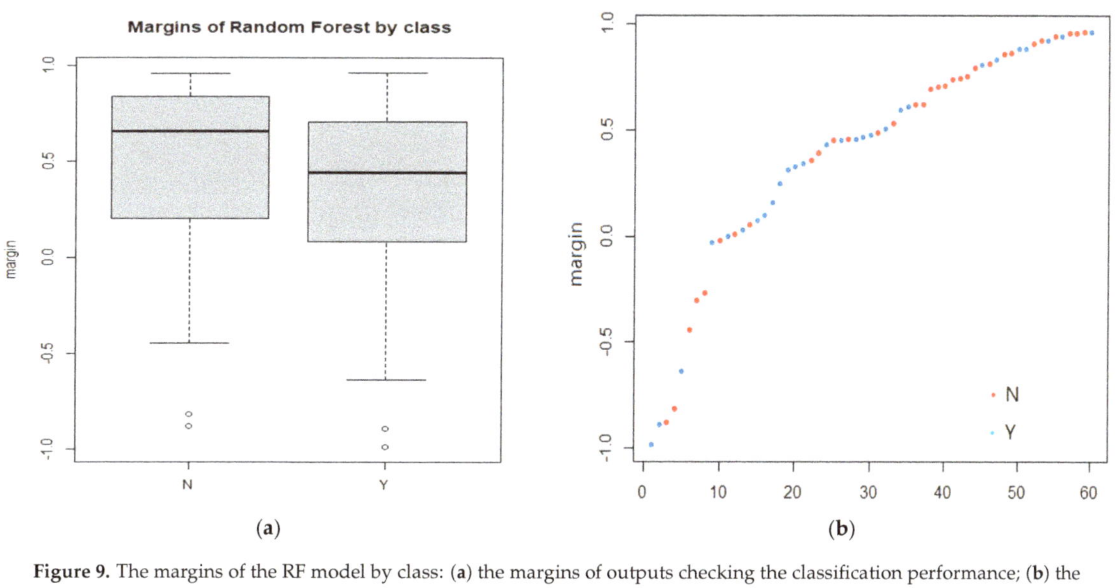

Figure 9. The margins of the RF model by class: (**a**) the margins of outputs checking the classification performance; (**b**) the outputs mapping from multiple dimensional data to 2 dimensions.

In another aspect, the MDS plot is made by R language (Figure 10). The positions of the predictions are marked in this 2D map. The predictions, especially the main bodies of Y and N predictions, have a clear boundary to each other. However, also, some points are mixed up with others, because of some inevitable abnormal points. The next step is to improve the accuracy of the detection in practice. Moreover, here, the predictions are classified into two groups, showing the classification ability of the RF model.

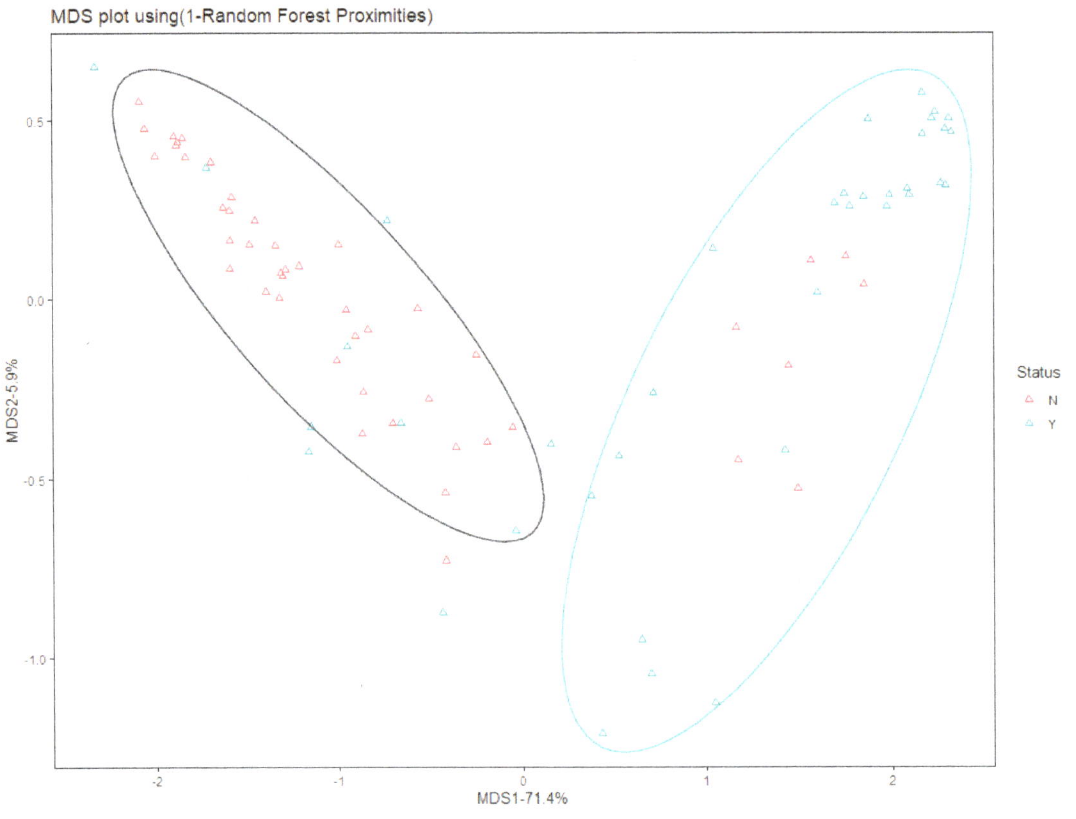

Figure 10. The MDS of the RF model.

6.5. Factor Importance

The importance of the variables is evaluated by a mean decrease in the accuracy index and a mean decrease in the Gini index for the explanation for the RF model, as shown in Figure 11. The more decrease the indexes have, the more important the factor is. The RI20, i.e., the increase in rutting over 20 days, is the most important factor, that is, if a point is rutting deeply in a short time, it is most likely to obtain moisture damage under the surface or surface failure soon. Following the RI20, i.e., the increase in deflection in the third test stage, the rutting at the 20th day and 5th day, and the position are outstanding in all variables. In these five factors, the most interesting thing is that three factors are linked to rutting, which directly presents the state of a road; a highly important factor is linked to deflection, which presents the strength of the road; and the point position is effective for predicting the project (i.e., the different construction method), which companies or materials adapted to in this road, leading this phenomenon. Based on the main important factors, it can be asserted that, before moisture damage occurs, there must firstly be a significant increase in rutting and deflection detected. Some original properties of the road, such as original rutting, original deflection, and surface splitting strength, have little weight in the model. This means that the moisture damage matters for a cumulative effect rather than the initial properties.

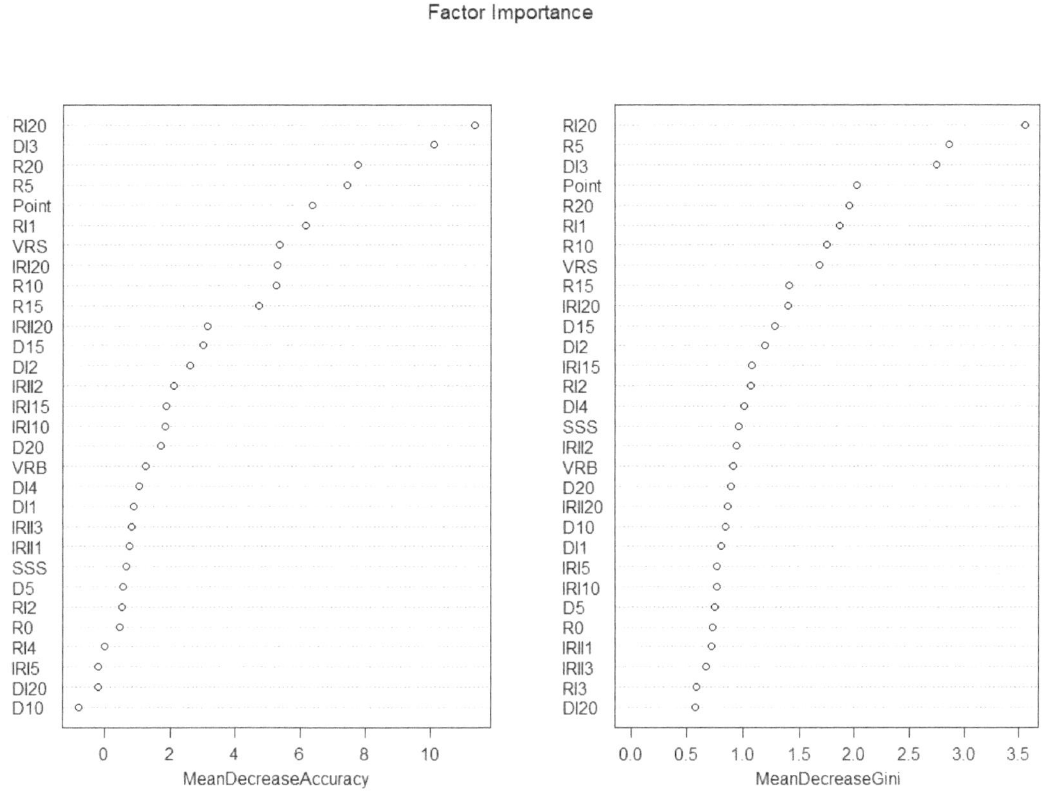

Figure 11. Factor importance.

These specified values of variables and their relationships are analyzed by density curve plots. For these important variables, the overlap section of the Y and N area is smaller, which has a higher classification strength (Figure 12). Thus, it will hold a bigger weight in the model.

To compare the traditional analysis of properties, the three most important categories of factors from the RF model are selected for making graphics of their relationships. The void rate, the rutting increase over 20 days, and the increase in deflection in the third test stage are fitted in Figure 13.

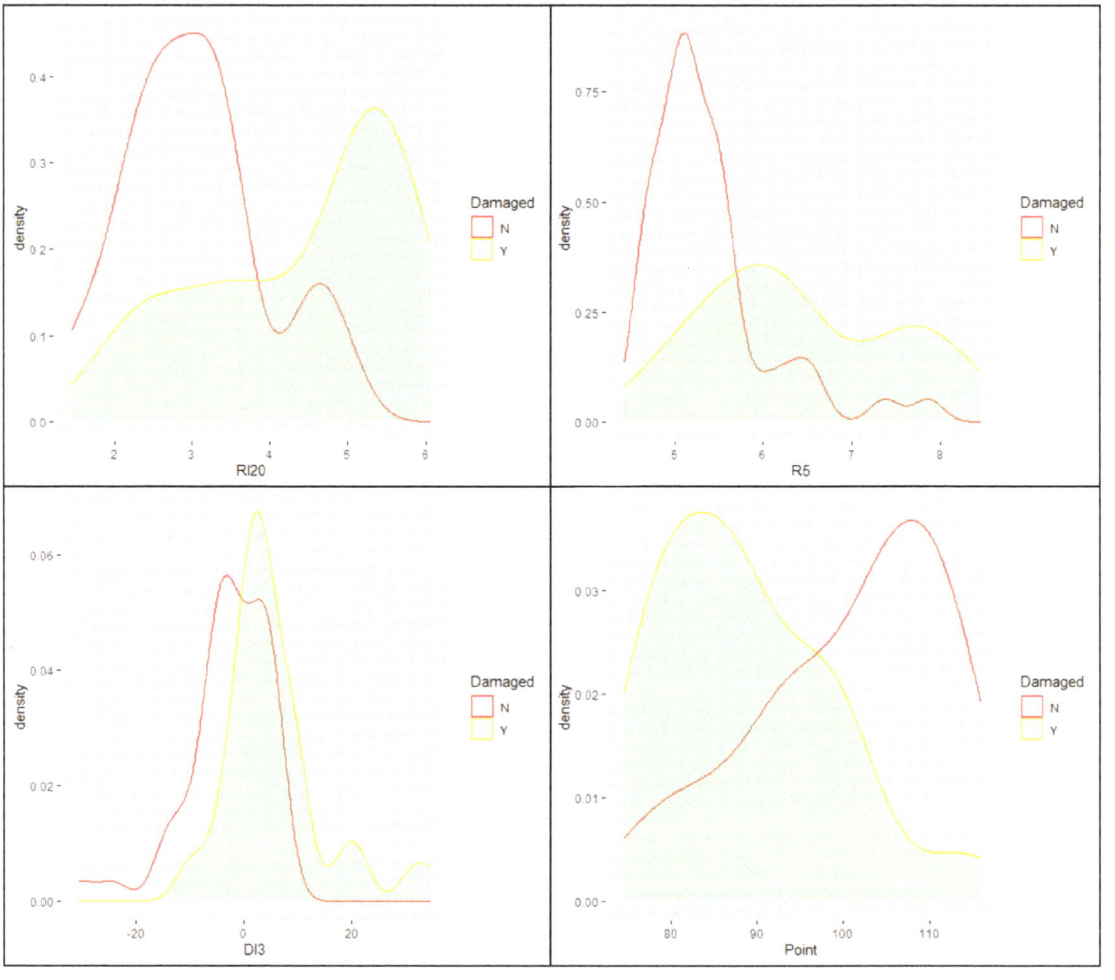

Figure 12. Specified values of variables and their relationships.

Another interesting phenomenon is found. The void rate between 4% and 6% has the lowest probability to drop in rutting and deflection [43,44]. This finding is very similar to the Superpave construction principles. Therefore, to control the moisture damage at an early age of a road, the most important thing is to minimize construction segregation. The result proves the interpretability of the RF model, which can be easily connected to the practice work. Furthermore, the prediction of the model is rational and logical.

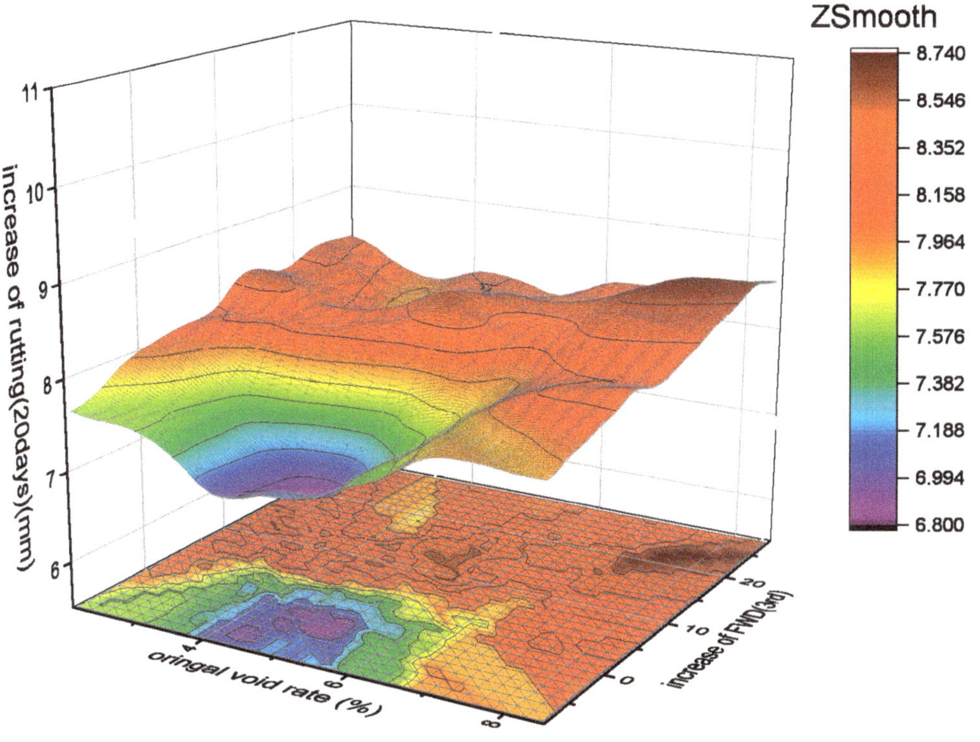

Figure 13. The relationships amongst the three important factors.

7. Conclusions

According to the construction and the evaluation of the RF model, a new method for predicting the potential damage is built in this article. Compared to the typical method, interrelationships of variables are analyzed through RF data mining. The excellent performance of the model is achieved with high accuracy and good interpretability by model optimization. It is concluded that:

- The RF model is suitable for the prediction of in-field properties in highway and road projects under multiple variables;
- The RF model has a good prediction accuracy and interpretability in this study;
- The optimized model can achieve a low OOB error rate of 16.67%, which can be further improved by the enhanced validity of detection data;
- The RF model and traditional method for analysis in road performance can be bridged together to obtain a more comprehensive and consistent understanding;
- The early moisture damage can be alleviated by controlling the void rate in the surface course, which is considered as a main factor by RF mining from another aspect;
- The RF model can be applied in nondestructive examination in roads for potential failure detection. In addition, by inputting more other roads' data sets into the training, the model will become more powerful which can fit almost all situations and properties.

Author Contributions: Conceptualization, X.G. and P.H.; methodology, X.G.; validation, X.G. and P.H.; formal analysis, X.G. and P.H.; investigation, X.G.; resources, X.G.; data curation, X.G.; writing—original draft preparation, X.G.; writing—review and editing, X.G.; visualization, X.G.; supervision, P.H.; project administration, P.H.; funding acquisition, P.H. All authors have read and agreed to the published version of the manuscript.

Funding: This research received no external funding.

Institutional Review Board Statement: Not applicable.

Informed Consent Statement: Not applicable.

Data Availability Statement: The data used to support the findings of this study are available from the corresponding author upon request.

Acknowledgments: This work was supported by my supervisor, Peiwen Hao. The authors would like to thank Peiwen Hao for excellent technical support and Jinghui Hou for critically helping in my life.

Conflicts of Interest: The authors declare no conflict of interest.

References

1. Transportation Research Board. Moisture Sensitivity of Asphalt Pavements: A National Seminar. In Proceedings of the TRB Committee on Bituminous–Aggregate Combinations to Meet Surface Requirements, San Diego, FL, USA, 4–6 February 2003.
2. Hicks, R.G.; Program, N.C.H.R. *Moisture Damage in Asphalt Concrete*; Transportation Research Board: Washington, DC, USA, 1991; ISBN 978-0-309-04924-5.
3. Sharaf, E.A.; Sinha, K.C. Estimation of Pavement Routine Maintenance Costs. *Transp. Res. Rec.* **1984**, *951*, 55–58.
4. Underwood, B.S.; Guido, Z.; Gudipudi, P.; Feinberg, Y. Increased Costs to US Pavement Infrastructure from Future Temperature Rise. *Nat. Clim. Chang.* **2017**, *7*, 704–707. [CrossRef]
5. Meneses, S.; Ferreira, A. Pavement Maintenance Programming Considering Two Objectives: Maintenance Costs and User Costs. *Int. J. Pavement Eng.* **2013**, *14*, 206–221. [CrossRef]
6. Kim, S.-H.; Jeong, J.-H.; Kim, N. Use of Surface Free Energy Properties to Predict Moisture Damage Potential of Asphalt Concrete Mixture in Cyclic Loading Condition. *KSCE J. Civ. Eng.* **2003**, *7*, 381–387. [CrossRef]
7. Cheng, D.; Little, D.N.; Lytton, R.L.; Holste, J.C. Use of Surface Free Energy Properties of the Asphalt-Aggregate System to Predict Moisture Damage Potential (with Discussion). *J. Assoc. Asph. Paving Technol.* **2002**, *71*, 59–88.
8. Aschenbrener, T. Evaluation of Hamburg Wheel-Tracking Device to Predict Moisture Damage in Hot-Mix Asphalt. *Transp. Res. Rec.* **1995**, *1492*, 193.
9. Hamzah, M.O.; Kakar, M.R.; Hainin, M.R. An Overview of Moisture Damage in Asphalt Mixtures. *J. Teknol.* **2015**, *73*, 125–131. [CrossRef]
10. Omar, H.A.; Yusoff, N.I.M.; Mubaraki, M.; Ceylan, H. Effects of Moisture Damage on Asphalt Mixtures. *J. Traffic Transp. Eng. Engl. Ed.* **2020**, *7*, 600–628. [CrossRef]
11. Ksaibati, K.; Armaghani, J.; Fisher, J. Effect of Moisture on Modulus Values of Base and Subgrade Materials. *Transp. Res. Rec.* **2000**, *1716*, 20–29. [CrossRef]
12. Behiry, A.E.A.E.-M. Laboratory Evaluation of Resistance to Moisture Damage in Asphalt Mixtures. *Ain Shams Eng. J.* **2013**, *4*, 351–363. [CrossRef]
13. Saarenketo, T.; Scullion, T. Road Evaluation with Ground Penetrating Radar. *J. Appl. Geophys.* **2000**, *43*, 119–138. [CrossRef]
14. Aavik, A.; Talvik, O. *Use of Falling Weight Deflectometer (Fwd) Measurement Data for Pavement Structural Evaluation and Repair Design*; Federal Highway Administration: Washington, DC, USA, 2021.
15. Chen, D.-H.; Xie, J.; Scullion, T. Using GPR and FWD to Assist in Selecting the Optimal Pavement Rehabilitation Strategy. In Proceedings of the GeoHunan International Conference 2011, Hunan, China, 9–11 June 2011; pp. 63–70. [CrossRef]
16. Khan, Z.; Ahmed, M.; Tarefder, R. Evaluation of Pavement Performance under FWD Test through Instrumented Pavement Section. In Proceedings of the NDE/NDT for Highways & Bridges: SMT 2018, New Brunswick, OH, USA, 27–29 August 2018; pp. 192–197.
17. Hoang, N.-D.; Nguyen, Q.-L. A Novel Method for Asphalt Pavement Crack Classification Based on Image Processing and Machine Learning. *Eng. Comput.* **2019**, *35*, 487–498. [CrossRef]
18. Witten, I.H.; Frank, E.; Hall, M.A.; Pal, C. *Mining: Practical Machine Learning Tools and Techniques*; Elsevier: Amsterdam, The Netherlands, 2011; ISBN 978-0-12-374856-0.
19. Zhang, X.-D. Machine Learning. In *A Matrix Algebra Approach to Artificial Intelligence*; Springer: Singapore, 2020; pp. 223–440. ISBN 9789811527692.
20. Mahesh, B. Machine Learning Algorithms—A Review. *Int. J. Comput. Sci. Inf. Technol.* **2018**, *9*, 7.
21. Naqa, E.I.; Murphy, M.J. What Is Machine Learning? In *Machine Learning in Radiation Oncology*; El Naqa, I., Li, R., Murphy, M.J., Eds.; Springer International Publishing: Cham, Switzerland, 2015; pp. 3–11. ISBN 978-3-319-18304-6.

22. Tsai, Y.-C.; Zhao, Y.; Pop-Stefanov, B.; Chatterjee, A. Automatically Detect and Classify Asphalt Pavement Raveling Severity Using 3D Technology and Machine Learning. *Int. J. Pavement Res. Technol.* **2021**, *14*, 487–495. [CrossRef]
23. Cheng, D.; Little, D.N.; Lytton, R.L.; Holste, J.C. Moisture Damage Evaluation of Asphalt Mixtures by Considering Both Moisture Diffusion and Repeated-Load Conditions. *Transp. Res. Rec. J. Transp. Res. Board* **2003**, *1832*, 42–49. [CrossRef]
24. Sebaaly, P.; Tabatabaee, N.; Bonaquist, R.; Anderson, D. Evaluating Structural Damage of Flexible Pavements Using Cracking and Falling Weight Deflectometer Data. *Transp. Res. Rec.* **1989**, *1227*, 44–52.
25. Karballaeezadeh, N.; Ghasemzadeh Tehrani, H.; Mohammadzadeh Shadmehri, D.; Shamshirband, S. Estimation of Flexible Pavement Structural Capacity Using Machine Learning Techniques. *Front. Struct. Civ. Eng.* **2020**, *14*, 1083–1096. [CrossRef]
26. Jia, X.; Woods, M.; Gong, H.; Zhu, D.; Hu, W.; Huang, B. Evaluation of Influence of Pavement Data on Measurement of Deflection on Asphalt Surfaced Pavements Utilizing Traffic Speed Deflection Device. *Constr. Build. Mater.* **2021**, *270*, 121842. [CrossRef]
27. Marcelino, P.; de Lurdes Antunes, M.; Fortunato, E.; Gomes, M.C. Machine Learning Approach for Pavement Performance Prediction. *Int. J. Pavement Eng.* **2021**, *22*, 341–354. [CrossRef]
28. Gong, H.; Sun, Y.; Mei, Z.; Huang, B. Improving Accuracy of Rutting Prediction for Mechanistic-Empirical Pavement Design Guide with Deep Neural Networks. *Constr. Build. Mater.* **2018**, *190*, 710–718. [CrossRef]
29. Gong, H.; Sun, Y.; Huang, B. Gradient Boosted Models for Enhancing Fatigue Cracking Prediction in Mechanistic-Empirical Pavement Design Guide. *J. Transp. Eng. Part B Pavements* **2019**, *145*, 04019014. [CrossRef]
30. Jia, X.; Woods, M.; Gong, H.; Hu, W.; Huang, B.; Faraj, B. Utilization of State Performance Indices to Correlate National Performance Measures for Asphalt Pavements in Tennessee. *Transp. Res. Rec. J. Transp. Res. Board* **2019**, *2673*, 379–388. [CrossRef]
31. Ham, J.; Yangchi, C.; Crawford, M.M.; Ghosh, J. Investigation of the Random Forest Framework for Classification of Hyperspectral Data. *IEEE Trans. Geosci. Remote Sens.* **2005**, *43*, 492–501. [CrossRef]
32. Paul, A.; Mukherjee, D.P.; Chintha, R.; Kundu, S. Improved Random Forest for Classification. *IEEE Trans. Image Process.* **2018**, *27*, 4012–4024. [PubMed]
33. Svetnik, V.; Liaw, A.; Tong, C.; Culberson, J.C.; Sheridan, R.P.; Feuston, B.P. Random Forest: A Classification and Regression Tool for Compound Classification and QSAR Modeling. *J. Chem. Inf. Comput. Sci.* **2003**, *43*, 1947–1958. [CrossRef] [PubMed]
34. Biau, G.; Scornet, E. A Random Forest Guided Tour. *TEST* **2016**, *25*, 197–227. [CrossRef]
35. Speiser, J.L.; Miller, M.E.; Tooze, J.; Ip, E. A Comparison of Random Forest Variable Selection Methods for Classification Prediction Modeling. *Expert Syst. Appl.* **2019**, *134*, 93–101. [CrossRef]
36. Gong, H.; Sun, Y.; Hu, W.; Polaczyk, P.A.; Huang, B. Investigating Impacts of Asphalt Mixture Properties on Pavement Performance Using LTPP Data through Random Forests. *Constr. Build. Mater.* **2019**, *204*, 203–212. [CrossRef]
37. Mohamed, O.A.; Ati, M.; Najm, O.F. Predicting Compressive Strength of Sustainable Self-Consolidating Concrete Using Random Forest. *Key Eng. Mater.* **2017**, *744*, 141–145. [CrossRef]
38. Gong, H.; Sun, Y.; Shu, X.; Huang, B. Use of Random Forests Regression for Predicting IRI of Asphalt Pavements. *Constr. Build. Mater.* **2018**, *189*, 890–897. [CrossRef]
39. Oshiro, T.M.; Perez, P.S.; Baranauskas, J.A. How Many Trees in a Random Forest? In *Machine Learning and Data Mining in Pattern Recognition*; Perner, P., Ed.; Lecture Notes in Computer Science; Springer: Berlin/Heidelberg, Germany, 2012; Volume 7376, pp. 154–168. ISBN 978-3-642-31536-7.
40. Breiman, L. Random Forests. *Mach. Learn.* **2001**, *45*, 5–32. [CrossRef]
41. Parmar, A.; Katariya, R.; Patel, V. A Review on Random Forest: An Ensemble Classifier. In Proceedings of the International Conference on Intelligent Data Communication Technologies and Internet of Things, Coimbatore, India, 7–8 August 2018; Springer: Berlin/Heidelberg, Germany, 2018; pp. 758–763.
42. Liaw, A.; Wiener, M. Classification and Regression by RandomForest. *R News* **2002**, *2*, 18–22.
43. Grant, T.P. *Determination of Asphalt Mixture Healing Rate Using the Superpave Indirect Tensile Test*; University of Florida: Gainesville, FL, USA, 2001.
44. Christensen, D.W.; Bonaquist, R.F. *Volumetric Requirements for Superpave Mix Design*; Transportation Research Board: Washington, DC, USA, 2006; Volume 567, ISBN 0-309-09867-X.

applied sciences

MDPI

Article

Numerical Simulation on Reflective Cracking Behavior of Asphalt Pavement

Houzhi Wang [1,2], You Wu [1,2], Jun Yang [1,2,*] and Haopeng Wang [3,*]

[1] School of Transportation, Southeast University, Nanjing 211189, China; houzhi_wang@seu.edu.cn (H.W.); 213161518@seu.edu.cn (Y.W.)
[2] National Demonstration Center for Experimental Road and Traffic Engineering Education, Southeast University, Nanjing 211189, China
[3] Section of Pavement Engineering, Faculty of Civil Engineering & Geosciences, Delft University of Technology, 2628CN Delft, The Netherlands
* Correspondence: yangjun@seu.edu.cn (J.Y.); haopeng.wang@tudelft.nl (H.W.)

Abstract: Cracks are one of the main problems that plague road workers. A correct understanding of the internal crack propagation mechanism of asphalt pavement will help road workers evaluate the road's working status more comprehensively and make more reasonable decisions in design, construction, and maintenance work. This paper established a three-dimensional asphalt pavement layered model using the software ABAQUS and fracture mechanics theory and the extended finite element method were used to explore the mechanical response of the pavement base layer's preset reflective cracks. This paper investigated the influence of the modulus of each layer, vehicle load on the principal stress, shear stress, J-integral, and two stress intensity factors (K1, K2) during the predetermined crack propagation process of the pavement base layer, and the entropy method was used to analyze the above-mentioned mechanical response. The results show that the main factor affecting the propagation of reflective cracks on asphalt pavements is the modulus of the bottom surface layer. However, from a modeling perspective, the effect of increasing load on crack growth is obvious. Therefore, in terms of technical feasibility, the prevention of reflective cracks should still be achieved by controlling the driving load and prohibiting overloading.

Keywords: reflection crack; numerical analysis; extended finite element method; J-integral; stress intensity factors

Citation: Wang, H.; Wu, Y.; Yang, J.; Wang, H. Numerical Simulation on Reflective Cracking Behavior of Asphalt Pavement. *Appl. Sci.* **2021**, *11*, 7990. https://doi.org/10.3390/app11177990

Academic Editors: Amir Tabakovic, Jan Valentin and Liang He

Received: 17 August 2021
Accepted: 27 August 2021
Published: 29 August 2021

1. Introduction

Cracking is a challenging topic which the researchers face. Insufficient and improper treatment of cracks often negatively impacts the pavement structure and its performance. Numerous studies on cracks' behavior have been conducted to reveal their influences on pavement performance. Initial cracks usually exist in the pavement structure in the form of micro-cracks, and these cracks grow due to temperature change, traffic loads, and other environmental factors. So, research on the mechanism of cracking growth will help monitor the cracks' behaviors from the beginning of their development and offer feasible ways of manufacturing and maintenance. Traditional fatigue methods explain the initiation and growth of cracks [1], but this only provides a rough approach that performs poorly while calculating longitudinal cracks. To solve this kind of problem, fracture mechanism theories were established. Paris established the approximate equation of crack growth rate under repeated load [2]. Majidzadeh expanded this theory, and normal forms of crack growth were obtained [3]. However, this equation needs four parameters related to material properties, which decrease its reliability and limit its usage. A life-related index was established based on engineering data, including traffic load, environment conditions, and embankment conditions [4]. Conditions included by this index are also limited, and it will take a long time to perform road surveys, which can be inconvenient.

These methods are always based on numerous engineering data and cannot involve many working conditions. Correction factors are always acquired when using these methods, so a convenient way of researching crack behavior is needed.

Conventional research on crack behavior heavily relies on laboratory-scale experiments, which require considerable time and effort. These experiments always focused on the final status of cracks and pavement, and the processes of cracks' growth were ignored due to testing methods. Over these years, the finite element method has become a popular approach to simulate the cracks' behavior during their life cycle. During their growth, every kind of cracks' behavior can be calculated using finite element software, requiring much less time than local experiments. The two-dimensional plane strain model was established by Myers using finite element software ABAQUS [5]. This model revealed the growth of pre-set cracks under traffic loads and the direction of cracks' growth, which offered a mature way of pavement simulation. García used the finite element method combined with fracture mechanics theory to investigate the cracking behavior of the orthogonal layered pavement, and uses the method of preset interlayer cracks to predict the cracking of each layer of the pavement during use. The results show that the cracking performance of the orthogonal layered pavement is largely affected by the initial shape and location of the cracks. These initial cracks mostly occur at the joints between layers, thinner layers and the free boundaries of each layer [6]. Results show that behaviors of crack growth are related to the viscoelasticity and creep properties of asphalt deeply. However, compared with two-dimensional models, three-dimensional models always perform better, especially when the process of crack growth was taken into consideration [5]. Parameters of crack growth and their time history curve can be calculated using the finite element method and predictions of pavement life, degree of damage can be drawn. Road workers can also choose the appropriate asphalt mixture based on the finite element calculation results, thus promoting the development of the asphalt mixture.

J-integral and stress intensity factors are often used to describe the behaviors of crack growth, from which researchers can have a direct understanding of the process. Okada developed a 3D model to compute the J-integral of large deformation solids [7]. Results show that J-integral is unconditionally path independent, reflecting that it is not very important to set a specific path while simulating cracks using J-integral, thus decreasing the difficulties. Moreover, energy can be used to calculate the J-integral. Okada added the strain energy density into the J-integral formulation to calculate the deformation histories of the specimen [7]. Yu proposed a new model of calculating J-integral based on the energy density equivalence by introducing 3D constraint functions to describe the relationships between J-integral and load in mode-I cracks [8]. Sasan compared two criteria of stress intensity factor and fracture energy to investigate the behavior of asphalt mixtures under combined tensile shear loadings [9]. The SCB (Semi-Cycle Bending Test) fracture test results show that the fracture behavior of asphalt mixtures is significantly dependent on testing temperature and loading rate, which can reflect the working conditions of asphalt mixtures during engineering construction. Moreover, finite element calculation was conducted to verify the correctness of SCB tests. As for pavement structure, J-integral and stress intensity factors can be different from theoretical calculations and mechanical experiments due to the structural differences between theoretical calculation models or mechanical specimens and pavement structure. Researchers often use elastic layered continuum theory to describe the structure of asphalt pavement. Alae established a three-layer 3D model of asphalt pavement with top-down cracks [10]. Working temperature and vehicle speed were assumed to be the changing working conditions of the pavement structure. Results show that top-down cracks in pavement structure can be broken down into I + II fracture, and J-integral can be used to describe stress intensity on the crack tip. Ma developed a wave propagation-based analytical solution to calculate the mechanical responses of transversely isotropic viscoelastic multi-layered asphalt pavement subject to moving harmonic load [11]. This solution can be used for asphalt pavement design and analysis with consideration of realistic load and material parameters. However, a specimen of asphalt mixtures can only reveal the crack resistance of the material, but the mixtures' behaviors in pavement structure are not widely studied. J-integral and stress intensity factors on the crack tip in pavement structures have not been researched systematically. Mechanical responses during the crack growth need to be calculated using the finite element method.

The research described in this paper is aimed to reveal the cracks' behaviors during the pavement's life cycle and under different working situations. The effects of traffic loads and Young's modulus of layers on the crack extension were investigated. This paper uses finite element software ABAQUS to simulate the pre-set crack's growth during the pavement's life cycle. J-integral, stress intensity factors, absolute stress, and strains were calculated to describe how the cracks extend while working. Moreover, the entropy method was used to analyze the relationship between these influencing factors aiming to help to deal with cracking problems efficiently.

2. Methods

2.1. Method

Fracture mechanics were widely used in crack analysis for their high accuracy, and the development of computer technology offers the convenience of calculating. Thus fracture mechanics are commonly used in pavement engineering. Using the finite element method for digital modeling can calculate the behavior of cracks. Responses of cracks could be summarized by modifying mechanical parameters of finite element model such as working conditions, structure combination, material composition.

Fracture mechanics divide cracks into three species (open crack—type I, sliding crack—type II, tear-open crack—type III) and stress intensity factors (K_I, K_{II}, K_{III}) were used to describe the tendency of cracking [12]. Linear superposition of three kinds of cracks can be used to describe the stress field at the tip of cracks. Normal stress and shear stress of the tip of cracks can be calculated using the following equations.

$$\sigma_x = \frac{K_I}{\sqrt{2\pi r}} cos\frac{\theta}{2}\left(1 - sin\frac{\theta}{2} sin\frac{3\theta}{2}\right) + \frac{K_{II}}{\sqrt{2\pi r}} sin\frac{\theta}{2}\left(-2 - cos\frac{\theta}{2} cos\frac{3\theta}{2}\right) \quad (1)$$

$$\sigma_y = \frac{K_I}{\sqrt{2\pi r}} cos\frac{\theta}{2}\left(1 + sin\frac{\theta}{2} sin\frac{3\theta}{2}\right) + \frac{K_{II}}{\sqrt{2\pi r}} sin\frac{\theta}{2} cos\frac{\theta}{2} cos\frac{3\theta}{2} \quad (2)$$

$$\tau_{xy} = \frac{K_I}{\sqrt{2\pi r}} sin\frac{\theta}{2} cos\frac{\theta}{2} cos\frac{3\theta}{2} + \frac{K_{II}}{\sqrt{2\pi r}} cos\frac{\theta}{2}\left(1 - sin\frac{\theta}{2} sin\frac{3\theta}{2}\right) \quad (3)$$

$$\tau_{zx} = -\frac{K_{III}}{\sqrt{2\pi r}} \sin\frac{\theta}{2} \quad (4)$$

$$\tau_{yz} = \frac{K_{III}}{\sqrt{2\pi r}} cos\frac{\theta}{2} \quad (5)$$

where r represents the distance from the calculation point to the crack tip. θ represents the angle between the calculation point and the crack tip in the polar coordinate system and the polar axis. Analysis of the above formula shows that the key to determining the stress field at the crack tip is the stress intensity factor. These factors are mainly determined by the properties of external load, the shape of cracks, and the geometry of elastomer, which can be expressed by the equation below.

$$K_m = \lim_{|r| \to 0} \sqrt{2\pi r} Z_m(r) \quad (6)$$

where Z_m (r) is an analytic function related to the boundary conditions, which can be calculated using finite element method software ABAQUS. The stress intensity factor at the ultimate fracture of a material is called fracture toughness, usually expressed by K_{IC}, which can be determined by experiments.

Compared with the stress intensity factor, J-integral is more suitable for analyzing fracture problems under elastoplastic conditions [13]. The J-integral is used to solve the situation that a certain range of plastic zone appears at the tip of the crack under the elastoplastic condition, which makes the problem very complicated. The J-integral can describe the stress and strain field strength in the crack tip area, and it can be easily determined by experiment. Use stress intensity factor and J-integral can reveal the crack behavior in pavement engineering. A plane crack problem was assumed when calculating J-integral. The integral is performed around the crack tip, starting from the lower crack surface, and stopping counterclockwise to the crack's upper surface, thus forming an integral loop. This integral value obviously has nothing to do with the integral path, so J-integral can reflect the intensity of the stress field near the crack tip. Figure 1 is a schematic diagram of the calculation of J-integral.

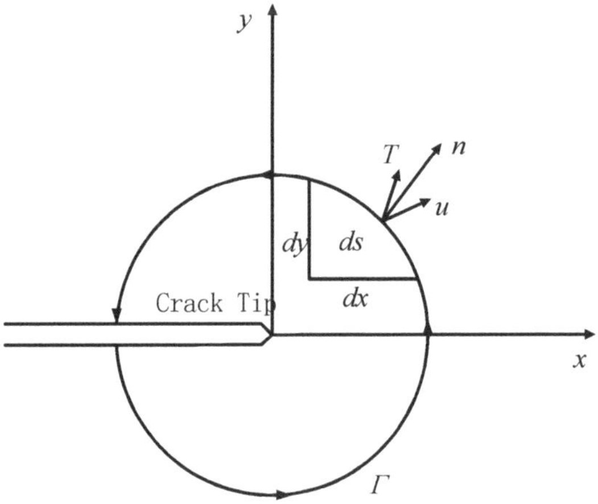

Figure 1. Calculation of J-integral.

J-integral is defined as:

$$J = \int_{\Gamma} \left(\omega dy - \vec{T} \cdot \frac{\partial \vec{u}}{\partial x} ds \right) \tag{7}$$

where ω represents the strain energy density on the loop. $\vec{T}$ is the stress vector at any point on the loop. $\vec{u}$ is the displacement vector at any point on the loop. In the case of linear elasticity (plane strain), the strain energy density ω is

$$\omega = \frac{1}{2}\sigma_{ij}\varepsilon_{ij} = \frac{1+\mu}{2E}\left[(1-\mu)\left(\sigma_{11}^2 + \sigma_{22}^2\right) - 2\mu\sigma_{11}\sigma_{22} + 2\sigma_{12}^2\right] \tag{8}$$

where σ_{ij} and ε_{ij} respectively represent the stress tensor and strain tensor at the crack. Substitute (8) into (7), we have

$$\omega = \frac{K_1^2}{2\pi r}\frac{1+\mu}{E}\left[cos^2\frac{\theta}{2}\left(1 - 2\mu + sin^2\frac{\theta}{2}\right)\right] \tag{9}$$

If we take a loop Γ with the crack tip as the center and a radius of r,

$$\int_{\Gamma} \omega dy = \int_{-\pi}^{\pi} \omega r cos\theta d\theta = \frac{K_1^2(1+\mu)(1-2\mu)}{4E} \tag{10}$$

As for stress vector $\vec{T}$

$$\vec{T} = \sigma_{ij}\eta_j \tag{11}$$

So

$$\int_{\Gamma} T_i \frac{\partial u_i}{\partial x_i} ds = \int_{-\pi}^{\pi} \left(T_1 \frac{\partial u_1}{\partial x_1} + T_2 \frac{\partial u_2}{\partial x_2} \right) r d\theta = -\frac{-K_1^2(1+\mu)(3-2\mu)}{4E} \tag{12}$$

Substituting (10) and (12) into (7),

$$J = \frac{1-\mu^2}{E} K_1^2 = \frac{K_1^2}{E'} = G_1 \tag{13}$$

That is, the J-integral is the crack propagation energy release rate G in the linear elastic state. This is a constant and can be obtained through experiments or theoretical calculations. Therefore, the J-integral is used to establish the basis for the determination of cracks and is consistent with the establishment of the K value.

When using the traditional finite element method to simulate static non-propagating cracks, it is necessary to consider the singularity of the crack tip stress field according to the principles of fracture mechanics. Singular meshes were used to solve the problem, such as setting the meshes of the crack tip into M font or refining meshes in the crack tip area. Difficulties in computing were generated due to a large number of meshes, especially in three-dimensional models. It is hard to compute in crack tip areas due to the low quantity of meshes. Finite element method software ABAQUS offered Extended Finite Element Method (XFEM) to solve the problem, which uses shape functions to simulate crack areas in the pavement structure. Thus, it does not need to remesh and improves the efficiency of calculating. Based on the thought of unit decomposition, XFEM added functions that can reflect the discontinuous characteristics of the crack area [14,15]. The thought of unit decomposition thinks that any functions $\psi(x)$ can be expressed using the equations below.

$$\psi(x) = \sum_I N_I(x) q_I \Phi(x), \ \sum_I N_I(x) = 1 \tag{14}$$

where q_I is a parameter to be adjusted to make the expression reach the best approximation, $N_I(x)$ is a function that satisfies the element decomposition, and $\Phi(x)$ is an extended function. As for crack simulation, XFEM uses the equations below to simulate.

$$u^k = \sum_I N_I(x) u_I + \sum_J N_J(x) q_J \Phi(x) \tag{15}$$

where q_J is the newly added degree of freedom and it has no meaning in physics. The only usage of q_J is to adjust the function $\Phi(x)$ to achieve the best approximation. Compared with the traditional finite element method, XFEM added more freedom degrees, thus improving its accuracy.

XFEM and fracture mechanics were used in finite element method software ABAQUS, which can simulate the crack behavior during its growth with considerable accuracy. Crack behavior can be calculated by modifying properties of models and materials, which can make predictions on pavement life and different working conditions, thus saves much time and cost compared with local experiments.

2.2. Simulation

The finite element model was established using the software ABAQUS. Previous studies show that 6 m × 6 m × 5 m cube specimen was the most suitable kind of simulating the actual pavement structure [16,17]. C3D8R was selected as the grid type, and no lateral movement is chosen as the boundary condition to reveal the actual conditions of cracking. A three-dimensional model of pavement was established, and the pavement structure was assumed to be composed of five layers, from top to bottom as SMA asphalt mixture, AC-20, asphalt treated base (ATB), cement stabilized base (CTB), and soil ground (SG). The specific layer composition is shown in Figure 2, and the properties of different layers are shown in Table 1. The initial location of the crack was set in CTB is shown in Figure 2.

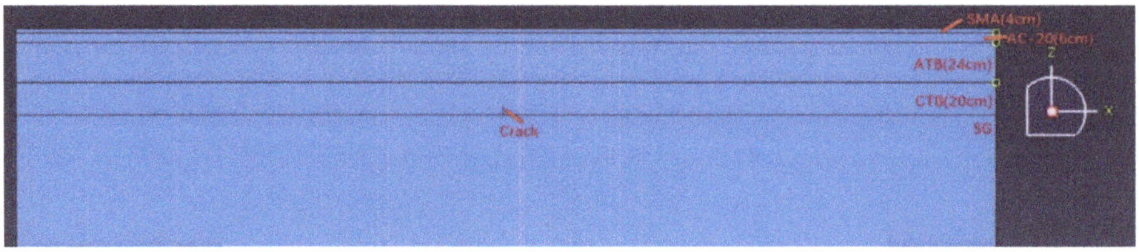

Figure 2. Schematic of pavement structure and crack location.

Table 1. Properties of layers.

Layer	Thickness (cm)	Young's Modulus @ 20 °C (MPa)	Poisson's Ratio
SMA	4	1400	0.35
AC-20	6	1200	0.3
ATB	24	1000	0.3
CTB	20	1500	0.25
SG	-	40	0.4

The initial time of the analysis step was set to 0.1 s on the basis of previous studies, and the total time of the analysis step was 25 s in order to control the solution time and take the calculation accuracy into account. The load area of traffic was set as two rectangular areas with a width of 0.213 m, a length of 6m, and an interval of 0.1065 m based on Chinese standard traffic load. In the static analysis of the structure, the load changes with the type of vehicle load. This paper adopts 1.0~2.6 times of the standard axle load specified in the Chinese highway pavement design code for analysis. When it is 1 time, the load should be 700 kPa. Boundary conditions were set as no stress, no displacement at the edge of the side and bottom of the model based on elastic layered system theory. In this paper, the pavement structure load and boundary conditions were set as shown in Figure 3.

Stress intensity factor and J-integral of the crack can reveal the tendency of crack growth. In addition, the principal stress and shear stress intensity at the crack tip can also reflect the cracking situation. Grids at the crack tip were subdivided in order to guarantee the accuracy of simulation, and four typical grids were selected to investigate the principal stress and shear stress at the crack tip. The location of the grids was shown in Figure 4.

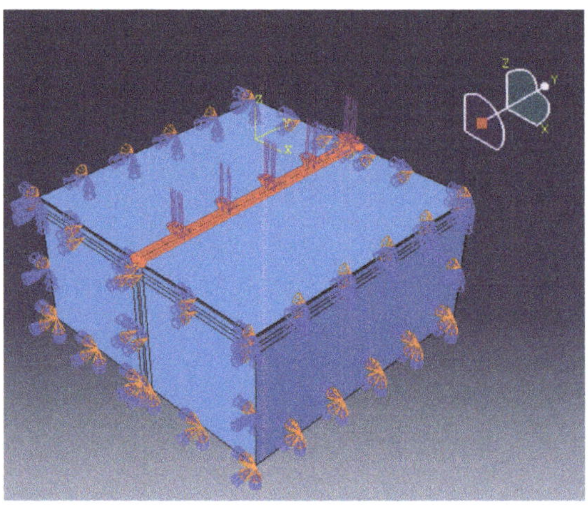

Figure 3. Schematic of load and boundary conditions.

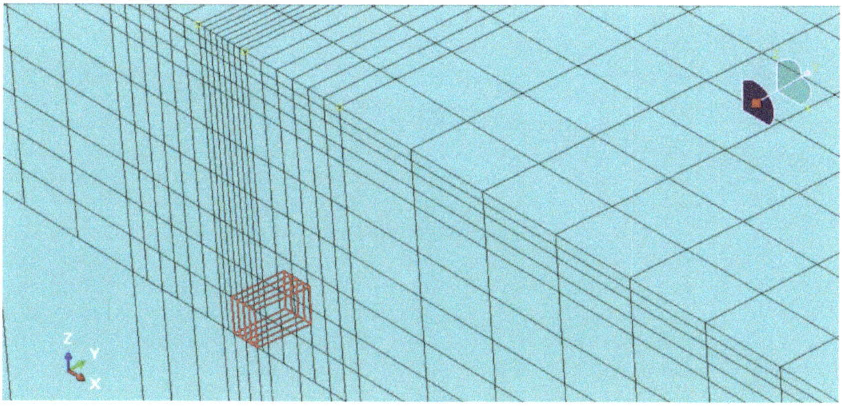

Figure 4. Selection of typical grids.

3. Results and Discussion

3.1. Behavior of Crack Extension

The relevant Chinese regulations stipulate that the design of asphalt pavement adopts a sing-axle-two-wheel set axle load with an axle load of 100 kN as the design axle load. When the vehicle is parked on the road surface, the pressure on the contact area of the road surface is 0.7 MPa. Overloading and over-limit often occur while working, both of which will have an adverse effect on the pavement structure. This paper uses 0.2 times as the axle load gradient to explore the propagation of reflection cracks on the road surface under the action of 1.0 to 2.8 times axle load. There are 10 working conditions in total. The principal stress (S33 in ABAQUS) cloud diagram and the shear stress (S23 in ABAQUS) cloud diagram of the asphalt pavement under the action of 1.0 times the standard axle load are shown in Figure 5.

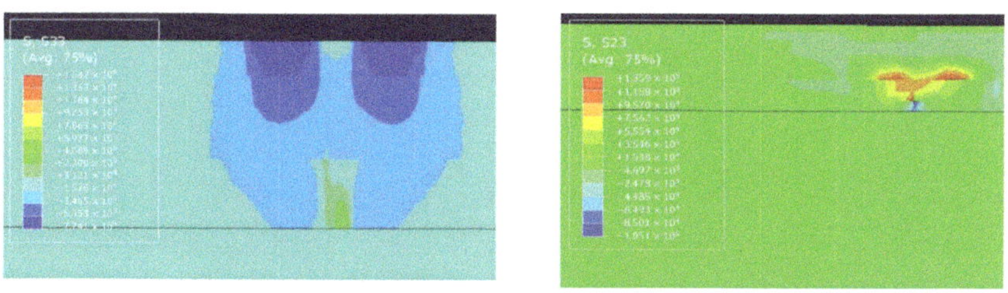

(a) (b)

Figure 5. (a) S33 diagram of crack area (b) S23 diagram of crack area.

Figure 5a indicates that the asphalt pavement structure bears vertical compressive stress under the vehicle load. The entire road surface is subject to greater compressive stress in the area of the surface layer directly in contact with the wheel under normal circumstances, which means that most areas of the pavement structure are compressed in the +z direction. However, due to the existence of reflective cracks in the base layer, an area with positive normal stress appears in the predetermined crack area, which indicates that the pavement structure is compressed in the −z-direction in the crack area due to the existence of crack. In addition, it can be seen that from the stress contour that the cracks do not grow strictly in the preset vertical direction during the expansion process. The normal stress of the crack area is obviously greater than the right on the left side of the preset vertical direction, and the crack continues to grow on the left. The growing process is affected by many factors, such as load, the thickness of each layer, the modulus of each layer, and the shape function of the crack area in ABAQUS. Figure 5a shows that there is a type I cracking in the crack growth process, and Figure 5b shows that there is also type II cracking in the process. Figure 5b indicates that there is a sliding trend between the different layers. The sliding trend between adjacent layers is caused by the existence of reflective cracks, the contact between the pavement structure layers is no longer stable. With the further action of the load, pavement cracks will spread around the entire pavement structure, causing serious damage to the pavement structure.

Crack development is a process, and ABAQUS believes that only when the mechanical parameters of the crack tip meet certain conditions, the cracks begin to develop. Meanwhile, the size of the three-dimensional layered structure of the pavement is considered to be infinite during the simulation process, and only the stress intensity factors and J-integral of the preset crack tip are calculated, that is, the destruction of the whole structure is not displayed for it was replaced by these mechanical factors. The J-integral and stress intensity factor time history curves of the crack tip were shown in Figure 6.

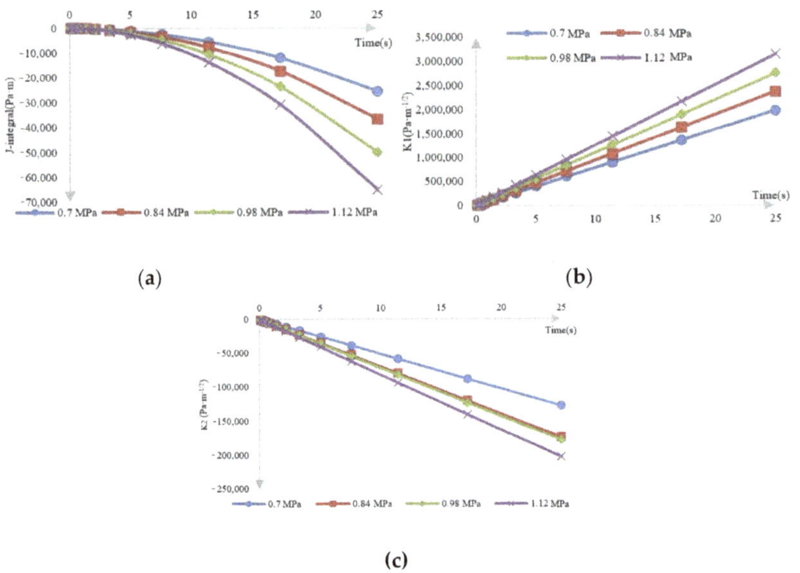

Figure 6. (**a**) J-integral time history curve (**b**) K1 time history curve (**c**) K2 time history curve.

Figure 6 indicates that under the action of static load, the curve of J-integral is close to half of the quadratic curve, while the changing trend of the stress intensity factor is a straight line. The time history curve shows that as the load time increases, the values of the three are increasing, and the crack propagation trend is also increasing, and the crack propagation rate will increase at a fixed growth rate before the structure is completely destroyed. This is because pavement damage will continue to accumulate under the action of traffic load, but the load will not disappear. As the crack continues to grow, the rate of crack propagation will gradually become faster and eventually grow at a stable speed, indicating that the crack has fully cracked in the middle and late stages of crack propagation, which behave as mesh failure in ABAQUS, and gradually expand around until the structure is completely destroyed. If the size of the simulating specimen is infinite, the crack continues to grow at a steady growth rate. In addition, it is not difficult to know that the accumulation of road damage will accelerate under the action of overload, and the development trend will increase.

3.2. Traffic Loads

Pavement may bear different loads during usage, and different traffic loads can generate different stress in the pavement structure. The principal stress and shear stress of the crack tip under different loads are shown in Figure 7.

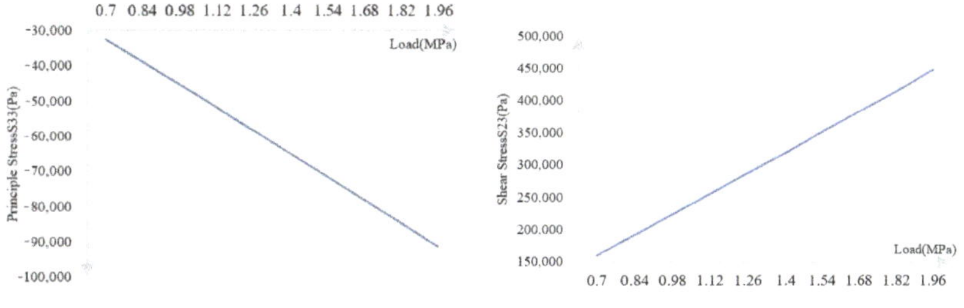

(a) (b)

Figure 7. (**a**) Principal stress change curve with load (**b**) Shear stress change curve with load.

The curve in Figure 7a satisfies Equation (16), where the value of R2 is 0.9999:

$$y = -46,682.23x + 273.52 \tag{16}$$

The curve in Figure 7b satisfies Equation (17), where the value of R2 is 0.9999:

$$y = 228,567x - 849.93 \tag{17}$$

The two curves indicate that the stress in pavement structure increases sharply with the increment of traffic loads. It can be seen that in the process of increasing the axle loads to 2.8 times the standard axle load, the principal stress increased by 181.29% and the shear stress increased by 180.82%, indicating that for every doubling of the load, the pavement structure will bear one more than the original standard axle load, especially in crack tips. The higher the stress, the faster the crack propagates, which leads to serious damage. The foregoing analysis also shows that the accumulation of road damage will accelerate under overloading, and the development trend will increase, which can be described using J-integral and stress intensity factors. Figure 8 shows the numerical changes of these mechanical factors related to crack growth under different overloads when the step time was set as 25 s.

Consistent with the stress response, the J-integral and the stress intensity factors continue to increase when the traffic load increases. The J-integral increases by 681.83%, K1 increases by 179.43%, and K2 increases by 177.11%. The above curve and increase indicate that when the pavement load gradually increases to 2.8 times of the standard axle load, the mechanical factors related to the growth rate of pavement reflective cracks are also greatly enlarged, and the reflective pavement cracks develop more rapidly, indicating that overloading affects the pavement structure. The impact of this is huge, and road traffic loads should be strictly restricted.

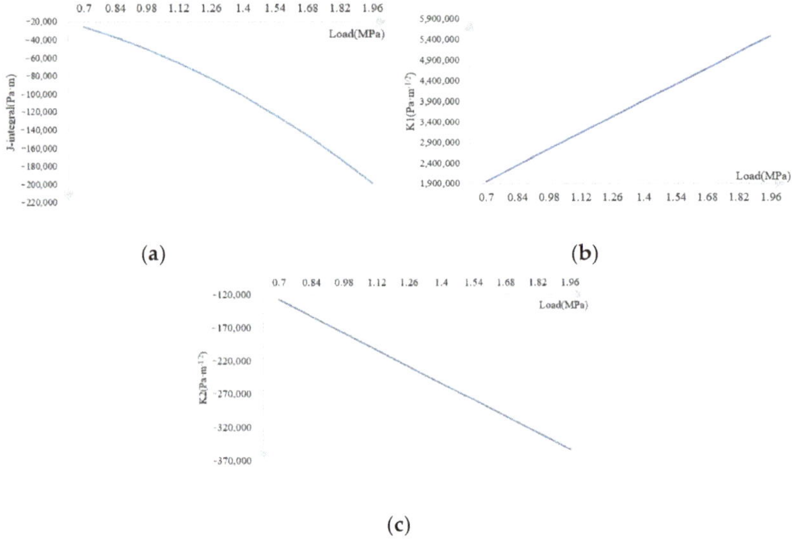

Figure 8. (**a**) J-integral curve with load (**b**) K1 curve with load (**c**) K2 curve with load.

3.3. Layers with Different Young's Modulus

The mechanical response of pavement materials to traffic load is not the same, and its influence on crack growth is also different. In the pavement structure simulation process in ABAQUS, the difference in properties of various materials is mainly reflected in the difference in modules. This paper simulated the influence of different modulus of surface layer, middle surface layer, and bottom surface layer on the crack growth. Both the three-layer take 500 MPa as the modulus gradient, and the mechanical responses of the surface layer under 10 working conditions from 1400 MPa to 5900 MPa, the mechanical responses of the middle surface layer under 10 working conditions from 1200 MPa to 5700 MPa is explored, the mechanical responses of the bottom surface layer under 10 working conditions from 1000 MPa to 5500 MPa were analyzed. The influence of the modulus of the surface layer, middle surface layer, and the bottom surface layer of asphalt pavement on the principal stress S33 and shear stress S23 under the action of standard axle load were shown in Figure 9.

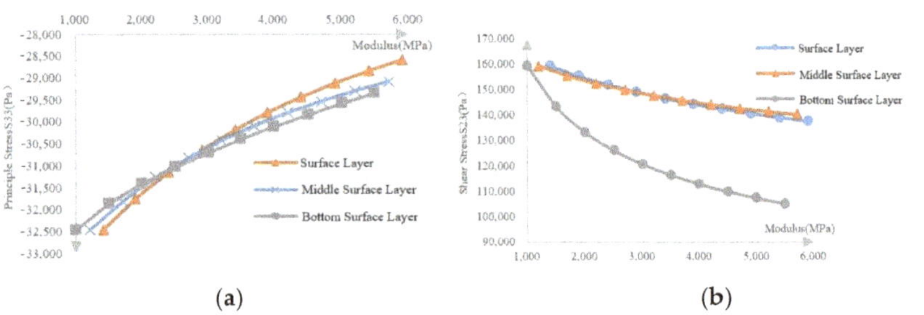

Figure 9. (**a**) Influence of the modulus of each layer on the principal stress (**b**) influence of the modulus of each layer on the shear stress.

The simulation results show that the influence of the modulus of each layer of the road surface on the stress of the crack tip during the growth of the reflective crack satisfies the inverse proportional function. The surface layer modulus has the greatest influence on the principal stress, and the bottom surface layer modulus has the greatest influence on the shear stress. It can be seen from Figure 9a that the principal stress on the crack area decreases with the increase in the modulus of each surface layer, indicating that the increase in the modulus of each layer has a certain inhibitory effect on the crack growth. This is because as the structural layers gradually harden and their strength increases, the stress that they can withstand progressively increases. The working area of the pavement structure moves upward, and more traffic loads will be borne by the surface layer structure. For the reflective cracks of the base layer, its cracking behavior will be suppressed. In terms of shear stress, it can be seen from Figure 9b that as the layer structure of the road surface gradually hardens, the shear stress on the crack area becomes smaller. This is also because the structure layers above the crack become harder, which leads to the upward movement of the pavement work area. The surface structure bears more load, and less load is transmitted to the crack area, which has a certain restraining effect. When the bottom surface layer becomes hard, the energy required for cracking the bottom surface layer increases correspondingly, and the initial energy required for slipping becomes larger. From Figure 9b, it can be seen that the layers slide mutually, that is, the distribution of shear stress. The most concentrated area is between the bottom surface layer and the base layer, so the change of the bottom surface layer modulus has a relatively large influence on the shear stress in the crack tip. However, it should be noted that this suppression effect is not obvious. In the process of changing the modulus of each layer by 4500 MPa, the largest reduction in normal stress is only 11.93% of the surface layer, and the value is only about 40,000 Pa. The largest reduction in shear stress is the bottom surface layer, the amplitude is only 34.13%, and the value is about 400,000 Pa. Compared with the damage caused by overloading the pavement structure, such an improvement appears inadequate. In addition, the production of high modulus asphalt mixture often means discarding other aspects of the mixture, such as anti-rutting performance and water damage resistance, and the production cost is correspondingly increased. Therefore, it is not feasible to prevent reflection cracks by increasing the modulus of each structural layer of the pavement. It can only be used in areas with less overload and over-limit phenomena and low performance in other areas of the road.

In addition, the values of J-integral and stress intensity factor both decrease with the increase in the modulus of each layer, and during the change of the modulus of the three-layer pavement structure, the change of the modulus of the middle surface layer has the least influence on the three parameters and the modulus of the bottom surface layer has the greatest influence. Figure 10 shows the curves of J-integral and stress intensity factors while modulus is changing.

The J-integral is a parameter related to crack initiation, and the K value is a quantity related to the rate of crack propagation. The constant increase in the modulus of each layer of the pavement will make it difficult for the crack to initiate. It will take a longer load time to reach the crack initiation criterion. Specified maximum stress and crack initiation energy. After the stress and energy in the crack area reach the crack initiation criterion, the further development of the crack in the structure still requires greater stress and energy than when the modulus is lower. Therefore, with the gradual increase in the modulus of each layer, the crack propagation rate, and the severity are suppressed to a certain extent. In addition, the interfaces with the bottom surface layer are the first and the easiest to reach when the reflection cracks develop upward due to they are the closest to the base layer, where the reflective crack was set, so the effect of change of the bottom surface layer modulus on the crack growth rate is relatively more obvious. Same as the effect on stress, even if the bottom surface layer with the greatest impact becomes hard, its J-integral was reduced by 31.3%, and the absolute value was about 8000 Pa·s; the reduction of K1 is 17.26%, and the absolute value is about 0.34 $MPa \cdot m^{-1/2}$; the reduction degree of K2 is 15.60%, and the absolute value is about 20,000 $Pa \cdot m^{-1/2}$. The above value is too small compared with the effect of overload on J-integral and stress intensity factors. In actual road engineering, the method of increasing the modulus of each layer to suppress the development of reflection cracks does not have a good effect. Therefore, the vehicle load should be strictly limited.

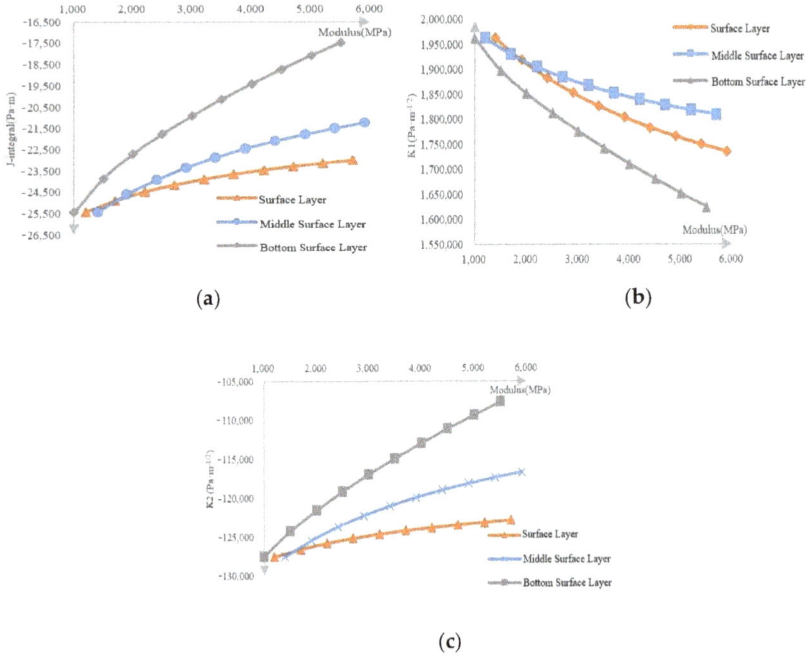

Figure 10. (**a**) J-integral curve with modulus (**b**) K1 curve with modulus (**c**) K2 curve with modulus.

3.4. Sensitivity of Various Influencing Factors on Asphalt Pavement

The entropy method was used to analyze the influence data of various working conditions on the mechanical response of the pavement crack tip in the foregoing three sections to obtain the primary and secondary relationship of the influence of various factors on the different mechanical responses. In the entropy analysis of this paper, the schemes to be evaluated are load effect, surface layer modulus, middle surface layer modulus, and bottom surface layer modulus. Each scheme has 10 working conditions. This paper counted the values of S33, S23, J-integral, K1, K2 under each working condition, taking the working condition as the vertical column and the evaluation plan as the horizontal row, and calculating the influence of each factor on them under 10 working conditions, and determined the weight of these influencing factors at the same time. The weight of the influence of various factors on S33, S23, K1, and K2 is shown in Table 2.

Table 2. Weight of various factors.

Mechanical Response	Surface Layer Modulus	Middle Surface Layer Modulus	Bottom Surface Layer Modulus	Traffic Load
J	0.2444	0.2365	0.2558	0.2633
K1	0.2443	0.2402	0.2872	0.2283
K2	0.1846	0.1817	0.4262	0.2076
S23	0.2335	0.2297	0.2791	0.2577
S33	0.2528	0.2533	0.2568	0.2370

It can be seen from the table above that among the factors that affect the J-integral, the modulus of the bottom surface layer of the surface layer is the most important factor, and the modulus of the surface layer and the middle surface layer modulus have the least influence. The factor that has the greatest influence on the four mechanical responses is the modulus of the bottom surface layer. This is because the bottom surface layer is the closest to the crack area. When the crack propagates to the bottom surface layer, the stress intensity and energy required for further expansion need to be accumulated at the contact between the bottom surface layer and the base layer. When the modulus of the bottom surface layer increases significantly, the cracking tendency becomes smaller, the cracks need to accumulate on the contact surface for a longer time, and the load needs to continue to act. Therefore, the most direct factor affecting the crack propagation process is the bottom surface layer modulus. However, the effects of several other factors need to be transmitted to the fracture area through the bottom surface layer, and the effect on the fracture area is not very direct. In addition, the analysis results also show that loads are not the most important factor affecting the mechanical response of the crack area, but from the foregoing analysis, it can be seen that they have a significant impact on the numerical value of the mechanical response of the crack area, far exceeding other working conditions. Therefore, special attention should be paid to controlling the traffic load.

4. Conclusions

Crack prevention and control is a long-term research field of road engineering. A correct understanding of the internal crack propagation mechanism of asphalt pavement will help road workers to evaluate road working conditions more comprehensively, and make better decisions in design, construction, and maintenance. In this study, the FEM software ABAQUS was used to establish a three-dimensional asphalt pavement layered model. Fracture mechanics theory and the extended finite element method were used to investigate the expansion behavior of preset cracks in the pavement base layer, and the following conclusions were obtained.

- A feasible three-dimensional pavement layered structure model was established, and numerical analysis was conducted using fracture mechanics theory and finite element software ABAQUS. The normal stress S33, shear stress S23, J-integral, and stress intensity factor K1 and K2 of the crack tip under different working conditions were calculated. Due to the existence of cracks, the difference between the positive and negative stresses between the base layer reached 2.209×10^5 Pa, and the difference between shear stress reached 4.018×10^4 Pa, indicating that the cracks showed a strong development trend both in the vertical and horizontal directions. The mechanical response of the crack tip is approximately a straight line when the vehicle load changes, and it is approximately an inverse proportional function when the modulus of each layer change. The R values of the fitted curves all reached over 0.99. In addition, the mechanical response of the stress and the stress intensity factor in some particular areas is high-order polynomial, generally second or third.
- Cracks will not strictly follow the preset path but will deviate from the preset vertical direction due to various factors. Therefore, in the actual monitoring of crack behavior, the normal stress and shear stress changes in the pavement structure should be monitored simultaneously in a larger area, and should not be limited to the crack tip.
- The load has a significant impact on the stress in the crack area. Under other conditions, the value of the stress change in the crack area is much smaller than under the load. In the process of increasing the load by 2.8 times, the principal stress increased by about 181.29%, and the shear stress increased by about 180.82%. In the process of increasing the modulus of each layer by 4500 MPa, the maximum reduction of normal stress is 11.93% of the surface layer, and the maximum shear stress is 34.13% of the ground layer. From a numerical point of view, the effect of increasing the modulus on crack propagation is relatively small compared with the effect of increasing load on cracking. Therefore, the vehicle load should be strictly limited to prevent road cracks.

- The entropy method was used to analyze the primary and secondary effects of various indicators in the process of reflecting crack propagation in asphalt pavement. The most important factor is the modulus of each layer, and the modulus of the bottom surface layer is the largest, followed by vehicle load. This shows that it is feasible to increase the modulus of each layer, especially the modulus of the bottom surface layer, to suppress the development of reflection cracks, but the impact of increasing the modulus on other road performance should not be ignored.

Author Contributions: Conceptualization, J.Y., Y.W., H.W. (Haopeng Wang) and H.W. (Houzhi Wang); methodology, J.Y., Y.W., H.W. (Haopeng Wang) and H.W. (Houzhi Wang); validation, Y.W. and H.W. (Houzhi Wang); formal analysis, Y.W. and H.W. (Houzhi Wang); investigation, Y.W. and H.W. (Houzhi Wang); data curation, Y.W. and H.W. (Houzhi Wang).; writing—original draft preparation, Y.W. and H.W. (Houzhi Wang); writing—review and editing, J.Y., Y.W., H.W. (Haopeng Wang) and H.W. (Houzhi Wang); visualization, Y.W. and H.W. (Houzhi Wang); supervision, J.Y. and H.W. (Houzhi Wang); project administration, J.Y. and H.W. (Houzhi Wang); funding acquisition, J.Y. and H.W. (Houzhi Wang). All authors have read and agreed to the published version of the manuscript.

Funding: This research was funded by the National Natural Science Foundation of China (No. 51778140, No. 52078130), Technology Research and Development Program of China State Railway Group Co., Ltd. (P2019G030), Postdoctoral Research Fund Program at Southeast University (1121002107).

Institutional Review Board Statement: Not applicable.

Informed Consent Statement: Not applicable.

Data Availability Statement: Not applicable.

Conflicts of Interest: The authors declare no conflict of interest.

References

1. Nemati, R.; Haslett, K.; Dave, E.V.; Sias, J.E. Development of a rate-dependent cumulative work and instantaneous power-based asphalt cracking performance index. *Road Mater. Pavement Des.* **2019**, *20*, S315–S331. [CrossRef]
2. Isied, M.M.; Souliman, M. Integrated predictive artificial neural network fatigue endurance limit model for asphalt concrete pavements. *Can. J. Civ. Eng.* **2019**, *46*, 114–123. [CrossRef]
3. Romeo, E.; Montepara, A. Characterization of Reinforced Asphalt Pavement Cracking Behavior using Flexural Analysis. *Procedia-Soc. Behav. Sci.* **2012**, *53*, 356–365. [CrossRef]
4. Hu, S.; Basheer, I.; Leidy, J.; Zhou, F. Development of a Universal Cracking Amount Model for Fatigue and Reflective Cracking of Asphalt Pavements. *J. Perform. Constr. Facil.* **2017**, *31*, 04017109. [CrossRef]
5. Ann Myers, L.; Roque, R.; Birgisson, B. Propagation Mechanisms for Surface-Initiated Longitudinal Wheelpath Cracks. *Transp. Res. Rec.* **2001**, *1778*, 113–122. [CrossRef]
6. García, I.G.; Carter, B.J.; Ingraffea, A.R.; Mantič, V. A numerical study of transverse cracking in cross-ply laminates by 3D finite fracture mechanics. *Compos. Part B* **2016**, *95*, 475–487. [CrossRef]
7. Okada, H.; Ishizaka, T.; Takahashi, A.; Arai, K.; Yusa, Y. 3D J-integral evaluation for solids undergoing large elastic–plastic deformations with residual stresses and spatially varying mechanical properties of a material. *Eng. Fract. Mech.* **2020**, *236*, 107212. [CrossRef]
8. Yu, S.-M.; Cai, L.-X. Analytical J-integral model for mode-I cracks in ductile materials with three-dimensional constraints. *Eng. Fract. Mech.* **2021**, *252*, 107801. [CrossRef]
9. Mehdinejad, S.; Fazaeli, H.; Moniri, A.; Dabiri, A.S. Comparison of two criteria of stress intensity factor and fracture energy to investigate the behavior of asphalt mixtures under combined tensile-shear loading modes-A statistical approach. *Constr. Build. Mater.* **2021**, *290*, 123230. [CrossRef]
10. Alae, M.; Ling, M.; Haghshenas, H.F.; Zhao, Y. Three-dimensional finite element analysis of top-down crack propagation in asphalt pavements. *Eng. Fract. Mech.* **2021**, *248*, 107736. [CrossRef]
11. Ma, X.; Quan, W.; Si, C.; Dong, Z.; Dong, Y. Analytical solution for the mechanical responses of transversely isotropic viscoelastic multi-layered asphalt pavement subjected to moving harmonic load. *Soil Dyn. Earthq. Eng.* **2021**, *147*, 106822. [CrossRef]
12. Gdoutos, E.E. *Fracture Mechanics: An Introduction*; Gewerbestrasse, Springer Nature: Heidelberg, Germany, 2020.
13. Rice, J.R. A path-independent integral and the approximate analysis of strain concentration by notches and cracks. *J. Appl. Mech.* **1968**, *35*, 379–386. [CrossRef]

14. Melenk, J.M.; Babuska, I. The partition of unity finite element method: Basic theory and applications. *Comput. Methods Appl. Mech. Eng.* **1996**, *139*, 289–314. [CrossRef]
15. Melenk, J.M.; Babuska, I. Approximation With harmonic and generalized harmonic polynomials in the partition of unity method. *Comput. Assist. Mech. Eng. Sci.* **1997**, *4*, 607–632.
16. Deng, Y.; Luo, X.; Gu, F.; Zhang, Y.; Lytton, R.L. 3D simulation of deflection basin of pavements under high-speed moving loads. *Constr. Build. Mater.* **2019**, *226*, 868–878. [CrossRef]
17. Sui, X.; Cao, L.; Ma, X.; Wang, H.; Dong, Z. Research on transversely isotropic permeability of asphalt pavement: Laboratory tests and computational simulation. *Constr. Build. Mater.* **2020**, *251*, 118958. [CrossRef]

applied sciences

MDPI

Article

Artificial Neural Network-Based Method for Real-Time Estimation of Compaction Quality of Hot Asphalt Mixes

Zhichao Xue [1], Weidong Cao [2,*], Shutang Liu [2], Fei Ren [3,*] and Qilun Wu [2]

[1] Shandong Hi-Speed Group Co., Ltd., Jinan 250101, China; 36186216@163.com
[2] School of Qilu Transportation, Shandong University, Jinan 250002, China; gtreesl@sdu.edu.cn (S.L.); qilun.wu@mail.sdu.edu.cn (Q.W.)
[3] School of Mechanical and Automotive Engineering, Qilu University of Technology, Jinan 250353, China
* Correspondence: cwd2001@sdu.edu.cn (W.C.); ren87@outlook.com (F.R.)

Abstract: With the advancement of intelligent compaction technology, real-time quality control has been widely investigated on the subgrade, while it is insufficient on asphalt pavement. This paper aims to estimate the real-time compaction quality of hot mix asphalt (HMA) using an artificial neural network (ANN) classifier. A field experiment of HMA compaction was designed. The vibration patterns of the drum were identified by using the ANN classifier and classified based on the compaction levels. The vibration signals were collected and the degree of compaction was measured in the field experiment. The collected signals were processed and the features of vibration patterns were extracted. The processed signals were tagged with their corresponding compaction level to form the sample dataset to train the ANN models. Four ANN models with different hidden layer setups were considered to investigate the effect of hidden layer structure on performance. To test the performance of the ANN classifier, the predictions made by ANN were compared with the measuring results from a non-nuclear density gauge (NNDG). The testing results show that the ANN classifier has good performance and huge potential for estimating the compaction quality of HMA in real-time.

Keywords: hot mix asphalt; compaction quality control; artificial neural networks; degree of compaction; vibration

Citation: Xue, Z.; Cao, W.; Liu, S.; Ren, F.; Wu, Q. Artificial Neural Network-Based Method for Real-Time Estimation of Compaction Quality of Hot Asphalt Mixes. *Appl. Sci.* **2021**, *11*, 7136. https://doi.org/10.3390/app11157136

Academic Editors: Amir Tabakovic, Jan Valentin and Liang He

Received: 25 June 2021
Accepted: 30 July 2021
Published: 2 August 2021

1. Introduction

The compaction of hot mix asphalt (HMA) is the last and most critical process during asphalt layer construction, which is critical for the safety and durability of an asphalt pavement. Improper compaction can cause a negative effect on the road surface performance of HMA [1]. The compacted layer usually has a high percentage of air voids if under-compaction. This may cause the pavement to be more vulnerable to oxidation and moisture infiltration [2]. Conversely, over-compaction will lead to a low percentage of air voids, which is the main cause of asphalt bleeding in high-temperature weather [3]. Therefore, quality control during compaction is a must for long-term performance assurance of asphalt pavement [4]. Currently, the most reliable method of measuring pavement density is to evaluate air voids of the extraction gained by field cores at limited locations. This test method is time-consuming, costly, destructive, and not indicative of the overall pavement quality [5]. Alternative ways to obtain the density of HMA layers in the field include nuclear density gauges and non-nuclear density gauges. However, these methods provide only point-wise measurements of density and cannot reflect the overall quality of the HMA layer in real time during compaction.

To solve these issues, the intelligent compaction (IC) technique has been evolved for more than 20 years to evaluate real-time compaction quality in the process of construction. So far, several IC measurement values (ICMVs) are proposed to indicate the compaction quality, such as some harmonic-based indicators including Compaction Meter

Value (CMV), Compaction Control Value (CCV), Vibrating Compaction Value (VCV), and some mechanical-based indicators including vibration modulus (E_{vib}) and soil stiffness (Ks) [6–10]. Among them, the CMV is calculated by the first and fundamental harmonic content, which is widely used for quality evaluation in the compaction of roadbed materials. At present, most research and application of these IC technologies have been conducted in highway and railway subgrades and earth dams [11–15]. However, some variability such as operational parameters of the compactor, characteristics of subgrade materials, and mix properties, make the above-mentioned IC measurement values difficult to accurately predict compaction quality, and thus not yet widely recognized [16]. In particular, since the strong nonlinear interaction between the roller drum and the pavement layer, and the complexity and flexibility of various factors as mentioned above, few studies on IC of asphalt layers are reported [17–19].

In recent years, a novel IC system based on the artificial neural network (ANN), named the IC analyzer (ICA), was proposed. It provides another way to estimate the stiffness of subgrade in real-time and forms an extension for estimating the quality of asphalt pavement during construction [16,20–22]. To verify the validity of the neural network-based IC method, a case study of the vibrating compaction process for the asphalt layer was conducted in this paper. The vibration patterns of the drum were identified and classified by using an ANN classifier. A field experiment of HMA compaction was designed. The field experiment was performed in a construction site to collect the vibration signals, while the degree of compaction was measured. The collected vibration signals were processed and the frequency features of vibration patterns were extracted. The processed signals were tagged with their corresponding compaction level to form the sample dataset for training the ANN model. At last, the predictions made by ANN were compared with the measuring results from a non-nuclear density gauge (NNDG) to test the performance of the ANN classifier.

2. Experimental Program and Signal Processing

A coupled system is assumed to be composed of the vibratory roller and the pavement underneath during the compaction. The response of the coupled system is influenced by the varying degree of compaction, causing the different vibration patterns of the drum. A mapping relationship should exist between vibration pattern and degree of compaction. The vibration signals and the degree of compaction were measured in a field compaction experiment. Then, the mapping relationship could be obtained by the ANN.

2.1. Experimental Program

In this study, a field test was performed in a construction site of Ning-Liang expressway project located in Shandong Province, China. Table 1 shows the pavement structure in the project, including the subgrade, one subbase layer, two semi-rigid base layers, one flexible base layer, and three hot-mix asphalt (HMA) surface layers. The compaction test was performed at the AC-25 layer with a thickness of 8 cm, as is shown in Figure 1. A Bomag roller with a single drum was used in this test. The weight of the drum is 13,000 kg, exerting a vertical force to the pavement. Other main operating parameters, such as speed and frequency, were selected by the paving crew based on their experience. The roller was operated at a speed of about 1 m/s, a frequency of 50 Hz, and an amplitude of approximately 0.35 mm in the vibrating compaction process.

Two accelerometers were used to collect the vertical vibration acceleration signals: one was carefully mounted on the drum frame and another one was mounted on the floor of the cab, as shown in Figure 2. The compaction was performed on two test lanes. The lengths of "test lane 1" and "test lane 2" were 50 m and 60 m, respectively. The width of the test lane was 2.13 m. The roller pass trajectory on each test lane is shown in Figure 3. The number of compaction passes performed for "test lane 1" and "test lane 2" were 14 and 16, respectively. In this study, TransTech's Pavement Quality Indicator Model 380 (PQI380) was employed in in situ tests, as shown in Figure 4. PQI380 uses impedance spectroscopy to measure the electrical response of asphalt from which the density is calculated. The density determined by the PQI380 is highly material-dependent, so it is required that the mix information for each mix design is input accurately into the gauge. The mix information includes stone sizes, depth, and maximum theoretical density (MTD). For one pass, the degree of compaction was measured about every 10 m and the mean value is used to indicate the degree of compaction of this pass.

Table 1. Pavement structure in the project.

HMA Surface Layers	4 cm SMA-13 6 cm AC-20 8 cm AC-25
Flexible base layer	12 cm LSPM-25
Semi-rigid base layer	2 × 18 cm Cement stabilized aggregate
Subbase layer	18 cm Cement stabilized gravel
Subgrade	Compacted natural soil

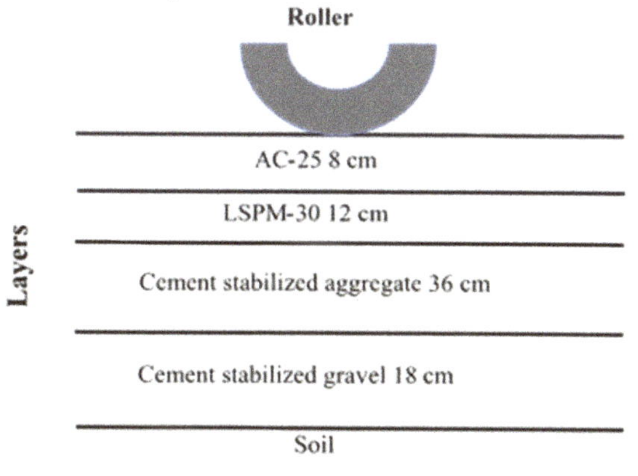

Figure 1. Pavement structure cross-section of compaction test.

Figure 2. Accelerometer setup.

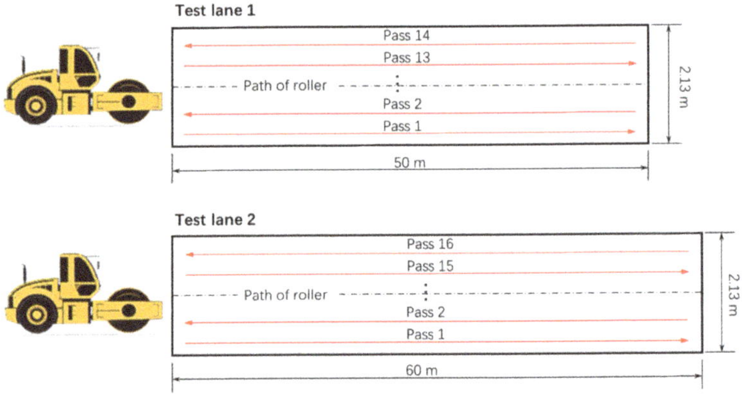

Figure 3. Test lanes.

Figure 4. PQI380 used in in-situ test.

Figure 5 shows the measuring results of the degree of compaction. The open circles denote the degree of compaction measured in the in-situ tests, the solid circles denote the average degree of compaction for each pass. It can be seen from Figure 5 that the degree of compaction increased with the increasing number of passes. For AC-25 asphalt mixture, it is generally required that the degree of compaction is at least 97%. In this compaction experiment, about 10 passes needed to be performed to satisfy the requirement.

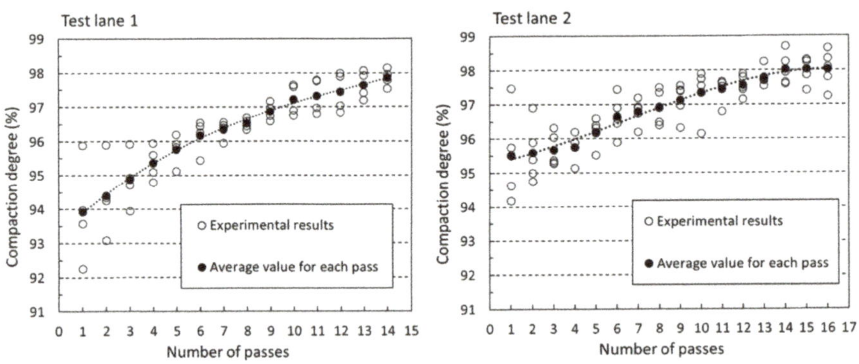

Figure 5. Relationship between the number of passes and degree of compaction.

2.2. Signal Processing

When the vibratory roller moved on the test lanes, the vibration signals measured by two accelerometers (in the cab and on the drum frame) were continuous. Since the velocity of the roller was about 1 m/s, it took about 60 s for one pass. Therefore, a 60-s long vibration signal was obtained from one accelerometer in one pass. A total of 60 long signals were collected since a total of 30 passes were performed on two test lanes. Theoretically, the frequency components of the vibration signals from the cab and drum frame should be the same, even though the amplitudes may be different.

The vibration was sampled at 2 kHz. The 60-s long signal (collected in one pass) was successively split into many 0.5-s short signals. A short signal had 1000 contiguous data samples, the first 500 values of a short signal overlapped with the last 500 values of the previous short signal. The short signals were converted into the frequency domain by using a fast Fourier transform (FFT). The Nyquist frequency was 1 kHz since the vibration was sampled at 2 kHz. The Nyquist frequency is half of the sampling rate of a discrete signal processing system. According to the Nyquist–Shannon sampling theorem, if a system uniformly samples an analog signal at a rate that exceeds the signal's highest frequency by at least a factor of two, the original analog signal can be perfectly recovered from the discrete values produced by sampling. In other words, the Nyquist frequency should be higher than the signal's highest frequency to express the signal's features accurately. The system in this study is a low-frequency system, 1 kHz Nyquist frequency is enough. By using the single-sided FFT, each short signal was converted to a frequency spectrum distributed between 0 and 1 kHz, expressed as an array of 500 elements, $a = (a_1, a_2, \ldots, a_{500})$. Therefore, the frequency features of the vibration signal were extracted by using FFT. In this study, the amplitude was not considered as a kind of feature. To eliminate the effects of amplitudes, array a should be normalized. The normalized array x was obtained as follows.

$$\begin{aligned} \bar{a} &= \ln(a) \\ x &= \frac{\bar{a}}{|\bar{a}|_{max}} \end{aligned} \tag{1}$$

In Equation (1), some inconspicuous frequency components can be amplified by the logarithmic operation. The signal processing procedure is summarized as the following three steps:

1. The 60-s long vibration signal collected in one pass is successively split into many 0.5-s short signals.
2. The short signals are converted into the frequency domain by using a single-sided FFT.
3. The short signals are normalized by using Equation (1).

The signal processing method is also diagrammed in Figure 6. The input arrays, x, were the samples used to train the ANN model, the elements of x were the features of the sample. Therefore, in this paper, 500 features were considered for a sample. Before the training, every sample was tagged with a target class, the signal sample's corresponding compaction level. In this study, four compaction levels were considered and represented by four target classes, as is shown in Table 2. Each sample was tagged according to the data of degree of compaction measured in the in-situ tests shown in Figure 5, then the tagged samples formed the dataset to train the ANN model.

Table 2. Target class and compaction level.

Target Class	Level No.	Degree of Compaction (%)	Target Array
Class 1	1	<95	(1, 0, 0, 0)
Class 2	2	95~96	(0, 1, 0, 0)
Class 3	3	96~97	(0, 0, 1, 0)
Class 4	4	>97	(0, 0, 0, 1)

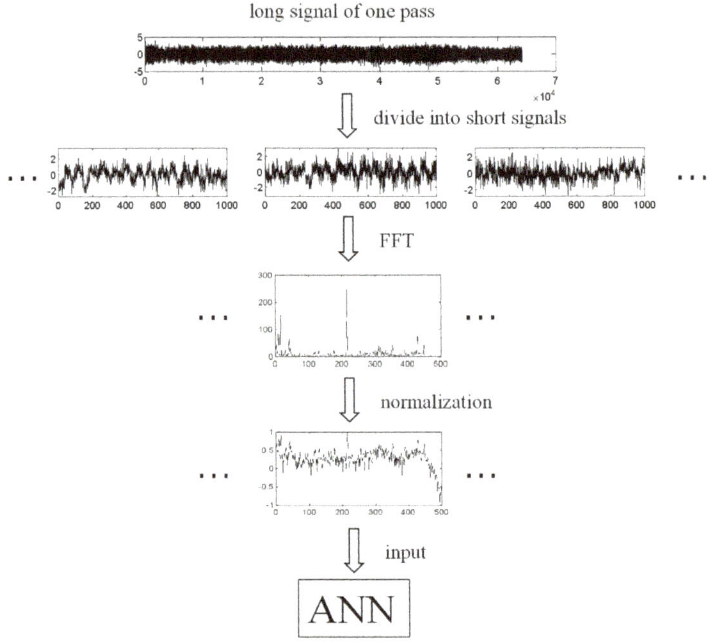

Figure 6. Signal processing.

3. Development of ANN Model

The multi-layer perceptron (MLP) feed-forward neural network was considered. The network structure shown in Figure 7 consisted of an input layer, several hidden layers, and an output layer. The input layer included 500 nodes since each sample had 500 features. There were four nodes in the output layer, representing four classes of compaction levels. Four networks with different hidden layer setups are considered in this paper, as is shown in Table 3. In Table 3, Network 4 has three hidden layers, and the first, second, and third hidden layers contain 20, 20, and 10 nodes, respectively. For convenience, the hidden layer setup of Network 4 is signified by an array [20, 20, 10], the other three networks follow this rule of expression.

Table 3. Four networks with different hidden layer setups.

	Network 1	Network 2	Network 3	Network 4
Number of hidden layers	1	1	2	3
Number of nodes	[40]	[20]	[40, 20]	[20, 20, 10]

The schematic of a single neuron is shown in Figure 7b. The governing equation of each node is shown as follows,

$$x_{s,j}^{(l+1)} = f\left(\sum_{i=1}^{k} x_{s,i}^{(l)} w_{i,j}^{(l)} + b_{0,j}^{(l)}\right) \tag{2}$$

In Equation (2), the subscript s denotes the sth sample, the number of nodes is k in the lth layer. $x_{s,i}^{(l)}$ denotes the ith input of the lth layer. $x_{s,j}^{(l+1)}$ denotes the jth output of the lth layer, meanwhile, $x_{s,j}^{(l+1)}$ also denotes the jth input of the $(l + 1)$th layer. $w_{i,j}^{(l)}$ denotes the weight from the ith input to the jth output. $b_{0,j}^{(l)}$ denotes the weight from the bias term of the lth layer to the jth output. $f()$ denotes the activation function. In this study, all the bias terms were set as 1. A softmax function, $Sof(z_j) = e^{z_j} / \sum_{i=1}^{4} e^{z_i}$, was adopted for the output layer, and a sigmoid function, $Sig(z) = (1 + e^{-z})^{-1}$, was used in the rest of the layers.

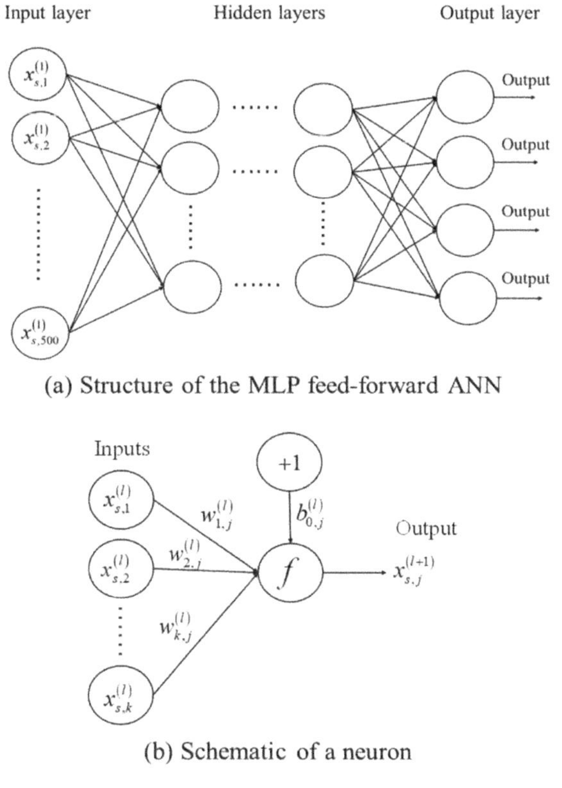

Figure 7. Structure of the MLP feed-forward ANN: (**a**) Structure of the MLP feed-forward ANN. (**b**) Schematic of a neuron.

4. Training of the ANN Model

A supervised learning method was adopted for the ANN training. The network was trained for vibration pattern recognition. The targets were expressed as the four-dimensional arrays of all 0 values except for a 1 in the cth element as shown in Table 2, where c denoted the compaction level they were to represent.

The network's performance was measured by the cross-entropy loss function. The loss of the sth prediction would be

$$CE_s = -\sum_{c=1}^{4} y_{s,c} \ln(\hat{y}_{s,c}) \quad (3)$$

In Equation (3), y denotes the target, $\hat{y}$ denotes the output from the output layer. The subscript c denotes the cth element of y or $\hat{y}$, and s denotes the sth sample. The entire loss of the dataset used for training is represented by the average of all samples' loss.

The scaled conjugate gradient (SCG) algorithm was adopted to perform training in this study; the derivatives of the loss with respect to the weights were obtained by backpropagation. SCG is an optimization algorithm based on the conjugate gradient method (CGM), it avoids the line-search per learning iteration by using a Levenberg-Marquardt approach to scale the step size [23].

The dataset used for training was divided randomly into three subsets: training set, validation set, and test set. The training set was used to calculate the gradient to update the weights. The overfitting phenomenon during the training was monitored by using the data of the validation set. Generally, the errors on the training and validation sets decrease together during the initial phase of training. When the overfitting occurs, the error on the validation set starts to increase. In this study, the training would be terminated when the error on the validation set kept increasing for six iterations, and the weights corresponding to the minimum validation error would be returned. The test set was used to test the network's performance after training. In this study, the training set, validation set, and test set were set to 70%, 15%, and 15% of the original data, respectively. Moreover, it was considered that the samples should be independent of each other, the samples were therefore arranged in a random order in the matrix of training data.

The training performances of the four networks are shown in Figure 8. Network 3 had the best performance 0.062. The performance of Network 3 is visualized in a confusion matrix in Figure 9. Each column of the matrix denotes the samples in an actual class while each row denotes the samples in a predicted class. The cells on the diagonal line denote the correctly classified samples. The off-diagonal cells denote the incorrectly classified samples. In each cell, the upper number denotes the number of samples and the lower number denotes the percentage of the total number of samples. The far-right column shows the accuracy of classification in each predicted class. The bottom row shows the accuracy of classification in each actual class. The cell in the bottom right shows the overall accuracy.

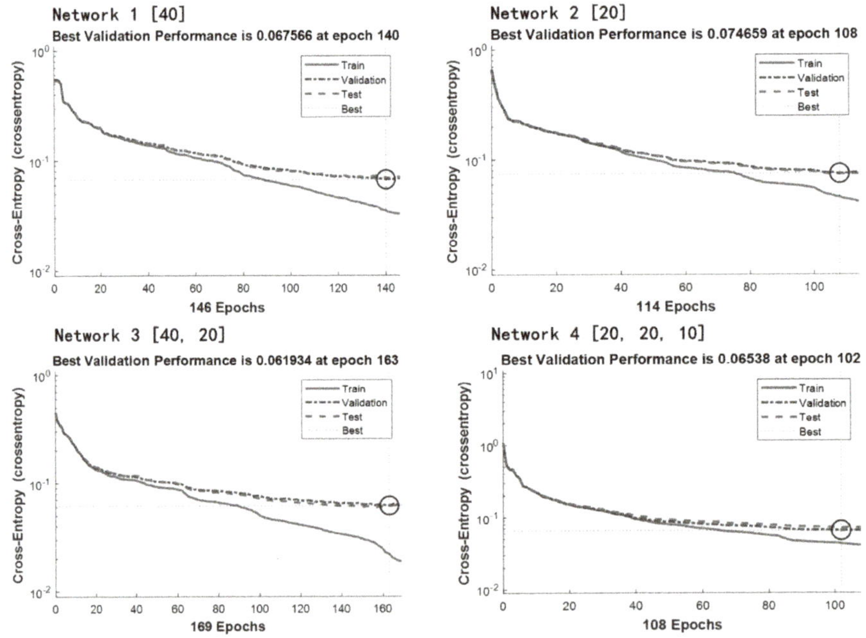

Figure 8. Output prediction error after each training cycle.

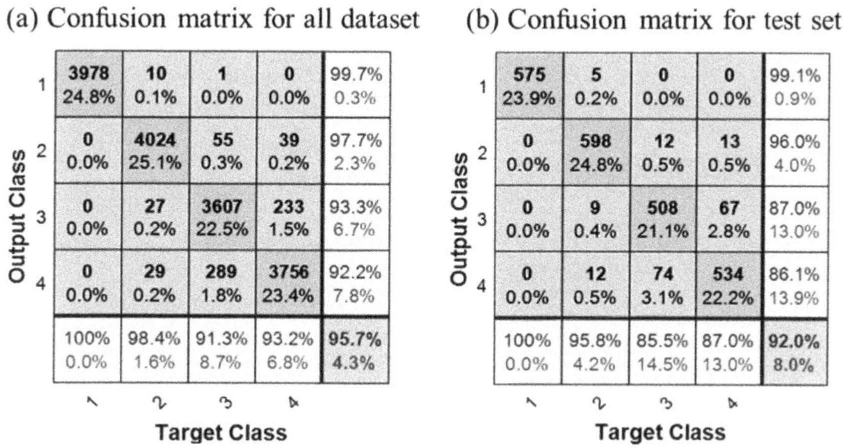

Figure 9. Confusion matrix of training performance of Network 3: (**a**) Confusion matrix for all dataset. (**b**) Confusion matrix for test set.

5. Validation of ANN Classifier

As shown in Figure 10, an ANN classifier can be gotten by integrating the signal processing procedure and the ANN model. Once inputting a 0.5-s signal, the ANN classifier will output an estimated compaction level. Network 3 was used as the ANN model in the ANN classifier since Network 3 had the best training performance.

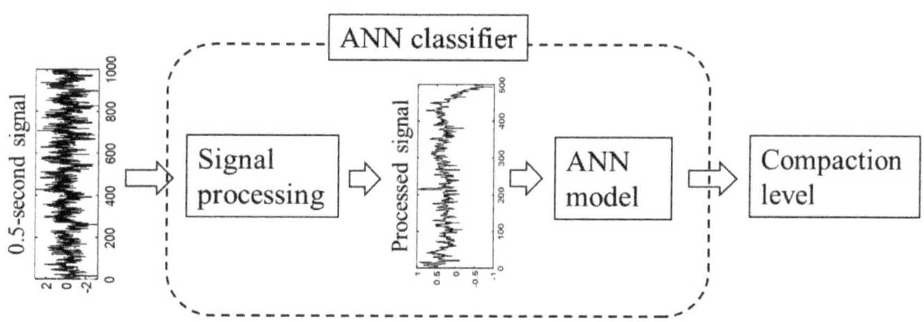

Figure 10. ANN classifier.

As is mentioned in Section 2.1, the data used to train the ANN models were collected during the compaction of two test lanes. The validation test of the ANN classifier was performed on the other 50-m lane in the construction site of the Ning-Liang expressway project. During the compaction of this validation lane, the degree of compaction was measured by NNDG for each roller pass. The number of the collected signals in each compaction level is shown in Table 4. The output accuracy of the ANN classifier is shown in Figure 11 in the form of a confusion matrix. The overall accuracy was 91.5%, and the ANN classifier had the worst performance on the recognition for the signals of level 3 with an accuracy of 82.9%.

Table 4. Number of the collected signals for the validation test.

Target Class	Degree of Compaction (%)	Number of Signals
class 1	<95 (Level 1)	850
class 2	95~96 (Level 2)	1000
class 3	96~97 (Level 3)	1040
class 4	>97 (Level 4)	1040

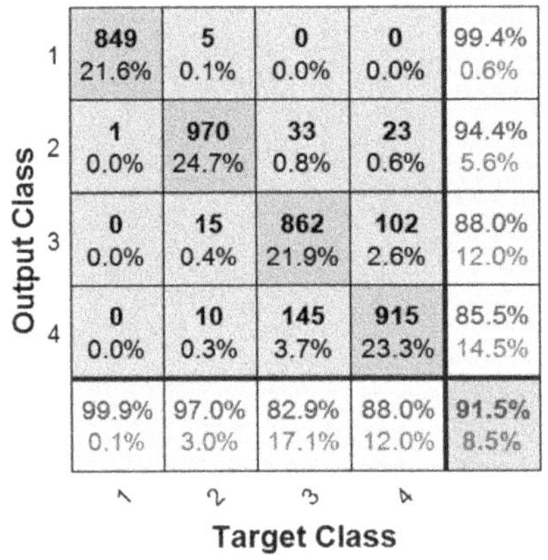

Figure 11. Output accuracy of ANN classifier.

6. Conclusions and Findings

In this study, an ANN classifier was developed for the real-time quality estimation of HMA compaction. Some conclusions and findings are listed as follows.

(1) To investigate the effect of hidden layer structure on the performance of the ANN model, four ANN models with different hidden layer setups were considered. The double-hidden-layer networks showed the best performance.

(2) The compaction quality was estimated by the ANN classifier based on the vibration signals. The results showed the validity of the ANN classifier for the real-time quality estimation of HMA compaction.

(3) A classifier was developed in this study, but not an estimator. This is due to the limitation of the accuracy of the non-nuclear density gauge. The range of degree of compaction was considered in this study, this was a kind of compromise. To develop the ANN estimator, an accurate measurement of the degree of compaction is necessary.

(4) It should be noted that although this study focused on the compaction of AC-25 asphalt mixture, it is not clear whether the ANN classifier developed in this study can be used for other asphalt mixes, such as AC-20 and SMA-13. Different materials may lead to different vibration patterns, thus ANNs may need to be re-designed and re-trained for different specified materials. Furthermore, different materials and different training dataset sizes may lead to different optimal network designs; the network in this study (Network 3) may not be suitable if the material is not AC-25, or if the amount of training data is very large.

In addition, the amount of training data is limited in this study. Collecting more training data may improve the performance of the ANN model. However, ANN still shows a huge potential for the problem of compaction quality control.

Author Contributions: Conceptualization, W.C. and F.R.; methodology, F.R.; formal analysis, W.C. and S.L.; investigation, W.C., F.R. and Q.W.; resources, Z.X. and S.L.; data curation, S.L. and Q.W.; visualization, F.R.; validation, W.C. and Z.X.; writing—original draft preparation, F.R., W.C. and Q.W.; writing—review and editing, W.C., S.L. and F.R.; project administration, Z.X.; supervision, W.C.; funding acquisition, Z.X. All authors have read and agreed to the published version of the manuscript.

Funding: This work was supported by the Science and Technology Plan of Shandong Transportation Department (No. 2019B48), and the authors would like to acknowledge their financial support.

Institutional Review Board Statement: Not applicable.

Informed Consent Statement: Not applicable.

Data Availability Statement: The data presented in this study are available on request from the corresponding author. The data are not publicly available due to them forming part of an ongoing study.

Conflicts of Interest: The authors declare no conflict of interest.

References

1. Wróbel, M.; Woszuk, A.; Franus, W. Laboratory methods for assessing the influence of improper asphalt mix compaction on its performance. *Materials* **2020**, *13*, 2476. [CrossRef] [PubMed]
2. Arambula, E.; Masad, E.; Martin, A.E. Influence of air void distribution on the moisture susceptibility of asphalt mixes. *J. Mater. Civ. Eng.* **2007**, *19*, 655–664. [CrossRef]
3. AASHTO; FAA; FHWA; NAPA; USACE; APWA; NCAE. *Hot Mix Asphalt Paving Handbook*; FAA: Washington, DC, USA, 2000.
4. Beainy, F.; Commuri, S.; Zaman, M. Asphalt compaction quality control using artificial neural network. In Proceedings of the 49th IEEE Conference on Decision and Control, Atlanta, GA, USA, 15–17 December 2010.
5. Beainy, F. Non-Contact Sensor for the Real-Time Measurement of Quality of Asphalt Pavements during Compaction. Ph.D. Thesis, University of Oklahoma, Norman, OK, USA, 2011.
6. Jeongho, O. A review on intelligent compaction techniques in railroad construction. *Int. J. Railw.* **2014**, *7*, 80–84.
7. Liu, D.; Wang, Y.; Chen, J.; Zhang, Y. Intelligent compaction practice and development: A bibliometric analysis. *Eng. Constr. Arch. Manag.* **2019**, *27*, 1213–1232. [CrossRef]
8. Mooney, M.; Adam, D. Vibratory roller integrated measurement of earthwork compaction: An overview. In Proceedings of the 7th International Symposium on Field Measurements in Geomechanics, Boston, MA, USA, 24–27 September 2007. [CrossRef]
9. Zhang, J.; Xu, G.; Cai, Y. An investigation on quality inspection and control for continuously compacting subgrade. *Rock Soil Mech.* **2015**, *36*, 1141–1146.
10. Xu, G.; Gao, H.; Wang, Z. Continuous dynamic monitor technology on subgrade compaction quality. *China J. Highw. Transp.* **2007**, *20*, 17–22.
11. Meehan, C.L.; Cacciola, D.V.; Tehrani, F.S.; Baker, W.J., III. Assessing soil compaction using continuous compaction control and location-specific in situ tests. *Autom. Constr.* **2017**, *73*, 31–44. [CrossRef]
12. Liu, D.; Gao, L. Analysis and improvement of roller vibration behavior-based indexes for monitoring compaction quality of earth-rock dams. *J. Hydroelectr. Eng.* **2018**, *37*, 111–128.
13. Liu, D.; Li, Z.; Lian, Z. Compaction quality assessment of earth-rock dam materials using roller-integrated compaction monitoring technology. *Autom. Constr.* **2014**, *44*, 234–246. [CrossRef]
14. Nie, Z.; Jiao, T.; Wang, X. Study on continuous compaction indicator of railway subgrade based on harmonic balance identification method. *China Railw. Sci.* **2016**, *37*, 1–8.
15. Nie, Z.; Wang, X.; Jiao, T. Anomalous data detection for roller-integrated compaction measurement. *Int. J. Geomech.* **2016**, *16*, 1–6. [CrossRef]
16. Commuri, S.; Mai, A.T.; Zaman, M. Neural network–based intelligent compaction analyzer for estimating compaction quality of hot asphalt mixes. *J. Constr. Eng. Manag.* **2011**, *137*, 634–644. [CrossRef]
17. Xu, Q.; Chang, G.K. Evaluation of intelligent compaction for asphalt materials. *Autom. Constr.* **2013**, *30*, 104–112. [CrossRef]
18. Hu, W.; Jia, X.; Zhu, X.; Gong, H.; Xue, G.; Huang, B. Investigating key factors of intelligent compaction for asphalt paving: A comparative case study. *Constr. Build. Mater.* **2019**, *229*. [CrossRef]
19. Hu, W.; Huang, B.; Shu, X.; Woods, M. Utilising intelligent compaction meter values to evaluate construction quality of asphalt pavement layers. *Road Mater. Pavement Des.* **2017**, *18*, 980–991. [CrossRef]
20. Commuri, S.; Zaman, M. A novel neural network-based asphalt compaction analyzer. *Int. J. Pavement Eng.* **2008**, *9*, 177–188. [CrossRef]
21. Imran, S.A.; Barman, M.; Commuri, S.; Zaman, M.; Nazari, M. Artificial neural network–based intelligent compaction analyzer for real-time estimation of subgrade quality. *Int. J. Geomech.* **2008**, *18*. [CrossRef]
22. Barman, M.; Imran, S.A.; Nazari, M.; Commuri, S.; Zaman, M. Use of intelligent compaction in detecting and remediating under-compacted spots during compaction of asphalt layers. In *Solving Pavement and Construction Materials Problems with Innovative and Cutting-Edge Technologies. GeoChina 2018*; Sustainable Civil Infrastructures; Springer: Cham, Germany, 2019. [CrossRef]
23. Møller, M.F. A scaled conjugate gradient algorithm for fast supervised learning. *Neural Netw.* **1993**, *6*, 525–533. [CrossRef]

applied sciences

MDPI

Article

Comparison of Unbound Granular Materials' Resilient Moduli Determined by Cyclic Triaxial Test and Innovative FWD Device

Dusan Stehlik [1], Petr Hyzl [1], Ondrej Dasek [1,*], Ludek Malis [2], Robert Kaderka [2], Radek Komenda [1], Jiri Sachr [1], Petr Vesely [1], Karel Spies [1] and Michal Varaus [1]

1 Institute of Road Structures, Faculty of Civil Engineering, Brno University of Technology, 602 00 Brno, Czech Republic; stehlik.d@fce.vutbr.cz (D.S.); hyzl.p@fce.vutbr.cz (P.H.); radek.komenda@colas.cz (R.K.); sachr.j@fce.vutbr.cz (J.S.); vesely.p@fce.vutbr.cz (P.V.); spies.k@fce.vutbr.cz (K.S.); varaus.m@fce.vutbr.cz (M.V.)
2 PavEx Consulting, s.r.o., 612 00 Brno, Czech Republic; lma@pavex.cz (L.M.); rka@pavex.cz (R.K.)
* Correspondence: dasek.o@fce.vutbr.cz

Abstract: In addition to the conventional laboratory tests of input materials and construction mixtures, the technically correct design of pavement structure involves more frequent functional tests of the input materials and mixtures. Our study summarizes findings from a long-term monitoring of pavement on selected road sections with lower traffic intensity. The research involved the comparison of resilient moduli as a design parameter of unbound pavement base layers' unbound granular materials (UGMs), determined in a laboratory by a test device for a cyclical triaxial test and in situ by impact load test using a falling weight deflectometer (FWD). In particular, the resilient moduli determined in the laboratory by a cyclic triaxial test ($M_{r,CTT}$) were compared with the real values of resilient moduli measured by FWD ($M_{r,FWD}$). A new, unique, and innovative FWD device was used to determine the $M_{r,FWD}$ of the UGMs. The innovation lay in the dual-axis deflection measurement—measurement performed not just in the axis parallel to the road, but also in the axis in the transversal direction. It was found that, especially on roads with lower traffic load, such dual-axis (spatial) determination of $M_{r,FWD}$ moduli was of great importance because it often allowed the identification of local pavement sections with low bearing capacities, especially at the edges, which have a significantly lower value of $M_{r,FWD}$.

Keywords: resilient modulus; pavement design; unbound granular material (UGM); cyclic triaxial test (CTT); non-destructive testing; bearing capacity; falling weight deflectometer (FWD); base layers; subbase layers

Citation: Stehlik, D.; Hyzl, P.; Dasek, O.; Malis, L.; Kaderka, R.; Komenda, R.; Sachr, J.; Vesely, P.; Spies, K.; Varaus, M. Comparison of Unbound Granular Materials' Resilient Moduli Determined by Cyclic Triaxial Test and Innovative FWD Device. *Appl. Sci.* **2022**, *12*, 5673. https://doi.org/10.3390/app12115673

Academic Editors: Amir Tabakovic, Jan Valentin and Liang He

Received: 19 April 2022
Accepted: 1 June 2022
Published: 2 June 2022

1. Introduction

Currently, the most commonly used materials in the base and subbase layers of a pavement designed to support asphalt mix layers are known as the unbound granular materials (UGMs). The unbound granular layers and the subgrade of a pavement structure provide a significant support for the structure as a whole. Hence, the mechanical properties of these materials are important for the overall performance of the structure [1].

A proper characterization of UGMs and subgrade soils is essential in the design and rehabilitation of pavement structures. The characteristics of the individual materials are of extreme importance during the design of pavement structures, and this includes the determination of optimum moisture content, grain size distribution, resilient modulus, and permanent deformation of all the materials that are to be used during the construction. Pavement layers formed by granular materials demonstrate a non-linear, time-dependent, and elastoplastic response under traffic loading. On the other hand, traditional elasticity theories consider the response of granular materials as linear-elastic, which requires a resilient modulus and Poisson's ratio. The parameters that influence the behavior of UGMs and fine-grained soils under repeated loads are stress level, density, gradation, fines content, grain size, aggregate type, particle shape, and moisture content or matric suction. The resilient modulus is mostly influenced by the level of applied stresses and the amount of moisture content in the material [2–4].

The resilient modulus is an important mechanical characteristic widely used for the analysis and design of pavements. Therefore, the determination of the resilient modulus of pavement materials is of vital importance for any mechanistically based design/analysis procedure for pavements. A laboratory resilient modulus test and a falling weight deflectometer (FWD) test are usually used to obtain the resilient modulus of a subgrade. However, the difference between the resilient moduli obtained from these two methods is considerably large due to the fact that these tests are conducted under quite different conditions [5–8].

Comprehensive research in this field is provided in [9]. A comparison was made between resilient moduli obtained from (i) a conventional small-scale resilient modulus test, (ii) a large-scale model experiment (LSME), and (iii) a falling weight deflectometer (FWD) in the field. The elastic moduli were back-calculated from the FWD data using the program MODULUS. The reasonable correspondence between the elastic moduli measured at different scales was obtained when empirical corrections were made for strain amplitude using a backbone curve for granular materials and by matching stress levels.

The differences between the back-calculated MR values of various types of roadbed soils and those obtained in the laboratory were presented and discussed in [10]. It was shown that the laboratory test results corresponded very well to the back-calculated ones. It was also shown that the MR of the roadbed soils was more or less constant, whether the soils were supporting flexible or rigid pavements.

Planning a pavement rehabilitation is based on pavement structure diagnostics, by which it is possible to determine the extent of pavement disruption as well as the cause of the disruption and thus subsequently remove this cause [11,12].

Possible causes of disruption include:

- Overall pavement structure—good-quality layers of insufficient thickness;
- Material of pavement structure layers—unsuitable, damaged, or poor-laying [13].

The above-mentioned disruptions of existing pavements can be prevented in the case of unbound mixtures in base layers by the functional testing of the resilient moduli using a cyclical triaxial device ($M_{r,CTT}$). The most frequently used test to determine the bearing capacity of the entire pavement structure is the impact load test with determination of the resilient moduli ($M_{r,FWD}$) [14–16].

The optimal solution is to design unbound mixtures for base pavement layers by functional testing so that it subsequently fulfills the requirements for the entire structure during control tests performed later [17].

The objective of the research described in this study was to find utilization for the adjustment of the frame (holder of geophones) of FWD and to describe the optimal way of designing UGMs by performance testing, which more accurately simulates the load on the tested base layers. Another objective was to find correlations between bearing-capacity characteristics derived using:

- A cyclic triaxial test according to EN 13286-7 standard;
- Non-destructive testing of the bearing capacity using FWD from real pavement structures on test sections using the designed UGMs [18];
- A method of 3D-bearing capacity testing by the adjusted measuring frame of the FWD device [19].

2. Monitored Sections and Methods

Data regarding the determination of bearing capacity at the monitored sections were analyzed in order to compare $M_{r,CTT}$ and $M_{r,FWD}$ and the development of the measuring frame of FWD. The following monitored sections were selected for research purposes (see Figure 1 and Table 1):

- Test section 1 (road II/379), length 1639 m, two-line road of width 11.5 m.
- Test section 2 (road III/6401), length 751 m, two-line road of width 7.5 m.
- Test section 3 (road III/3785), length 364 m, two-line road of width 7.5 m.

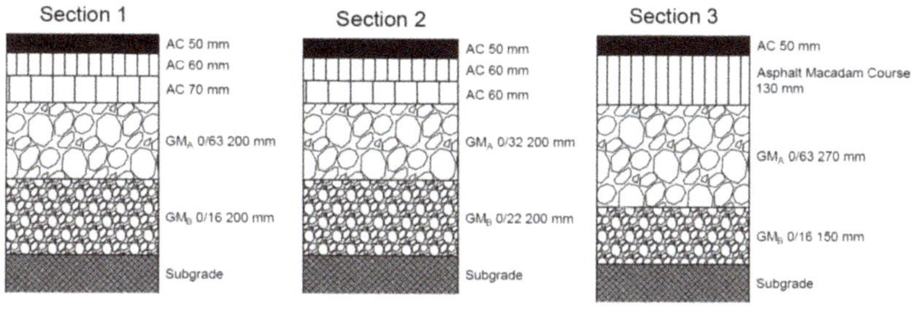

Figure 1. Pavement structure schemes.

Table 1. Pavement structures of the monitored roads.

Section	Surface of Pavement	Base Layer	Subbase Layer
1	180 mm asphalt concrete	200 mm GM_A [1] 0/63	200 mm GM_B [1] 0/16
2	170 mm asphalt concrete	200 mm GM_A [1] 0/32	200 mm GM_B [1] 0/22
3	AC 50 + 130 mm asphalt macadam course [2]	270 mm GM_A [1] 0/63	150 mm GM_B [1] 0/16

[1] $GM_{A,B}$: granular mixtures according to EN 13285; [2] asphalt macadam course: granular course 32/63 grouted with bitumen approximately 5 kg/m^2 and filled by finer fraction of aggregate.

The standard design and assessment of unbound mixtures do not commonly include functional (performance) testing of the pavement base and subbase layers. In addition to the common laboratory tests of gradation, compactability, and water and frost resistance, the individually designed mixtures were also subjected to experimental functional tests to determine the modulus in accordance with the EN 13286-7 standard. This European standard is intended specifically for testing UGMs. The resilient modulus $M_{r,CTT}$ was determined after a constant curing period of the testing specimens. Before the actual determination of $M_{r,CTT}$, the properties of the used building materials were determined.

The properties of UGMs were verified in a CTT. The $M_{r,CTT}$ of the compacted cylindrical testing specimens was determined, taking into account the assumed number of standard axles based on the conditions for designing flexible pavements [20].

The measurement of the pavement's bearing capacity was performed using the FWD deflectometer Sweco. The bearing capacity measurement was performed during different seasons and various weather conditions. Material samples taken from the probes focused especially on unbound base layers.

The $M_{r,CTT}$ determination was performed using CTT (see Figure 2). The course of the $M_{r,CTT}$ of the compacted UGMs was monitored and determined using FWD.

The bearing capacity was determined using FWDs. The dynamic tests (drop of weight through bumpers to loading plate) simulated the passing of a heavy vehicle at a speed of 60 km/h. At the same time, the values of deflection from sensors (geophones) deployed on the longitudinal frame to a distance of up to 2500 mm from the center of the loading plate were recorded.

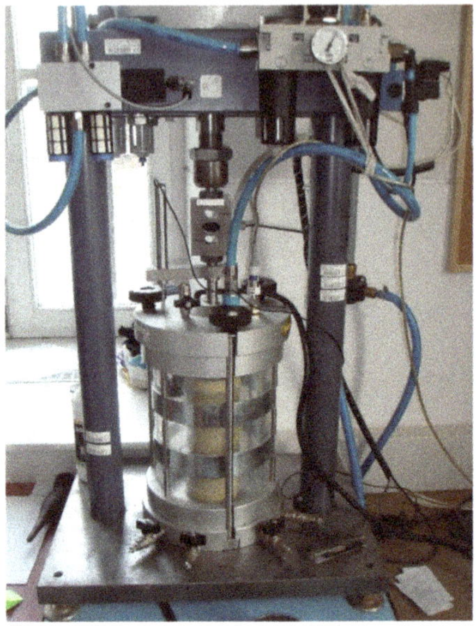

Figure 2. CTT device.

In order to be able to gather information about the transversal course of deformation as well, the standard FWD frame was equipped with a transversal frame with four additional sensors. These sensors recorded the values of deflection in the transversal direction and allowed us to make a comparison of the shape of deflection bowl in 3D and calculate the modulus of the structural layers or half-space in both directions to describe the pavement stiffness also close to the edge of the structure (see Figure 3).

Figure 3. Layout of an auxiliary cross frame.

The PRIMAX FWD trailer by Sweco was mounted on a double-axle trailer. It was independent of the towing vehicle. The operator controlled all FWD functions from the PC, which was placed in the towing vehicle.

The standard trailer-mounted FWD was supplied with a personal computer, Windows FWD software, time-history module, transport lock, DMI (Distance Meter Indicator) integrated in the software, three temperature sensors, four-split loading plate, nine geophones in the longitudinal direction, and four geophones in the transversal direction. The impact loading applied to the pavement was 50 kN. The scheme of the load frame, including the position of the geophones, is provided in Figure 4.

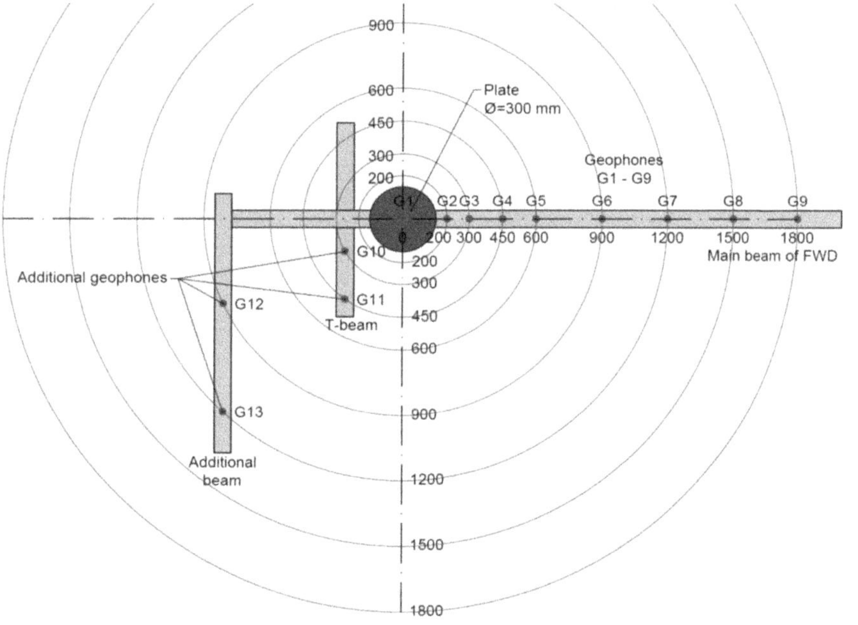

Figure 4. Scheme of the frame.

A back-calculation software for road data analysis was used for data treatment. However, the files generated from the equipment can be applied in any other back-calculation program.

We used the Primax design system from SWECO based on the following theory: When a force is transmitted to a load plate, this results in a deflection of the road surface—a "deflection bowl". The deflection is the largest at the center of the plate and decreases as the distance from the center becomes greater. The layer stiffness in a road structure is also called the E-modulus of a layer.

Based on this theory, the shape of a deflection bowl is symmetrical, and deflections measured using a standard beam normally positioned parallel with the horizontal axis of the communication should be equivalent also in the transversal direction in the case of ideal structures. If the measured deflections at geophones are larger than those in the transversal beam, it means that the structure is less durable in this direction, either due to a smaller thickness of the structure layers or their lower quality, or due to greater disruption of these layers. By comparing the deflections at the same distance from the loading plate in horizontal and transversal direction, one can determine the differences in pavement durability in that particular direction. We are currently in the process of verifying the adjusted setup of the evaluation software so that we can directly compare the elasticity moduli (E-moduli) of the structure layers of an n-layered system in transversal and horizontal direction.

The calculation of a road surface deflection is based on the theory of elasticity and the method of equivalent thickness, as framed by engineers Messrs. J. M. Kirk and N. Odemark on the basis of Boussinesq's equations. The deflection is the sum of the deformation in the layers and the subgrade. The deformation of one layer is linearly elastic, i.e., the deformation is directly proportional to the force and the thickness of the layer, but inversely proportional to the stiffness of the layer. The deformation of the subgrade is calculated from the stress and stiffness.

In accordance with the Czech national standards, sampling at national and regional communications is required every 25 m. During special measurements or in the case of severely damaged communications, the distance between two subsequent measuring points can be up to 5 m in a traffic lane.

3. Results

The obtained results can be divided into several focus sections.

3.1. Assessment of UGM Mixtures

A commonly used parameter for the evaluation of the bearing capacity is the deflection line parallel to the direction of traffic flow. This, however, might not provide ideal information about the bearing capacity of the entire pavement structure area. When evaluating the properties of the test sections, significant differences have been observed in the bearing capacity of the pavement. When considering the bearing capacity, the pavement structure is non-homogeneous in the transversal direction. For this reason, attention was paid to the development and modification of the test device. The deflectometer was upgraded by an auxiliary cross frame capable of measuring deflection also in the transversal direction from the measuring point. The following tables (Tables 2 and 3) describe the test results of functional tests, i.e., the cyclic triaxial test and CBR test. A comparison of $M_{r,CTT}$ was performed in relation to the changes in moisture of the unbound mixture tested (see Figure 5) since a variable moisture state was expected across the pavement width. Tables 2 and 3 therefore show the $M_{r,CTT}$ determined at various moisture contents. It was possible to make specific comparisons of the calculated and measured values of the elasticity moduli of the GMs. The results show that the proposed $M_{r,CTT}$ testing for unbound mixtures in base layers and the control of the non-destructive testing by impact load test with determination of $M_{r,FWD}$ are the most optimal systems for the assessment of suitability of used GMs.

Figure 5. Specimens for CTT: GM 0/22 (**left**); GM 0/32 (**right**).

Table 2. Average $M_{r,CTT}$ values and approximate values of CBR for subbase layers of the monitored road sections.

Test Section	$M_{r,CTT}$ [MPa] w_{opt} [1]	$M_{r,CTT}$ [MPa] w_{opt}-1% [1]	$M_{r,CTT}$ [MPa] w_{opt}-2% [1]	$M_{r,CTT}$ [MPa] w_{opt}+1% [1]	$M_{r,CTT}$ [MPa] w_{opt}+2% [1]	CBRsat [%] [2]
1	320–480	380–520	* [3]	280–330	200–300	124
2	180–250	220–260	* [3]	250–300	* [3]	139
3	250–350	280–450	300–420	250–300	* [3]	187

[1] Determined at optimal moisture content (w_{opt} = 4–5%) in accordance with the EN 13286-2 (Proctor modified). [2] Determinations of bearing ratio CBR in accordance with the EN 13286-47 was performed for the comparison of materials used in pavement subgrades. [3] Test samples for determination of $M_{r,CTT}$ disintegrated after compaction during manipulation.

Table 3. Average $M_{r,CTT}$ values and approximate values of CBR for base layers of the monitored road sections.

Test Section	$M_{r,CTT}$ [MPa] w_{opt} [1]	$M_{r,CTT}$ [MPa] $w_{opt}-1\%$ [1]	$M_{r,CTT}$ [MPa] $w_{opt}-2\%$ [1]	$M_{r,CTT}$ [MPa] w_{opt}+1% [1]	$M_{r,CTT}$ [MPa] w_{opt}+2% [1]	CBRsat [%] [2]
1	420–520	480–550	450–550	350–400	250–300	231
2	420–500	400–450	* [3]	450–500	380–420	189
3	300–350 [4]	280–350 [4]	* [3]	350–500 [4]	* [3]	117

[1] Determined at optimal moisture content (w_{opt} = 5–7%) in accordance with the EN 13286-2. (Proctor modified). [2] Determinations of bearing ratio CBR in accordance with the EN 13286-47 was performed for the comparison of materials used in pavement subgrades. [3] Test samples for determination of $M_{r,CTT}$ disintegrated after compaction during manipulation. [4] Values of elasticity moduli for GM at test Section 3 are only approximate because the gradation had to be adjusted for test purposes from 0/63 to 0/31.5 mm.

3.2. Relationship between $M_{r,CTT}$ and $M_{r,FWD}$

The relationship between $M_{r,CTT}$ and $M_{r,FWD}$ is given in Table 4 and Figure 6 at the current natural moisture content, sampled from the edge of the pavement where higher moisture content can be expected.

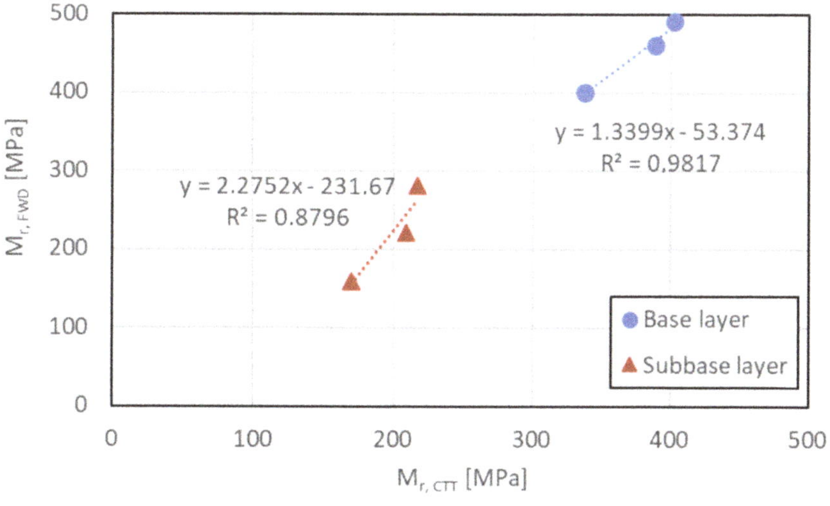

Figure 6. Relationship between $M_{r,CTT}$ and $M_{r,FWD}$ of materials from the base and subbase layer.

Table 4. Determined elasticity moduli of layers at the monitored sections at natural material humidity during the time of measurement.

	$M_{r,CTT}$ [MPa]		$M_{r,FWD}$ [MPa]	
Test Section	Base Layer	Subbase Layers (Subgrade)	Base Layers	Subbase Layers (Subgrade)
1	GM_A 0/63 403	GM_B 0/16 217	GM_A 0/63 492	GM_B 0/16 281
2	GM_A 0/32 338	GM_B 0/22 209	GM_A 0/32 401	GM_B 0/22 221
3	GM_A 0/63 389	GM_B 0/16 170	GM_A 0/63 461	GM_B 0/16 159

The average $M_{r,CTT}$ value for GM_A was 350 MPa. The measured values varied from 315 to 403 MPa. The bearing capacity of GM_A was especially affected by the gradation of the mixture with the fine particle content. The average elasticity modulus of GM_B was 200 MPa. Some values, however, were below 100 MPa. The reason for this rapid decrease in bearing capacity of GM_B was typically the increase in moisture content. Another reason was most likely the penetration of the lower unbound layer with a fine-grain base, especially in the case of pavements with a current lifespan longer than 5 years.

3.3. The Benefit of the Innovated FWD in Determination of $M_{r,FWD}$

The FWD device used for the determination of bearing capacity is capable of measuring in the axis of the traffic line or in the outer track of the wheels. In some cases of pavement disruptions, especially spatial deformations of pavement edges, the measured values of deflection determined in longitudinal direction do not correspond to the edges of the pavement. In situations where this is due to changes in the pavement structure (gradual historical widening) or changes of the parameters, specifically those of the base layer and subbase layer, there is a significant decrease or fluctuation of the bearing capacity of the structure layers. This phenomenon brings about the non-homogeneity of the pavement base layer or subbase layer characteristics, and this can directly affect the optimum planning of maintenance and rehabilitations or the reconstructions of the entire road section. An auxiliary cross frame added to FWD allows the gathering of information about changes of the parameters in transversal directions, in particular the bearing capacity of the structure layers. This allows for better description of the causes of disruption of pavement edges.

Searching for places with lower bearing capacity of the base layer and surface of a pavement allows, for example, a direct comparison of the measured pavement deflection by a sensor 900 mm away from the roadway center in the longitudinal and transversal directions. When a positive difference is found, the lower parameters of these layers can be expected, and closer attention should be paid in these cases during further diagnostic process (tests and laboratory material analysis) or special rehabilitation actions to be planned for these sections (see Figure 7).

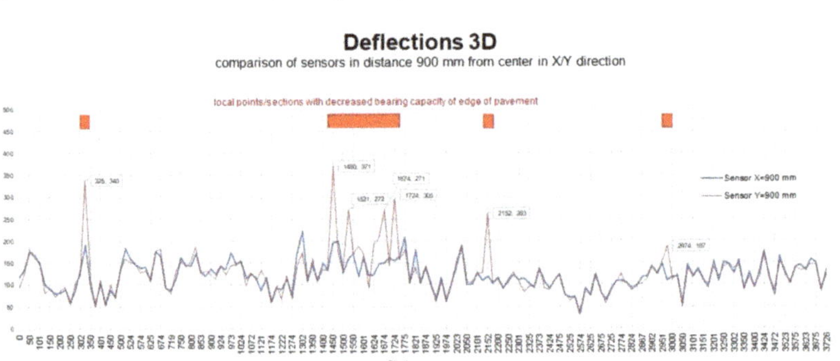

Figure 7. Comparison of sensors in distance 900 mm from center in X/Y direction.

Another possible use of the data from such modified FWD is to draw up a detailed analysis of the course of the deflection bowl and surface moduli at selected measuring points. Such data can then be used either to determine the homogeneity or non-homogeneity of the bearing capacity parameters of specific structure layers in the longitudinal and transversal directions or to allow focusing on layers which negatively affect the overall bearing capacity of the pavement (see Figure 8).

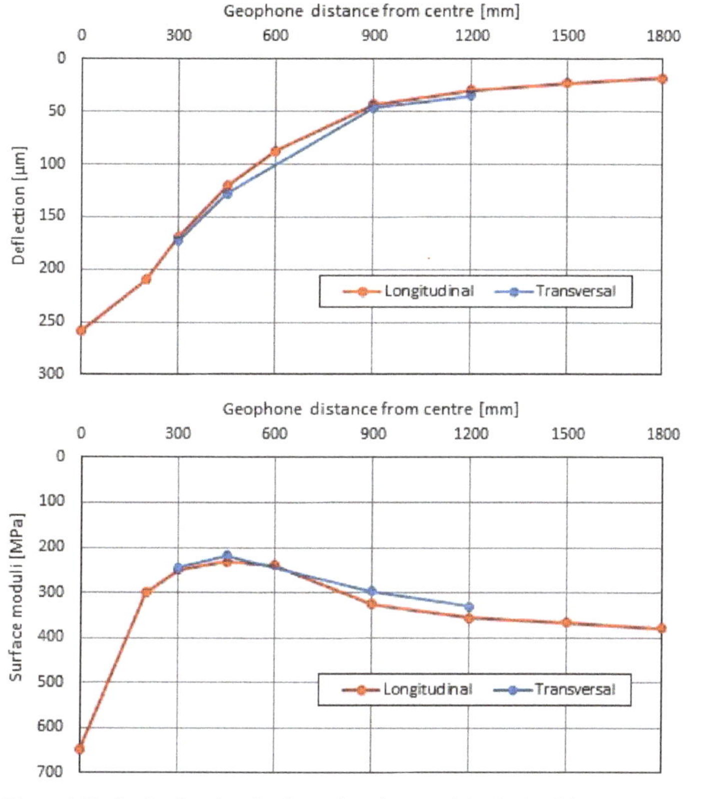

Figure 8. Deflection bowl and values of surface moduli obtained from innovative FWD device.

4. Discussion

Our results show that it was possible to compare the values of the determined $M_{r,CTT}$ with the real values of $M_{r,FWD}$ at selected road sections with a lower traffic load, which often suffer from the problem of low bearing capacity at their edges. Thanks to the use of this new, unique, innovative FWD device consisting of a dual-axis system of moduli measurements, it was always possible to find the spots with lower bearing capacity of the pavement edges. Materials sampled from these spots were then subjected to a cyclic compression test, and, indeed, lower parameters of $M_{r,CTT}$ were found. During subsequent testing of these samples, a higher level of moisture was also determined, and thus the cause of the pavement edge disruptions was revealed.

5. Conclusions

Based on the described research, it can be concluded that:

- A strong correlation was found between $M_{r,CTT}$ and $M_{r,FWD}$ for both base layers as well as subbase layers.
- The indicative values of CBR_{sat} were also determined. For the monitored unbound mixtures, they ranged between 117% and 231%.
- The values of $M_{r,CTT}$ and $M_{r,FWD}$ for the subbase layers are comparable.
- Using the innovative FWD device, it is possible to reveal locally non-durable points at the edge of the pavement during pavement diagnostics.
- It was verified that the innovative FWD device was usable specifically for pavements with lower traffic load, at which the problematic durability of the pavement edges can be assumed.
- Additionally, during the phase of deflections evaluation, it was possible to label a point as non-durable if there were differences in deflections in the horizontal and transversal directions larger than approximately 20%. This criterion can be used in technical regulations dealing with pavement diagnostics during their reconstructions.

An auxiliary cross frame can have various uses in the measurement of durability. It provides advantages when enlarging the width of roads with various pavement structures in the cross section or in pavements with neglected maintenance from the perspective of correct drainage function.

Author Contributions: Conceptualization, D.S. and L.M.; methodology, M.V., P.H. and O.D.; software, R.K. (Robert Kaderka), J.S. and P.V.; validation, D.S., P.H., O.D., L.M. and M.V.; formal analysis, M.V. and K.S.; investigation, K.S.; resources, P.H. and O.D.; data curation, M.V. and R.K. (Radek Komenda); writing—original draft preparation, D.S., P.H., O.D., L.M., R.K. (Robert Kaderka), R.K. (Radek Komenda), J.S., P.V., K.S. and M.V.; writing—review and editing, D.S., P.H., O.D., L.M., R.K. (Robert Kaderka), R.K. (Radek Komenda), J.S., P.V., K.S. and M.V.; visualization, R.K. (Radek Komenda), J.S., P.V. and K.S.; supervision, M.V.; project administration, O.D.; funding acquisition, D.S., P.H. and O.D. All authors have read and agreed to the published version of the manuscript.

Funding: The research was funded by the Technology Agency of the Czech Republic with Project CK01000158 "Use of higher amounts of reclaimed material in asphalt mixtures with PmB bitumen", Project CK03000169 "Development of road construction layers with optimized grain size replacing insufficient aggregate fractions", and Project TH04030342 "Recycled building materials for transport infrastructure construction".

Institutional Review Board Statement: Not applicable.

Informed Consent Statement: Not applicable.

Data Availability Statement: Not applicable.

Conflicts of Interest: The authors declare no conflict of interest.

References

1. Erlingsson, S.; Rahman, S.; Salour, F. Characteristic of unbound granular materials and subgrades based on multi stage RLT testing. *Transp. Geotech.* **2017**, *13*, 28–42. [CrossRef]
2. Cetin, A.; Kaya, Z.; Cetin, B.; Aydilek, A.H. Influence of laboratory compaction method on mechanical and hydraulic characteristics of unbound granular base materials. *Road Mater. Pavement Des.* **2014**, *15*, 220–235. [CrossRef]
3. Mousa, R.; Gabr, A.; Arab, M.; Azam, A.M.; Arab, M.G.; Azam, A.; El-Badawy, S. Resilient Modulus for Unbound Granular Materials and Subgrade Soils in Egypt Model for Prediction of Resilient Modulus Incorporating Matric Suction for Recycled Unbound Granular Materials. View Project 4th International Jordan Conference View project Resilient Modulus for Unbound Granular Materials and Subgrade Soils in Egypt. 2017. Available online: https://www.researchgate.net/publication/316521210 (accessed on 15 October 2021).
4. Adu-Osei, A. Characterization of Unbound Granular Layers in Flexible Pavements (Sponsored by the Aggregates Foundation for Technology, Research and Education). Available online: https://repositories.lib.utexas.edu/bitstream/handle/2152/35440/502-3.pdf?sequence=2&isAllowed=y (accessed on 15 October 2021).
5. Ji, R.; Siddiki, N.; Nantung, T.; Kim, D. Evaluation of resilient modulus of subgrade and base materials in indiana and its implementation in MEPDG. *Sci. World J.* **2014**, *2014*, 372838. [CrossRef] [PubMed]

6. Quinta-Ferreira, M.; Fung, E.; Andrade, P.S.; Branco, F.C. In-place evaluation of a limestone base course modulus, using a van-integrated falling weight deflectometer (FWD) and the GeoGauge (SSG). *Road Mater. Pavement Des.* **2012**, *13*, 817–831. [CrossRef]
7. Rahim, A.; Asce, A.M.; George, K.P.; Asce, H.M. Falling Weight Deflectometer for Estimating Subgrade Elastic Moduli. *Transp. Eng.* **2003**, *129*, 100–107. [CrossRef]
8. Saleh, M. Utilisation of the deflectograph data to evaluate pavement structural condition of the highway network. *Road Mater. Pavement Des.* **2016**, *17*, 136–152. [CrossRef]
9. Tanyu, B.F.; Kim, W.H.; Edil, T.B.; Benson, C.H. Comparison of Laboratory Resilient Modulus with Back-Calculated Elastic Moduli from Large-Scale Model Experiments and FWD Tests on Granular Materials. *Symp. Pap.* **2003**, *1437*, 191–208. Available online: https://www.astm.org/stp12532s.html (accessed on 15 October 2021).
10. Dawson, T.A.; Baladi, G.Y.; Sessions, C.P.; Haider, S.W. Backcalculated and Laboratory-Measured Resilient Modulus Values. *Transp. Res. Rec.* **2009**, *2094*, 71–78. [CrossRef]
11. Stehlik, D.; Varaus, M.; Hyzl, P. Monitoring of experimental sections with recycled waste materials. In Proceedings of the 5th International Conference on Structural Engineering, Mechanics and Computation, Cape Town, South Africa, 2–4 September 2013.
12. Stehlik, D.; Dasek, O.; Hyzl, P.; Coufalik, P.; Krcmova, I.; Varaus, M. Pavement construction using road waste building material—From a model to reality. *Road Mater. Pavement Des.* **2015**, *16*, 314–329. [CrossRef]
13. Salour, F.; Erlingsson, S. Investigation of a pavement structural behaviour during spring thaw using falling weight deflectometer. *Road Mater. Pavement Des.* **2013**, *14*, 141–158. [CrossRef]
14. Elshaer, M.; Ghayoomi, M.; Daniel, J.S. The role of predictive models for resilient modulus of unbound materials in pavement FWD-deflection assessment. *Road Mater. Pavement Des.* **2020**, *21*, 374–392. [CrossRef]
15. Rocha, S.; Tandon, V.; Nazarian, S. Falling Weight Deflectometer Fleet: Repeatability and Reproducibility. *Road Mater. Pavement Des.* **2004**, *5*, 215–238. [CrossRef]
16. Mehta, Y.; Asce, A.M.; Roque, R.; Asce, F. Evaluation of FWD Data for Determination of Layer Moduli of Pavements. *J. Mater. Civ. Eng.* **2003**, *15*, 25–31. [CrossRef]
17. Islam, K.M.; Gassman, S.; Rahman, M.M. Field and Laboratory Characterization of Subgrade Resilient Modulus for Pavement Mechanistic-Empirical Pavement Design Guide Application. *Transp. Res. Rec. J. Transp. Res. Board* **2020**, *2674*, 921–930. [CrossRef]
18. Horak, E.; Maina, J. Structural number determined with the falling weight deflectometer and used as benchmark methodology Advanced numerical modeling View project Permeability of Hot Mix Asphalt View project. *IEEE Proc. Hong Kong* **2014**. [CrossRef]
19. Pożarycki, A.; Górnaś, P.; Wanatowski, D. The influence of frequency normalisation of FWD pavement measurements on backcalculated values of stiffness moduli. *Road Mater. Pavement Des.* **2019**, *20*, 1–19. [CrossRef]
20. Hyzl, P.; Varaus, M.; Stehlik, D.; Zdralek, P. Experience with triaxial loading systems for the testing of road construction materials. In Proceedings of the 7th International Rilem Symposium Atcbm09 on Advanced Testing and Characterization of Bituminous Materials, Rhodes, Greece, 27–29 May 2009; pp. 909–926.

MDPI
St. Alban-Anlage 66
4052 Basel
Switzerland
Tel. +41 61 683 77 34
Fax +41 61 302 89 18
www.mdpi.com

Applied Sciences Editorial Office
E-mail: applsci@mdpi.com
www.mdpi.com/journal/applsci

www.ingramcontent.com/pod-product-compliance
Ingram Content Group UK Ltd.
Pitfield, Milton Keynes, MK11 3LW, UK
UKHW060122010826
14089UKWH00006B/108

9 783036 554938